AF522148

ENCYCLOPAEDIA
OF
ASIA

TIBETAN YAK-DRIVER, WITH PRAYER-WHEEL.

ENCYCLOPAEDIA
OF
ASIA

Compendium of Geography
with Maps and Illustrations

Volume I

Northern and Eastern Asia

By

A.H. KEANE, F.R.G.S.

DISCOVERY PUBLISHING HOUSE PVT. LTD.
NEW DELHI-110 002

Published by:

DISCOVERY PUBLISHING HOUSE PVT. LTD.
4383/4B, Ansari Road, Darya Ganj
New Delhi-110 002 (India)
Phone : +91-11-23279245; 23253475; 43596065
E-mail : discoverybooksindia@gmail.com
discoverypublishinghouse@gmail.com
namitwasan9@gmail.com
web : www.discoverypublishinggroup.com

Reprinted: 2021

ISBN: 978-81-7141-405-5 (Set)

Encyclopaedia of Asia

Printed at:
Infinity Imaging Systems
Delhi

CONTENTS

CHAPTER I

INTRODUCTION : GENERAL SURVEY

NORTHERN ASIA: CAUCASIA, RUSSIAN TURKESTAN, SIBERIA

CHAPTER II

CAUCASIA

CHAPTER III

RUSSIAN TURKESTAN

CHAPTER IV

SIBERIA

COMPENDIUM OF GEOGRAPHY

EASTERN ASIA: CHINESE EMPIRE, JAPAN

CHAPTER V

CHINESE EMPIRE

CHAPTER VI

JAPAN

senting many striking points of analogy with Spain, Italy, and Greece, the three corresponding peninsulas of South Europe. Arabia, like Spain, forms a vast tableland with a monotonous coast-line, unvaried by any deep inlets. Like Italy, India is sheltered from the north by a great alpine region, is traversed by a mountain range running north and south, and terminates at its southern extremity with a large and fertile island. In the same way the Eastern Archipelago, continuing the Indo-Chinese peninsula towards Australasia, answers to that of the Ægean Sea, serving to connect the Hellenic peninsula with Asia Minor. Asia Minor, itself the westernmost projecting peninsula of Asia, may be compared with Brittany, the westernmost promontory of Central Europe, and the analogy is completed by the peninsulas of Korea and the Crimea, both projecting into narrow inland seas, and by the great archipelagoes of Japan and the British Isles, nearly equal in extent and even in population, but with their positions towards the mainland necessarily reversed.

Along the whole northern section of the continent there stretches a boundless lowland region, which, for hundreds of miles inland, is covered with the so-called *tundra*—dreary and almost uninhabitable wastes, exposed to the full fury of the fierce Arctic gales, ice-bound for nine months in the year, and in many places permanently frozen to a considerable depth. Farther south the land ascends gradually to the south Siberian highlands, whence flow the Ob, Yenisei, Lena, and other great streams, which during the short open season roll their sluggish waters northwards to the Arctic basin. This great polar sea washes the whole of the flat and low-lying North Asiatic seaboard, the exploration of which has been completed by the Swedish navigator Nordenskjöld, who for the first time made the north-east passage in 1878-79, and determined the northernmost point of the continent, at

Cape Severo, close to Cape Chelyuskin, in 78° 20′ N. lat. and 104° E. long. These bleak Northern shores, facing the neighbouring archipelago of New Siberia, and the more distant Wrangel Land, form a true Arctic region, lying entirely within the Arctic Circle, and sparsely inhabited by a few nomad Samoyede, Yakut, Yukaghir, and Chukchi tribes. Its low level and exposed northern aspect, combined with its high latitude and enormous extension southwards, are the chief reasons which cause the climate of this region to be the most "continental" —as it is technically termed, that is, subject to the greatest extremes of cold and heat—of any on the globe. "Siberian" winters have become proverbial, but the summers are almost equally intense; and while the mercury becomes frozen to a hard malleable mass during the clear Arctic nights in midwinter, it will occasionally rise to above 100° F. at midday in June. The most unfavourably situated tracts are undoubtedly those which stretch along the Frozen Ocean, from the Taimur peninsula to the River Kolima, which in many places are permanently frozen for some distance below the surface. But farther east also, and generally speaking throughout the whole of the north-east, the aspect of the land is extremely dreary, especially in the Chukchi country, which reaches quite to Bering Strait. Here is developed the great peninsula of Kamchatka, which stretches southwards, and is continued across the Kurile archipelago as far as the northern extremity of the large Japanese island of Yesso. Igneous agencies, elsewhere all but exhausted or long quiescent on the Asiatic mainland, are still active in Kamchatka, whose eastern seaboard is traversed by an imposing line of burning mountains. These volcanoes are continued across the barren Kurile group, which, with the peninsula, encloses a deep inlet of the Pacific Ocean known as the Sea of Okhotsk, along whose desolate shores dwell a few scattered Lamut and other tribes of Tungus stock. More inviting and more favourable for agricultural life becomes the region where the mighty Amur rolls its waters to the sea over against the long and narrow island of Sakhalin. This metalliferous island, which is at one point almost connected with the continent, marks the extreme eastern limit of the Czar's authority, but since its cession to Russia by Japan it has been chiefly used as a convict station for political offenders.

Sakhalin is separated by the narrow Strait of La Perouse from the Japanese group of islands which stretch, in a slightly curved arch, southwards to the Korean peninsula. Thus is formed the almost landlocked Sea of Japan, which communicates through the Gulf of Tartary northwards with the Sea of Okhotsk, and through the Strait of Korea southwards with the Yellow and Eastern Seas. East of these waters, and along the east coast of Japan, flows the Kuro Siwo, or "Black Stream," which is situated nearly under the same parallels of latitude as the Gulf Stream, and which plays almost a more important part in the Pacific than that remarkable current does in the Atlantic. Favoured by the boundless extent of the Pacific Ocean, which is here encumbered by but few island groups, the Kuro Siwo finds far fewer obstacles to its full development than its Western rival, and is thus enabled to pursue a more decided course, attended by correspondingly greater influences on the climate and vegetation of the lands lying in its track. Its effects are especially visible in Japan itself, where everything reminds us that we have entered a mild, and, in some places, even a sub-tropical zone. Here a delightful climate, combined with a lavish display of grand natural scenery, unites all the conditions required for the development of that peculiar civilisation which cannot fail to excite the admiration of the Western world, and create a deep sympathetic feeling for the Japanese people, with their varied industrial pursuits, their populous cities lying at the foot of threatening volcanoes, their well-tilled lands, their many ingenious social and political institutions.

A southern continuation of the Japanese archipelago is formed by the much smaller group of Liu-kiu Islands, which have long constituted a subject of contention between the governments of China and Japan. This group forms a link in the chain of islands which are developed in a series of successive festoons, as it were, along the east Asiatic seaboard, between the Bering and China Seas. Another, and a still more important link, is formed by the extensive but little known island of Formosa, whence the transition is easily effected through the Batanes and Babuyan groups to the Philippines. Formosa occupies an important position, both physically and ethnically, for it is crossed, nearly in its centre, by the Tropic of Cancer. It thus stands on the verge of the torrid and temperate zones, marking the extreme northern extension of the Malay race, which here meets the Chinese on common ground. Beyond this point we pass with the Philippines into Australasia proper, and the great Archipelago of Malaysia, through which the south-eastern extremity of Asia merges imperceptibly with the continent of Australia.

Notwithstanding the labours of Ney Elias, Montgomerie, Forsyth, Margery, Gill, Prjevalsky, Krapotkin, Kostenko, Richthofen, Vambery, Schlagintweit, Desgodins, Nain Sing, The Mirza, Carey, Younghusband, Conway, Rockhill, Bower, Grombchevsky, Robarofsky, Dutreil de Rhins, Bonvalot, and many other illustrious modern explorers, much geographical work still remains to be done in various parts of the continent. British India, West Siberia, Palestine, and the Caucasus alone can be said to have been thoroughly surveyed, and India especially is certainly one of the best-known countries in the world;

but with these exceptions, and although nearly all great geographical problems have been solved, our knowledge of the Arabian and Korean peninsulas, of Tibet, the interior of Indo-China, Yun-nan, Kan-su, North-east Siberia, and several other regions is still far from complete, and often extremely inadequate.

2. *Relief of the Land: Plateaux and Highlands.*

The bold lines on which Asia has been framed are especially conspicuous, no less in its main political and social than in its physical features. The heart of the continent consists of a vast tableland, by far the most elevated and extensive on the globe, with an altitude ranging from 10,000 to 17,000 feet, above which tower the mighty Himalayan, Kuen-lun, Tian-shan, and Altai ranges. This tableland broadens out eastwards, and converges westwards in the nucleus of the Great Pamir, or "Roof of the World." A western extension of the same tableland is formed by the Iranian plateau, which stretches from the Hindu-Kush and Suliman Mountains, across Afghanistan, Baluchistan, and Persia, to the Persian Gulf and Mesopotamian lowlands. Culminating towards the north-west in the Kurdish and Armenian highlands, Irania merges westwards in the tableland of Asia Minor and the snowy crests of Lebanon, but falls abruptly northwards to the valley of the Kur. Beyond this historic stream the land again rises to the mighty barrier of the Caucasus, which is continued north-westwards through the Taman peninsula into the Crimea, and south-eastwards across the Caspian to the highlands separating Irania from the Turkestan lowlands.

The vast central plateau itself is enclosed on the south by the mighty barrier of the Himalayas, sweeping round from Afghanistan to Burma in a graceful curve, which presents its convex side towards the Indian Ocean. On the north the tableland is hemmed in by the Altai, with its eastern projections, the Sayan, Yablonovoi, and other Siberian ranges; on the east by the less continuous Yung-Ling, Inshan, and other Chinese ranges; while westwards the Himalayas and Altai, through the Karakorum, Hindu-Kush, Tian-shan, and Alai, close round the Great Pamir, here interlacing in the focus of the whole continental mountain system.

But within these stupendous rocky walls the central tableland, occupying an area of perhaps 3,000,000 square miles altogether, presents several clearly-defined divisions, differing greatly in their relief, and even in their physical aspect, one from the other. The great Tibetan plateau maintains, between the Himalayas and the Kuen-lun, a mean elevation of 14,000 to 16,000 feet. The Pamir steppe in the west, and the Koko-nor basin in the east, fall to 12,000 and 9000 feet respectively. But beyond the Kuen-lun and its possible eastern extensions, there is almost an abrupt descent to the vast region of the Gobi desert, which is scarcely more than 4000, and which sinks westwards in the Tarim or Lob-nor depression as low as 2000 feet above sea-level. Yet, notwithstanding these deviations, the enormous extent and great mean elevation of the whole region are sufficient to give to the entire continent an average altitude of no less than 1600 feet, or about 600 feet more than Europe, and 500 more than the estimate made by Humboldt, on insufficient data, early in the present century.

The mountain ranges intersecting the plateaux, mainly in the direction from the north-west to the south-east, but occasionally running nearly due west and east or north and south, consist chiefly of crystalline rocks, old schists, palæozoic and other primitive formations, in the Siberian, Kuen-lun, and Karakorum sections. But the Himalayas,

VIEW IN THE HIMALAYAS.

although resting on granite masses, which crop out in many of the highest peaks, are, to a very large extent, of comparatively far more recent formation, having been upheaved during the tertiary epoch, when the eocene strata in Ladak were raised to an elevation of nearly 12,000 feet.

Simultaneously with the tendency towards greater dryness in the interior of the continent, there is clear evidence to show that a process of slow upheaval has been going on, at least around most of the seaboard, throughout the present geological epoch. On the north coast, islands, which a hundred years ago stood at some distance from the land, are now connected with it by rocky isthmuses. The upheaved coral reefs skirting the west coast of Arabia show that here also the land is rising, and similar tendencies have been observed in the Euxine and Ægean in the extreme west; about the Amur delta, Kamchatka, and China, in the extreme east; along the shores of Burma, Ceylon, Malabar, and Baluchistan, in the extreme south. On the other hand, symptoms of subsidence have been detected at a few points on the coast of Syria, near the Indus delta, on the shores of Annam and Fo-kien over against Formosa, and especially in the Laccadive and Maldive islands, where the atolls or round coral reefs are disappearing, and where the Chagos bank has already vanished.

3. *Hydrography: Rivers and Lakes—Inland and Seaward Drainage.*

Several distinct systems of inland drainage are formed by deep depressions, partly within the tablelands themselves, partly in the plains by which they are nearly everywhere surrounded. Such is the depression of Eastern Turkestan, 2000 feet above sea-level, through which the Tarim and its tributaries drain eastwards to Lake Lob; that of Lake Sistan, which receives, through the Helmand and other streams, a great part of the Afghanistan drainage; and the remarkable trough of the Dead Sea, the deepest on the surface of the earth, fed mainly by the Jordan from the north. But by far the most important system of inland drainage is that of the Aral Sea, which comprises the whole of the western Turkestan lowlands, and which was formerly even still more extensive. At present it drains the Great Pamir, Alai, and western Tian-shan highlands alone, through the twin rivers Oxus (Amu-darya) and Jaxartes (Sir-darya). But it seems to have at one time stretched eastwards to Lake Balkhash, westwards to the Caspian, southwards to the north Iranian highlands, and northwards to the low range of hills forming the water-parting between the Ob and Aral basins. Altogether, the area of all the lands which have no present outflow seawards is estimated at about 4,000,000 square miles, or nearly one-fourth of the whole continent. The significance of this fact will be best realised when it is added that both Europe and America are almost destitute of an inland drainage, while that of Africa seems limited mainly to the Chad, Ngami, and Rudolf basins.

The seaward drainage of the continent is determined only to a very small extent by the lofty ranges enclosing the great central and western tablelands. These ranges form scarcely anywhere true water-partings; for, except where they converge about the Pamir, they are everywhere pierced by the great continental rivers, which rise, not on their outer flanks, but within the plateaux, and which have thus to force their rocky barriers to reach the surrounding oceans. Thus, of the three great Siberian rivers flowing north to the Arctic, both the Ob and Yenisei have their farthest head-streams south of the mountains fringing the Kobdo and Mongolian plateaux; and even the Lena, now rising on the outer slopes, seems to have formerly been connected with the Angara (Yenisei basin), in the neighbourhood of Irkutsk. So also the Amur, Hoang-ho, and Yang-tse-kiang, the three main streams flowing each to the Pacific, rise all of them far beyond the encircling ranges of the Mongolian, Koko-nor, and Tibetan tablelands. The great southern rivers, Me-khong, Salwin, Brahmaputra (San-po), and Indus, have also their sources behind the Himalayas on the Tibetan steppe. Here a solitary but important exception is the Ganges system, of which both the head-streams, the Ganges and the Jamna, rise on the outer or southern flanks of the Himalayas. The same remarkable phenomenon is presented in the extreme west of the continent, where the Tigris and Euphrates flow to the Persian Gulf and the Araxis to the Caspian, from the very heart of the Armenian and Kurdistan highlands. Here also the Kizil-Irmak has to force its way from the Anatolian tableland through the Anti-Taurus to the Euxine, while the Orontes reaches the Mediterranean from the Bekaa (Cœle-Syria), behind the Lebanon and Nusarieh coast ranges. The list of great Asiatic rivers is almost completed by those of Southern India, where the Nerbadda flows from the furthest extremity of the Vindhya hills westwards to the Arabian Sea, and where the Godavari and the Kistna, rising also on the plateau of the Deccan, trend in the opposite direction to the Bay of Bengal.

Compared with the other divisions of the globe, Asia is singularly deficient in large fresh-water lakes. Apart from the intensely salt Dead Sea and the salt or brackish Caspian, Aral, and Balkhash,—apparently remnants of a vast Asiatic Mediterranean, which at present communicates with the Euxine,—the only sheet of fresh water worthy of mention by the side of the great inland seas

WEST SIDE OF LAKE BAIKAL AT LISTVINICHNAYA.

of equatorial Africa and North America is Lake Baikal, which discharges its overflow through the River Angara to the Yenisei. The so-called lakes of the West Siberian steppes are little better than swamps, and no large bodies of water occur in West Asia except Gokcha, Van, Urmia, and the marshy Hamun or Sistan, none of which has any outflow seawards. In the whole of India and Indo-China the Tonle-sap of Cambodia is the only lake of any size; in China, the Tong-ting and Po-yang alone deserve mention; and even on the Tibetan uplands, although lakes are now found to be very numerous, few appear to be of large size except the Tengri-nor, Palti, Buka, Ike-Namur, Koko-nor, and the recently discovered Horpa, Charol, Aru, Ghalaring, Kyaring, Dangra Yum, Daru, and Sira Cho. The Kos-gol, Ubsa, and Kulon of North Mongolia, the Kenka of Manchuria, the Zaizan and Ulyungur of the Upper Irtish, the Issik-kul of the Tian-shan highlands, the Lob-nor of Eastern Turkestan, and the Kara-kul of the Pamir, almost complete the list of large Asiatic lakes.

4. *Main Political Divisions.*

While Europe may geographically be described as a dependency of Asia, politically Asia may almost be regarded as a dependency of Europe. Notwithstanding its vast extent and enormous population, this continent has very few independent States, and of these not one can be said to be entirely independent of European influences. The whole of the northern division, comprising nearly one-third of the mainland, may be regarded as practically a mere extension of European Russia eastwards to the Pacific seaboard. In the south the British Queen and Empress of India (" Kaisar-i-Hind ") is either the absolute sovereign or the suzerain of the Indian peninsula, together with the whole of Burma in Farther India, of the Malay Peninsula, of Ceylon, and most of the islands scattered over the Indian Ocean, besides possessing either treaty rights or a political status in Baluchistan and Afghanistan, which bring a large portion of the Iranian tableland within the British political system. In the west, the Muhammadan world, embracing the rest of Irania, Anatolia, Syria, and Arabia, is divided between the Turkish Sultan and the Shah of Persia. The Sultan has, by the Anglo-Turkish Convention of 1878, practically accepted the protectorate of England for his Asiatic possessions, while the Shah remains much under the influence of England and of Russia. Lastly, in the east the Buddhist world is divided between China, Japan, Siam, England, and France. Here also European influences are in many respects predominant. The "Middle Kingdom" has opened its ports to the trade of the world in virtue of treaties concluded after the close of military operations. Japan also has adopted the culture of the West, perhaps with the view of preserving its political independence. Lastly, Farther India, Tonking, Annam, Cochin-China, and Cambodia, together with a large slice of Siam, are directly controlled by France; the rest of Siam owes a precarious independence to British and French rivalries; Burma has been absorbed in the Anglo-Indian empire, and the petty Moslem States of the Malay Peninsula are distributed between Siam and Great Britain.

Thus we behold the Asiatic world mapped out into four political regions, which roughly correspond to four main natural divisions, and even to the four religious systems predominant in this quarter of the globe. The Russian possessions in the north, comprising Siberia, Caucasia, Western Turkestan, and part of Manchuria, have mainly an Arctic and inland drainage through the Ob,

Yenisei, Lena, Jaxartes, and Oxus, and here is the original home of Shamanism. In the west, still held by the two effete Moslem powers of Turkey and Persia, the drainage is chiefly through the Shat-el-Arab, the Orontes, the Kizil-Irmak, and other Anatolian streams, to the south-western landlocked basins of the Euxine, Mediterranean, and Persian Gulf. The southern, or British division, drains almost exclusively to the Indian Ocean through the Indus, Ganges, Brahmaputra, Narbada, Godavari, Kistna, and other streams of the Deccan, and here Brahmanism is the prevailing form of belief. Lastly, the Buddhist world, occupying the whole of the east, and comprising the Chinese Empire, Japan, and most of Farther India, drains partly through the Tarim to the inland basin of the Lob-nor, but mainly through the Hoang-ho, Yang-tse-kiang, Mekhong, and Menam, to the Pacific Ocean.

5. *Climate: Diminished Moisture—Rainfall.*

Although the great bulk of the land lies within the temperate zone, the climate is essentially continental, that is, characterised by the extremes of heat and cold, and by great dryness. Excluding the three southern peninsulas, which are mainly tropical, and China proper and Japan in the east, and parts of Persia, Syria, and Anatolia in the west, which are mainly temperate, the general climatic conditions are remarkably uniform, notwithstanding the great differences in the relief of the land. Thus, the Aral basin, which in many places is scarcely 200 feet above sea-level, and the Tibetan tableland, which is nowhere less than 10,000 feet, and occasionally attains an altitude of over 17,000 feet, are both subject to the same intense heat and long droughts in summer, followed in winter by almost equally intense cold. On the whole,

no longer reach the Oxus or the Caspian, and the map of East Persia is scored with many watercourses which seem to run nowhere, but which formerly combined to fertilise the now arid wastes of Kerman and Khorasan.

But while the inland and south-western plateaux are amongst the driest, the great southern and south-eastern peninsulas are perhaps the wettest regions on the globe. Over one-half of the total annual rainfall is said to be absorbed by India, Indo-China, and the neighbouring archipelagoes of the Philippines and Malaysia. The coasts of Malabar and Burma are deluged by the summer monsoons, which also discharge tremendous downpours on the advanced ramparts of the Himalayas. At the head of the Bay of Bengal the moisture-charged clouds from the Indian Ocean are almost completely arrested by the lofty ranges enclosing the lower Brahmaputra basin, and the annual rainfall, varying in the Indian peninsula from 240 to 480 inches, amounts on the Assam highlands in some years to no less than 600 inches. Hence arise the striking contrasts everywhere presented by the climate, flora, and fauna of the north Indian lowlands to those of the neighbouring Tibetan tablelands. On one side of the dividing range we have tropical heats, a magnificent southern vegetation, varied animal life, flourishing cities, and teeming populations; on the other Arctic winters, bleak and almost uninhabited steppes, stunted vegetable growths, a fauna restricted to a few hardy upland species. Such a contrast scarcely occurs elsewhere in regions separated from each other by forty or fifty degrees of latitude.

6. *Flora and Fauna.*

Within the vast limits of the Asiatic mainland, which almost touches the equator at Cape Romania, and advances to within twelve degrees of the North Pole at

Cape Chelyuskin, every variety of animal and vegetable life finds a congenial home. While the southern peninsulas abound in tropical and aromatic products, the northern tundras are almost destitute of vegetation. Cereals cease to be cultivated beyond the 62nd parallel; but, on the other hand, a tropical and sub-tropical flora prevails, even in the temperate zone on the eastern seaboard. Here, as well as in India and Indo-China, rice forms the staple food of many hundred millions of human

WILD YAK.

beings, whereas the nomad Kirghiz and Kalmuk tribes of the Mongolian and West Siberian steppes are limited almost exclusively to an animal diet. The tea plant flourishes in Japan, China, Annam, and has in recent years been successfully cultivated in Assam, along the slopes of the Himalayas, and in Ceylon. Coffee, indigenous in Arabia, has also been introduced into Ceylon and the uplands of Southern India. Opium is largely grown in India, and the area of its cultivation is yearly increasing in China. Cotton, indigo, and sugar flourish in the two eastern peninsulas; cinnamon in Annam and Ceylon;

aromatic plants in Arabia. Forest trees occupy, on the whole, a relatively limited area, being restricted mainly to the north coast of the Euxine, Caucasia, the southern shores of the Caspian, India, Indo-China, and the South Siberian uplands. The most useful species are the oak, walnut, pine, cedar, box, poplar, teak, bamboo, coco-nut

OVIS POLI.

and date palms, apricot, peach, and other stone-fruit trees. Central Asia appears to be the true home of the rose, whence it spread in remote times eastwards to China, westwards to Irania, and thence at an early period through Asia Minor to Greece and Italy.

Of the larger animals, the elephant, tiger, buffalo, and bear abound chiefly in India and Farther India; the yak in Tibet; the horse, wild ass, camel, and dromedary in Turkestan, the West Siberian steppes, Irania, and Arabia;

the reindeer in the extreme north and north-east. But the tiger has penetrated north as far as the Altai highlands, and the buffalo is indigenous in China. Characteristic of the central tablelands and Mongolia are the argali, ovis poli, and other large-sized wild sheep and goats, and the hair of the Kashmir and Angora breeds is unequalled for its delicate texture. The sable, civet, marten, blue and silver fox, and other valuable fur-bearing animals are widely diffused throughout Siberia and Manchuria, but are almost everywhere rapidly disappearing before the Russian, Ostiak, Tungus, and other trappers.

7. *Inhabitants: Social Culture—Religions.*

Asia, which is believed by some anthropologists to be the cradle of the human race, is still the home of nearly three-fifths of the inhabitants of the globe. But these teeming multitudes are far from being evenly distributed over its surface. While the bleak plateau of the Great Pamir, the frozen tundra stretching along the Arctic Ocean, the deserts of Gobi and Turkestan, are almost uninhabited, and the greater part of Siberia, Tibet, Persia, and Arabia occupied only by a scanty nomad population, the rich and well-watered alluvial plains of the Ganges, Yang-tse-kiang, and Hoang-ho are amongst the most densely-peopled regions in the world. On the whole the density of the population is in direct ratio to the abundance of the rainfall, and in the southern and eastern lands—India, Indo-China, China, and Japan—which are directly exposed to the moist winds from the Indian and Pacific Oceans, are concentrated over half of the human race. The popular views long entertained regarding the enormous masses of people occupying these countries had often been suspected of exaggeration, but they have been more than confirmed by the results of the official enumerations which are now regularly taken in India and Japan. The estimates for China are still matter of conjecture; but when the latest census for India with Ceylon reveals a total population of considerably over 290,000,000, strength is added to the generally received opinion that the inhabitants of the "Flowery Land" may number from 300,000,000 to 350,000,000, or even 400,000,000.

Almost every variety of physical types, of speech, social culture, and religion, finds representatives amongst the Asiatic peoples. A few specimens of the dark, woolly-haired Negrito stock occur not only in the Andaman Islands, but in the interior of Malacca, and possibly also amongst the low-caste hill tribes of Southern India. In the extreme north-east certain affinities have been traced between the Chukchis and the Eskimo of the North American seaboard. The relations of the neighbouring Yukaghirs, Kamchadales, and Koriaks, as well as of the Ainus and Giliaks of Yesso, Sakhalin, and the Amur delta, have not yet been satisfactorily determined. But, with the exception of these few outlying communities, all the inhabitants of the continent belong to the two great fair and yellow stocks, conventionally known as the Caucasic and Mongolic.

Throughout the historic period the Caucasic peoples have been mainly confined to the south-western region of Caucasia, which gives its name to the type, to Asia Minor, Syria, Arabia, Irania, parts of Central Asia, and Northern India. They are roughly cut off by the Himalayas, the Hindu-Kush, and its western extensions to the Caspian, from the Mongolic, sometimes collectively grouped as the "Turanian" races, which occupy nearly all the rest of the mainland. At the same time there have at all times been frequent crossings and overlappings, especially in Turkestan, Persia, and Asia Minor, which have presented many complicated problems to the student of ethnology in this division of the globe. Mongol and Caucasic tribes seem to be intermingled in the Cochin-Chinese and Yun-nan highlands; many of the low-caste tribes of the Deccan, if not all the Dravidian and Kolarian peoples, must be classed rather with the Mongol than with the Caucasic races; the Akkads, a people apparently of Turanian origin, were the founders of the earliest civilisation in Babylonia, and numerous members of the Mongol family have been so long settled in Irania and Anatolia that they have become largely assimilated in physique to the surrounding Caucasic peoples. On the other hand the Caucasic Tajiks have from the remotest times been settled amid the nomad Mongols of the Aral basin, and, according to Prjevalsky, have penetrated eastwards as far as the dreary shores of Lob-nor.

The various grades of human culture, broadly described as the hunting, pastoral, and agricultural states, depend in Asia rather on soil and climate than on race. Thus the Mongoloid Chinese and Japanese have for ages been settled agricultural peoples, while the Caucasic Arab tribes still remain mostly in the pastoral condition. The Tunguses, a large north-eastern branch of the yellow stock, follow the chase, tend their herds, or till the land, according to their position on the shores of the Arctic, in the Siberian steppes, or along the fertile banks of the Amur. Some of the Turki races also, such as the Usbegs and Osmanli, have formed settled agricultural communities in Bokhara, Khiva, and Asia Minor, while the kindred Kirghiz and Kara-Kirghiz hordes of the West Siberian steppes and the Tian-shan still dwell in tents, and migrate with the seasons between the lowlands and the uplands of Central Asia. But, speaking generally, the hunting and fishing state is confined to a northern zone, reaching from the Frozen Ocean southwards to about the 60th parallel. The nomad pastoral tribes occupy the heart of the continent as far south as the 35th parallel, besides the arid plains of Irania and Arabia. Elsewhere, and especially in Japan, China, India, Indo-China, and Anatolia, the populations have long formed settled and more or less civilised communities on an agricultural basis.

But if social culture is chiefly conditioned by the outward surroundings, religion, on the other hand, is still largely determined by race and nationality. Asia, the original home of monotheism, is also still the land of paganism in some of its crudest aspects, while the originally pure doctrines of Brahma and Buddha alike have almost everywhere degenerated to the grossest polytheism and superstition. Judaism has almost vanished from Palestine, but there are scattered Jewish communities in many parts. Christianity is spreading very gradually through missionary effort, but is at present professed only by the Hellenes of Anatolia, the Maronites of Mount Lebanon, many Syrians, the Armenians, the Georgians and kindred peoples of Caucasia, the so-called "Nestorians" or "Chaldeans" of the Upper Tigris and Lake Urmia, the Eurasians, and the more recently converted communities in India, some 500,000 converts in China, Farther India, and Japan; lastly, the Russians of Siberia and Caucasia, far more numerous than all the rest put together. Most of the hill tribes in India, in Kafiristan, on the Indo-Tibetan frontier, in Farther India, and in the western and southern highlands of China are still pagans. A few survivors of the old Iranian "fire-worshippers" still linger on in Persia; and their descendants, the flourishing tribe of Parsis in India, also follow the religion of Zoroaster. But, with these exceptions, the whole of the Asiatic populations belong either to the Shamanist, the Buddhist, the Brahmanic, or the Muhammadan religious world. Shamanism, openly or thinly disguised, is diffused throughout all the Finno-Tatar tribes of Siberia and Manchuria. Buddhism is the religion of fully two-fifths of mankind, nearly all of Mongol stock, and concentrated mainly in Japan, the Chinese Empire, Farther India, and Ceylon. Hinduism, or, as it should be more exactly termed, Brahmanism, is professed by 208 of the 287 millions of inhabitants of India. Muhammadanism, divided into the two great sects of the Sunnis and the Shiahs, prevails amongst the Tatar peoples of Turkestan, and South-West Siberia, and West China, in Irania, Anatolia, Syria, and Arabia, and has over 57,000,000 adherents in different parts of India. The tenacity with which the Asiatic peoples almost everywhere adhere to their particular forms of belief is curiously illustrated by the Kalmuk and Kirghiz nomad tribes settled side by side in the steppe lands of the Lower Volga. The Mongolian Kalmuks, like their remote kinsmen of Zungaria and Mongolia, are all still Buddhists. But the Kirghiz, like all the other Kirghiz hordes of the Siberian steppes and uplands, are all Muhammadans. In the same way, nearly all the Iranians of pure Persian blood belong to the Shiah sect, while the mixed Tajik communities of Western Turkestan and Afghanistan are almost invariably Sunnis. An exception, however, to this rule is presented by the Aimaks and Hazarahs of the North Afghan highlands, between Herat and Kabul, both of Mongol origin, but the former of whom, like those of Persia, are Sunnis, while the Hazarahs are of the Shiah sect. Besides these two great divisions, Muhammadanism embraces some other communities which are addicted to mysterious rites, and are consequently looked on with suspicion by their neighbours. Some of these, such as the Kizil-Bashis of Anatolia, Persia, and Afghanistan, are doubtless recent developments. But others, like the Druses and Nusarieh of Syria, and the Yezides, or so-called "Devil-worshippers" of Kurdistan, seem to date back from pre-Moslem and even pre-Christian times.

8. *Topography: Chief Towns.*

In reference to the aggregate of its population, Asia is not remarkable for the number of its large towns. A consideration of the present statistics relating to many Asiatic cities of historic renown, within comparatively recent times—such as Bagdad, Isfahan, Shiraz, Tabriz, Samarkand, Pekin, Ormuz, Goa—would disclose a remarkable decline in population, affording melancholy instances of the instability of material greatness. On the other hand, within the last century an equally striking progress is perceptible in many seaport towns. But that has generally arisen under European auspices, or, as it might be more accurately said, under British influence. Calcutta (with Howrah) having close on a million, Bombay considerably over three-quarters of a million, and Madras about half-a-million inhabitants, may be reckoned in the second rank of the cities of the world. Singapore, Penang, and Hong-Kong have also rapidly increased in population. The prosperity of Tientsin, Shanghai, and Tokyo—all cities of the second rank in the world—is largely due to British trade. But of cities in the first rank—like London, Paris, New York, Chicago, and others—almost the only instance now found in Asia is Canton, with an estimated population of 1,800,000.

9. *Highways of Communication.*

Except in India, where nearly 40,000 miles of railway are open to traffic, there are few railways worth mentioning in Asia. By far the most important are those of Caucasia, now continued by the Transcaspian line as far as Samarkand. The Russians have also begun at several points the great trunk line across Siberia; several railways have been opened in Asiatic Turkey, by one of which Jerusalem is connected with the coast at Jaffa, and over 1200 miles of the Japanese system was completed in 1893. Even China, hitherto opposed to railway enterprise, has opened a short mineral line in the north-east, and this is to be continued for strategical purposes into the heart of Manchuria.

India also can show several highways, each many hundred miles in length, which may bear comparison with the roads in Europe. Elsewhere in Asia there are few highways fit to be called such in the European sense of the term. Irrespective of highways, properly so termed, there are very few tracks easily passable in Asia; except in Siberia, there is not one such track traversing the continent from end to end. There is no through road from the British to the Russian dominions in Asia; no road from India to China, or to Tibet, or to Central Asia. The great central plateau, already described, interposes extraordinary obstacles in the way of such communication. The only instance of this nature is in the south. A horseman might, without meeting any real difficulty as regards ground, ride from any part of India through southern Afghanistan to Persia, and thence to the shore of the Euxine. Both China and India have magnificent rivers navigated by small craft. China also has navigable canals of great length. In Mesopotamia the two rivers are the natural highways. But extensive regions in Asia are destitute of water traffic.

Connected with communications is the subject of the electric telegraph. British India is the only Asiatic country which has telegraphic communication between all the principal towns. But some other countries in Asia have one or two through lines. From Constantinople there runs a telegraph line across Asia Minor, then down Mesopotamia to the head of the Persian Gulf. From Tiflis, in Russian territory, another line runs to Tehran, then southwards across Persia to the head of the Gulf. Both these lines are joined to the Indian system by a line along the shore of the Persian Gulf and of Baluchistan. A long line passes from European Russia, near the Ural Mountains, across Siberia to the Pacific Ocean, near the mouth of the Amur. All these land lines are supplemented by submarine cables, one of which runs from Egypt down the Red Sea to Aden, and thence across the Arabian Sea to India. Another has been laid from Madras across the Gulf of Bengal to the Straits of Malacca, and thence turning northwards passes near the Chinese coast to join the Japanese and Russian systems. The introduction of telegraphs is entirely due to the British and Russian Governments, and, in some degree, to the French Government. Japan is the only Asiatic country that has adopted the electric telegraph *ex proprio motu.*

10. *Administration.*

In all Asia, British India and Ceylon alone have an administration completely organised, in the European sense of organisation. The administration in Siberia doubtless approaches this standard so far as may be possible in a country thinly peopled and wild in parts.

In Central Asia—that is, Khokand, Bokhara, and Khiva —civilised principles are being gradually introduced under Russian auspices. The French have established their rule in Cochin-China and Cambodia. The Chinese management of affairs, while evincing an elaborate culture in some respects, is in other respects semi-barbaric. Japan has been remodelling all its institutions after the European example, but whether these multifarious reforms have really taken root is more than the best-informed authorities seem able to pronounce. Both Persia and Asiatic Turkey are unreformed, and have nothing commendable in their administration. Independent Arabia has scarcely an administration in the strict sense of the term; its political organisation is for the most part tribal.

Respecting geography, the administrative results are in this wise:—

"The greater part of Asia has not yet been touched by scientific operations on a complete scale. In the whole of Asia, only India, Ceylon, Cyprus, Western Palestine, Caucasia, the Caspian basin, part of Western Siberia, and part of Japan, also many points in the Asiatic coast-line, have been subjected to trigonometrical observation. The altitudes of mountains have been determined only in the Himalayas, the Caucasus, and the Urals by trigonometry. But in many ranges the heights have been approximately ascertained by the barometer. Professional surveys in detail have been completed only in India, Ceylon, Western Palestine, Caucasia, parts of Western and Eastern Siberia, the Tian-shan region, the greater part of Western Turkestan, Cambodia, parts of Cochin-China, parts of Afghanistan, also on certain lines of Persia, Mesopotamia, and Asia Minor. Even in the professionally surveyed territories many defects and imperfections are acknowledged to remain.

"Though the southern coasts of Asia have been surveyed in sufficient detail for geographical purposes, yet, according to the demands of a growing traffic and of maritime resort, these surveys need frequently to be amplified in detail. The old surveys by the Indian Navy were good in their day, reflecting honour on Moresby, Ross, and others; still the Government have ordered a new survey to be made for nautical purposes. A fresh survey, like that made by Nares for the Gulf of Suez, may have to be ordered one day for the whole of the Red Sea and the Persian Gulf. The British Admiralty are making yearly additions to the surveys of the Chinese coasts, of which the work done by St. John (R.N.) is an example. Whether the Russians will see fit to attempt a scientific survey throughout the Arctic coast of Siberia remains to be seen.

"Of geological surveys, the largest example is that in India, which, though far advanced, is far from complete. Very much remains to be done in this respect for the Himalayas. Geological surveys have been made in the Caucasus, the Urals, the Tian-shan and Altai ranges, Kamchatka, many parts of China and Japan, Cambodia, Ceylon, some parts of Arabia and Persia, much of Asia Minor and Palestine." [1]

In recent years official surveys have been extended to Upper Burma, a great part of Siam, and the Malay Peninsula, to Afghanistan, Baluchistan, and the Aralo-Caspian region. But few of the results have yet been published in accessible form.

11. *Statistics.*

The size of Asia, in comparison with that of the other

[1] Paper read before the British Association for the Advancement of Science, on 2nd September 1881, by Sir Richard Temple.

main divisions of the globe, may be seen by the following statement of the relative areas of the five continents:—

	Square miles.
Asia, including Malaysia	17,300,000
America	16,000,000
Africa	11,800,000
Europe	3,800,000
Australasia	3,400,000
Total	52,300,000

According to Mr. E. G. Ravenstein, Asia contains considerably more than half the population of the world, as shown in the subjoined estimate for the year 1890:—

	Population.	Per sq. mile.
Europe	360,200,000	101
Asia	850,000,000	57
Africa	127,000,000	11
Australasia	4,750,000	1·4
North America	89,250,000	14
South America	36,420,000	5
Polar Regions	300,000	—
	1,467,920,000	29

Of the four great political systems, already described as existing in Asia, the approximate areas and populations are as under:—

	Area in sq. miles.	Population.
Western Asia: Muhammadan States . .	2,200,000	30,000,000
Southern Asia: British and French Political System	2,700,000	300,000,000
Eastern Asia: Buddhist States	5,500,000	500,000,000
Northern Asia: Russian Political System .	6,730,000	20,000,000
	17,130,000	850,000,000

The direct and indirect European possessions in Asia are:—

	Area in sq. miles.	Population.
British Political System[1] .	2,500,000	290,000,000
Asiatic Russia . . .	6,730,000	20,000,000
French Territory . . .	200,000	20,000,000
Portuguese „ . . .	8,000	940,000
	9,438,000	330,940,000

1 Including Afghanistan, Baluchistan, and parts of the Malay Peninsula.

NORTHERN ASIA

CAUCASIA, RUSSIAN TURKESTAN, SIBERIA.

CHAPTER II

CAUCASIA

1. *Boundaries—Extent—Area.*

NORTHERN ASIA forms one vast political system, comprising nearly one-third of the whole continent, and, with a few trifling exceptions, directly administered by Russia. It embraces three distinct geographical regions—Caucasia, Turkestan, and Siberia—which will here be treated under three separate chapters.

Caucasia consists, broadly speaking, of the Ponto-Caspian isthmus—that is, of the narrow neck of land separating the Euxine (Pontus) from the Caspian Sea, and connecting the south-east corner of Europe with South-Western Asia. From its peculiar geographical position and intermingled ethnical elements, this region has been regarded as a sort of neutral or debatable border-land between the two continents. But one marked physical feature seems to be decisive in favour of its claim to be included within the limits of Asia. This is the deep depression of the Manich steppe river, which may be geologically looked upon as a survival of the broad strait formerly connecting the two seas. When flooded in spring by the swollen waters of the Kalaus from the northern slopes of the Caucasus, the eastern branch of this stream still finds its way to the Kuma delta on the Caspian, while the western branch reaches the left bank of the Lower Don at the head of the Sea of Azov.

The Manich depression thus clearly indicates the former direction of the Ponto-Caspian Strait, which in a not very remote geological epoch flowed in a broad channel between the two continents. And as this Ponto-Caspian Strait lay entirely to the north of the Caucasus, it follows that this great highland region belongs physically not to Europe but to Asia. Hence the Manich depression may now be taken as at once the parting-line between the two continents and the northern boundary of Caucasia. Its western and eastern limits are marked by the Euxine and Caspian respectively. But the southern frontier line between the Turkish and Persian States is somewhat irregular, and even arbitrary. The new boundary towards Turkey is so laid down as to include Kars and Batum, as described in vol. ii. ch. i., while the frontier towards Persia follows the windings of the River Aras from Mount Ararat to within a short distance of its junction with the Kura. Here the line is deflected south and east to the Caspian, thus including, in Russian territory, the hilly coast district of Lenkoran, which forms a part of the Iranian plateau.

At its narrowest point, between Poti on the Euxine and Derbent on the Caspian, the Ponto-Caspian isthmus is about 350 miles broad west and east. But from the Strait of Kerch to the mouths of the Kuma the distance is nearly 500 miles in a straight line, and from the low water-parting of the eastern and western Manich to Mount Ararat, on the Perso-Turkish frontier, a similar line will measure 420 miles north and south. But even these dimensions fall considerably short of the actual length of the Great Caucasus, whose axis stretches in an oblique line for 720 miles across the isthmus, from the Taman peninsula, between the Black and Azov Seas, to the Apsheron peninsula in the Caspian. By this central range the whole region is divided into two unequal parts —Cis-Caucasia and Trans-Caucasia, with a joint area of 186,000 square miles.

2. *Relief of the Land: The Great and Little Caucasus—Armenian Plateau—Ararat and Ala-goz.*

The Caucasian region has been during recent years visited by several English travellers—Moore, Grove, Bryce, and especially by Freshfield, who ascended apparently for the first time the summit of Elbruz, and has presented to English readers a charming narrative of his proceedings.

The Caucasus presents in its general outlines one of the best-defined mountain systems in the world. Approached from the northern steppes, it everywhere offers the appearance of an unbroken rocky barrier, rising rapidly from the plains, and surmounted all along the line by a series of magnificent snowy peaks. Southwards, also, it falls everywhere abruptly towards the valleys of the Rion and Kura, which form a nearly continuous trough or depression running from sea to sea between Poti and the Kura delta. This southern depression answers somewhat to that of the Manich on the north, the whole of the intervening highlands constituting the Caucasus proper, or the Great Caucasus. They take the latter name in contradistinction to the Little or Anti-Caucasus, which consists of the spurs rising in confused masses beyond the Rion-Kura depression. The connection between the two systems is effected by the Suram or Mesk range, which forms the Rion and Kura water-parting east of Kutais.

Except at this point the Great Caucasus is thus completely isolated from the southern Lazistan and Armenian highlands. The direction of its main axis, which is continued by a submarine ridge across the Caspian to the Balkan hills, about Krasnovodsk, also shows that the Great Caucasus forms the real north-western continuation of the north Iranian escarpment. This escarpment, which is itself a western continuation of the Hindu-Kush and Paropamisus, broken only by the Tajand (Hari-rud) valley, may now be regarded as stretching under the name of the Kuren-dagh along the northern frontier of Khorasan to within a short distance of the south-eastern shores of the Caspian. Here it ramifies into two branches, one sweeping round the south coast of the Caspian, as the Elburz range, and merging north-west in the Armenian highlands, while the other effects through the Little and Great Balkans and the already-mentioned submarine ridge a junction with the south-eastern extremity of the Great Caucasus in the Apsheron peninsula. We thus see that the Great Caucasus forms the direct continuation of the Central Asiatic systems north-westwards to the Taman peninsula, and beyond it to the South Crimean highlands. The continuity of the whole system is clearly shown by the underground fires, naphtha and oil wells, mud volcanoes, and other still active igneous agencies, occurring at intervals all along the line.

The Great Caucasus bears in many respects a striking resemblance to the Pyrenees. Both run between two marine basins, both are marked by the Sierra formation in their higher crests, and both are divided into two sections of unequal length. But in the Caucasus the break formed by the tremendous fissure of the Dariel Gorge lies almost exactly midway between the two seas.

Through this pass runs the great military highway from Vladikavkaz to Tiflis, the respective capitals of Cis- and Trans-Caucasia. It was also owing to this remarkable geological fault that the Russians were enabled to divide the Caucasus, so to say, into two military zones, preventing any possible combination of the western and eastern tribes, and thus effecting piecemeal the reduction of the whole region.

Most of the eastern section is comprised under the general name of Daghestan, a Turko-Persian compound meaning "Highlands." This region, which is politically included in Trans-Caucasia, is considerably lower and far more irregular than the western section. Here a uniform elevation of 10,000 to 12,000 feet is maintained north of the Rion basin, and above this there tower, besides Elbruz (18,526 feet), several other snowy peaks, all higher than Mont Blanc—the Koshtan-tau (17,096); Shkara, two peaks (17,036 and 16,590); Janga, two peaks (16,569 and 16,527); Kazbek (16,546); Kartan-tau (16,296); Ushba and Aghish, or Adish-tau, each considerably over 16,000 feet. Here also the mean altitude is so great that for 100 miles between the sources of the Kuban and Adai-Kokh (15,274) there are no passes lower than 10,000 feet. Even the Mamisson, about the head of the Rion near the Zikari ridge, is still 9352 feet; but east of this point openings are found from 6000 to 9000 feet high. Mr. Freshfield describes the Mamisson as superior in variety and beauty of scenery even to "the sometimes over-praised Dariel. The road is at first only a steppe track along the foot of the mountains, without bridges over the glacier streams descending from Kazbek and Gumaran Khokh, the peaks of which are in sight. It runs across a sea of wild-flowers, amongst which, in July, the pale mauve hues of the mallows predominate, for two stages to Alagyr—a large Ossete village, where a numerous population nestles in wattled huts shaded by orchard trees, and half hidden after midsummer by a luxuriant growth of sunflowers and other vegetables. At Alagyr a road in the western sense begins. It immediately enters a limestone gorge, where the crags are fringed

MOUNT ELBRUZ.

with ferns, and clothed in hanging woods of elm, lime, maple, oak, and alder."

West of Elbruz the western section assumes the character of a coast range skirting the Black Sea from the mouth of the Rion to the Taman peninsula. For some distance beyond Elbruz it retains a great altitude, with snowy peaks such as the Marukh, the Juman-tau, and the Oshten, rising far above the snow-line. On the coast the incline is also continued for a great depth below the surface, where depths of over 12,000 feet occur close in shore. But beyond Pitzunda the coast range falls rapidly towards the Idokopaz hills near the port of Novo-Rossiisk, after which the chain merges through a few low scattered hills in the alluvial plains of the Kuban delta.

The eastern section, or Daghestan, although crowned by no peaks equal to those of the Western Caucasus, exceeds it not only in breadth but also in its mean elevation. From Mount Borbalo (11,100 feet) at its western extremity, the main range and the Andi ridge diverge right and left towards the Apsheron peninsula and the Terek delta, thus enclosing, with the sea for a base, the triangular space entirely occupied by the irregular masses and upland valleys of Daghestan.

This intricate highland system, which still awaits thorough exploration, culminates with the Tebulos or Shebulos-mta (14,930 feet) in the Andi range, which also contains several other " mta " or peaks, such as the Kachu and Diklos-mta, considerably over 14,000 feet. The highest elevations are the Basarjusi (14,722), the Kumito-tau (14,140), the Tuga-mta (13,940), the Addala-shukchol (13,780), the Bogos-mir (13,610), the Belenki-mir (13,520), the Sari-dagh (12,160), and the Baba-dagh (12,080). Within the triangular space two crests, the Shah-dagh and the Shalbuz-dagh, exceed 14,000 feet, lying towards its south-eastern extremity, where it falls down to the hills of the Baku district in the Apsheron peninsula.

These romantic highlands were visited in 1892 by Herr Merzbacher for the purpose of surveying the lofty and somewhat isolated Tebulos and Bogos groups, between Daghestan proper and the southern territories of the Chevsur and Tush (Tushet) tribes, and extending our knowledge of the Bogos range which traverses the heart of Daghestan, but which had previously been explored by Dr. Radde only as far as the snow-line. Herr Merzbacher speaks in glowing terms of the magnificent scenery of the whole of this comparatively little known alpine region. "The characteristics of this Tebulos group are the jaggedness of the ridges and sharpness of the ice-crests. In the Donos group the variety and grandeur of the mountain forms exceed even those of the Tyrolese Dolomites. In the Bogos (Bochok) group, on the other hand, the extent and magnificence of the glaciers is the chief feature. In this respect they can compare with many parts of the Central Caucasus, as the inferior height of the peaks is compensated for by the low elevation to which the glaciers descend in the valleys. Nowhere have I seen more beautiful valleys, and I think my photographs will prove this. The mountains of Daghestan conceal in their recesses scenery the magnificence of which is still wholly unknown to the world, and beyond my powers of description. The character of the landscape differs entirely from that of the Central Caucasus and the Alps" (*Geo. Jour.*, January 1893, p. 63). The highest peak ascended by this explorer was the Tebulos-mta, which approaches 15,000 feet, and which appears to be the culminating point of Daghestan, unless it is overtopped by Basarjusi.

Although Dr. Radde had already visited the district, little was known of the Basarjusi and surrounding heights before the year 1890, when the group was explored by Mr. G. P. Baker and Mr. George Yeld, who followed the route from Nukha in the Kur basin over the Salawat Pass (9283 feet) eastwards to another pass 9250 feet high at the source of a head-stream of the Achti-chai tributary of the Samur, a Caspian coast stream. From this pass a superb panoramic view was commanded of all the surrounding heights—the twin-peaked Shalbuz (13,679) in the north, the tremendous Shakh-dagh mass (13,751) in the east, Messent in the south, and in the south-east the culminating ice-clad peaks of Basarjusi (14,722), said by some to be the highest in Daghestan. Both Shalbuz, legendary home of the Roc of the Arabian Nights, and Basarjusi, as well as the neighbouring Kishin-dagh (12,500), were ascended, Basarjusi for the first time, and the whole district carefully mapped. It lies 190 miles east by south of the Dariel gorge, and 60 miles from the Caspian Sea, which was faintly visible from the south peak of Shalbuz. The explorers returned to Nukha by a southern route, traversing the Bum district in the Kur basin, the vegetation of which is described as "almost tropical: at first fruit-trees, then a paradise of blackberries, shrubs with their brilliant bunches of red berries, and the tamarisk also in fruit; then majestic plane and chestnut trees of extraordinary girth, while bunches of purple grapes hung side by side with walnuts and chestnuts, and from the topmost boughs of the poplars drooped, in sweeping festoons, the same graceful tendrils of the vine, waving softly above our heads their luscious burdens" (G. P. Baker, *Geo. Proc.*, 1891, p. 327). The district is thinly inhabited by picturesque-looking Lesghians, "a wild and uncouth race, their shaggy black sheepskin busbies overhanging their eyes, and their large skin cloaks making them look most ferocious."

Few alpine regions present more numerous or extensive ice-fields than the Caucasus, where a few years ago scarcely any glaciers were supposed to exist. In Suanetia alone Mr. Freshfield enumerates the Kalde and Adish, both 4½ miles long and descending to nearly 7000 feet above the sea; the great Zanner, nearly 7 miles long, covering with its tributary from Tetnuld 20 square miles; the Thuber, 5½ miles long; the Leksur

(Gvalda), 8 or 9 miles long, and with its affluents covering over 18 square miles; the Salaam descending from Ushba down to 5180 feet, "the lowest level reached by ice in the Caucasus; it is seen like a silver staircase from the Latpari Pass, with Elbruz just visible beyond. We shall hardly be far wrong in setting down 100 square miles of ice and snow as the amount drained by the Ingur alone" (*Geo. Proc.*, 1892, p. 112).

The Adai Khokh heights are encircled by a magnificent series of ice-fields, such as the Ceja, Dargom, Songuta, Skatikom, Karagom, and Burjula, which flow down the flanks of this central group in all directions, feeding the head-streams of the Rion, Ardon, Urukh, and other rivers. Still more remarkable for its glacial streams is the Nalchik district in the province of Terek, where occur the Ulu-Chiran (Bezingi) glacier, largest in the Caucasus, nearly 11½ miles long, from 580 to 1170 yards wide and 214 feet thick at its lower end, 6538 feet above the sea. Here are also the Ulu-auz, nearly 5 miles long, and the Mishirgi-chiran, which at its lowest extremity (7422 feet) is no less than 425 feet thick, and which, like the Ulu-chiran, is a feeder of the Cherek Shkara River. Most of the Central Caucasus glaciers have been explored by Mr. Freshfield, who has spent many consecutive summers in the survey of this alpine region.

The northern and southern slopes of the Caucasus highlands differ greatly in their general aspect. The descent towards the Rion and Kura is everywhere far more abrupt than towards the Manich depression. Here the fall is broken first by a succession of nearly parallel ridges, and then by a series of upland limestone terraces sloping gently towards the steppe, but often presenting nearly vertical walls 3000 feet high towards the central range. This range consists mainly of crystalline schists, and it is remarkable that the same formation prevails in the transverse Mesk ridge, connecting the Great with the Little Caucasus. On both sides of the higher schists the chief rocks are eocene and other old limestones, which disappear northwards beneath the pliocene formations of the steppes.

Porphyries and other igneous rocks abound in the higher regions, where Elbruz was probably a still active volcano down to the close of the tertiary period, during which the Euxine and Caspian were connected by the Manich Strait. Its rival, the mighty Kazbek, overlooking the Dariel Gorge, together with the more northern crests, is also of volcanic origin. Underground forces are even still at work, not only in the mud volcanoes and slumbering fires at both ends of the range, but also in the numerous hot springs and naphtha wells which occur on both sides, but especially in the Terek and Kura valleys about Vladikavkaz and Tiflis. The Lower Kura and Aras valleys are, moreover, still subject to violent earthquakes, while clear traces of continuous upheaval on the Euxine coast, and of oscillations of level on the Caspian side, are visible at Sukhum-Kaleh, about Baku and elsewhere. On the other hand, there is a remarkable absence of large waterfalls and alpine lakes; the great reservoirs, which formerly studded the plains on both sides, having been drained since the glacial epoch.

The Little or Anti-Caucasus presents in almost every respect the most decided contrast to the great northern barrier. Instead of one sharply-defined system, rising somewhat rapidly from the plains, and with a single main axis running throughout in a given direction, we have here rather a rugged plateau formation, intersected by irregular masses, with axes running in all directions. So ill-defined is the whole system that it has no natural southern limits at all. It rises abruptly from the Rion and Kura valleys, but towards the south merges everywhere imperceptibly in the Lazistan and Armenian highlands. Hence the Little Caucasus forms the true north-western scarp of the Iranian tableland, with which the Great Caucasus is connected only by the narrow Mesk ridge intersecting the Rion-Kura valley.

But although narrow, the connection is complete; for the Mesk ridge, which maintains an elevation of 8000 feet, is continued south-westwards by the Ajara or Akhaltzikh range into Lazistan. This range, which skirts the Black Sea within a mile of the coast, rises gradually towards the Turkish frontier, where it culminates with the Karch-shall (11,410 feet), south-east of Batum.

The Ajara range is separated eastwards by the Upper Kura valley from the Akhalkalaki plateau, which is limited southwards by the Kars-chai valley beyond Lake Chaldir. This extremely irregular plateau, which has a mean elevation of 8000 feet, thus forms the water-parting between the Upper Kura and Aras basins. It is a rugged, bleak region, which seems to have been formerly flooded by an extensive lacustrine basin, of which the sole remnants are the Chaldir and other smaller lakes, draining some to the Kura, some to the Aras, while others are mere brackish tarns or marshes, without any outflow. Eastwards the plateau is limited by a double line of volcanic peaks, culminating in Mount Samsar (11,115 feet), with a crater nearly two miles long; and the Great and Little Abul (11,000), with their two cones springing, like those of Ararat, from a common base.

Southwards the Akhalkalaki plateau passes through rough transitions to the desolate tableland of Erivan and Kars, which has a mean elevation of nearly 3500 feet above the sea, and which continues to form the parting line between the Kura and Aras north and south. Here the great features are the basin of Lake Gok-cha and the mighty Ala-goz, lying about midway between the lake and the Russian fortress of Kars. Farther south the plateau has been politically extended across the Aras valley to include Ararat, which lies on the borders of the Russian, Turkish, and Persian frontiers.

Ararat and Ala-goz.

Although thus rising in apparently isolated grandeur at the converging point of three empires, Ararat really forms the eastern and culminating point of the volcanic range which here forms the water-parting between the Aras and Murad-chai, or eastern branch of the Euphrates. Its seemingly isolated position and imposing appearance have from the remotest times encircled it with a mysterious halo of legends and traditions. A number of places in the vicinity still betray in their very names the traces of the Noachian tradition. Thus the village of Aghurri lying on its slope means "he planted the vine"; Nakhichevan, the spot where the patriarch is said to have reached the valley, is interpreted, "here he first descended"; and Erivan itself indicates the place where he permanently settled. His grave is shown at Nakhichevan, and is held in great veneration both by the Armenians and Tatars. But while these Biblical traditions survived, the Biblical name itself of the mountain was forgotten in the neighbourhood. The Armenians know it only as the Masis Lern—that is, the "Grand or Sublime Mountain"—and the Tatars and Turks as the Agri-dagh, or "Steep Mountain," the Persians alone calling it the Koh-i-nuh, or "Noah's Mount."

Viewed from Nakhichevan, Ararat presents the appearance of a single cone bounding the horizon towards the north-west. But it really consists of two separate cones known as the Great and Little Ararat, resting on a common base, and separated by a deep intervening de-

MOUNT ARARAT.

pression. The higher cone consists itself of a double peak, and the whole mass, with its projecting spurs, covers a space of over 370 square miles between Erivan and the frontier Turkish town of Bayazid.

The Armenians have a firm conviction that the summit is altogether inaccessible, and received with absolute incredulity the statement that Parrot had for the first time succeeded in scaling the highest peak in 1829. Since then the exploit has been repeated, amongst others, by H. F. B. Lynch, his cousin Captain Lynch, and his Swiss guide Taugwalder, in September 1893.

On the northern slope there is a vast chasm where formerly stood the Convent of St. James and the village of Aghurri, both of which were overwhelmed by a terrific earthquake in the year 1840.[1] The upper portion of the chasm is filled by one glacier, while another glacier occupies a narrow channel towards the north-east.

Owing to the slight moisture there are no large forest trees on Ararat, which, however, is clothed with vegetation to an altitude of over 11,000 feet. Pasturage extends thence to nearly 13,000 feet, beyond which an alpine flora struggles up to 14,200, which marks the snow-line. Its fauna is also very poor, including on the higher grounds little beyond a mountain goat, a species of hare, and the polecat.

The chief elevations are—Summit of Great Ararat, 16,916 feet; Little Ararat, 12,840; connecting ridge, 8780; Upper Aras Valley, 2800. This valley, in which are situated Erivan and Echmiadzin, political and spiritual capitals of Armenia, separates Ararat from its northern rival, the magnificent Ala-goz, a truncated volcanic cone 13,436 feet high. With its advanced spurs, the Ala-goz,

[1] Some attribute the catastrophe to a landslip or avalanche, some to an earthquake, and others to the reopening of an old crater above the Convent of St. James.

or "Motley Mountain," so named from the various colours of its pumice, scoriæ, obsidian rocks, and foliage, covers a wider area than its southern rival.

3. *Hydrography: The Kalaus, Terek, Kuma, Ingur, Rion, Kura, and Aras Rivers—Lake Gok-cha.*

The Great Caucasus forms a clearly-defined water-parting between the Terek, Kuma, Kalaus, and Kuban, flowing from its northern slopes towards the former Ponto-Caspian Manich Strait and the Ingur, Rion, and Kura, draining from its southern slopes to the Black and Caspian Seas. South of the Rion-Kura depression the hydrographic system is far more intricate, comprising, besides an inland drainage, represented chiefly by Lake Gok-cha, the farthest sources of the Aras, Euphrates, and Chorukh, which flow east to the Caspian, south to the Persian Gulf, and north-west to the Euxine.

Since the disappearance of the Ponto-Caspian Strait, the Terek and Kuma find their way eastwards to the Caspian, the Kuban westwards to the Azov and Black Seas, while the Kalaus, a true steppe river, reaches the Manich only when swollen by the melting of the snows in spring. The Kalaus has its farthest head-streams in the advanced spurs of the Caucasus above Stavropol, and joins the Manich exactly at the water-parting between the Euxine and Caspian, 25 feet above sea-level. Its waters are thus divided into two channels, flowing during the floods one through the West Manich to the Azov Sea, the other through the East Manich to the Caspian.

The Terek, rising in a cirque 8000 feet above the sea, at the northern foot of the Kazbek, sweeps round through the Dariel Gorge and by Vladikavkaz northwards nearly to the 44th parallel. Above the Malka, its largest affluent, the discharge is over 17,000 cubic feet per second, and such a quantity of alluvia is washed down that the delta is encroaching on the Caspian at the rate of about 40 yards annually. Fishing hamlets which early in the present century stood on the coast are now 10 or 12 miles from the sea, and Baer asserts that the Terek is contributing even more than the Volga to the filling up of the Caspian. The waters brought down are doubtless considerable; but these are rapidly evaporating, while the sedimentary matter remains continually accumulating.

A combined system of canalisation and drainage has brought several hundred thousand acres under cultivation about the Lower Terek.

The Kuma has its source nearly under the meridian of Elbruz, and pursues a uniform north-easterly course towards the Caspian, which it occasionally enters through several small channels at a point between the Terek and Volga deltas. Although a considerable stream on emerging from the hills, it gradually contracts as it approaches the coast,- the diminished volume being due to evaporation, to the absence of any affluents during a sluggish course of 150 miles through the steppe, and to the irrigation works of the Kalmuk and Tatar stock-breeders on both sides of its banks. From these combined causes the Kuma is often entirely exhausted within 50 or 60 miles of the Caspian. Formerly the discharge was much greater, as indicated by the old channels and dried-up watercourses, some joining the Manich, some reaching the coast at Serebrakovskaya, and all still occasionally flooded.

The Kuban is the only river flowing to the Euxine basin from the northern slopes of the Caucasus. Its farthest head-stream rises on the west side of Elbruz, whence it flows north-west and west to its delta below Yekaterinodar. Here the main branch continues to flow westwards through the Taman peninsula to the Black Sea, while a considerable quantity of water is diverted through several smaller channels northwards to the Sea of Azov. During the spring, summer, and autumn floods the Taman is often swollen to the proportion of a large stream, from 300 to 400 yards wide and over 10 feet deep. But at other times it nowhere exceeds 4 feet, while the Azov channels sometimes run dry. It is ascended during the floods for some miles by the Kerch steamers, but it is permanently navigable only for flat-bottomed craft. The mean discharge is estimated at 40,000 cubic feet per second.

Two important rivers reach the Black Sea from the southern slopes of the Western Caucasus. These are the Ingur and Rion (Phasis), whose basins are completely enclosed north, east, and south by the Great Caucasus, the Mesk, and Ajara ranges. All the head-streams of the Ingur lie on the southern slopes of the Adish-tau and Ushba, two of the highest peaks in the Caucasus, whence it flows through mountain gorges and upland valleys down to the coast near Redut-Kaleh.

The Rion also and its chief tributary the Kvirila rise at great elevations, the former at the Pasis-mta near Mount Garibolo, the latter at the Mamisson Pass (9520 feet). But the Kvirila soon reaches the Mingrelian plains, where it joins the left bank of the Rion below Kutais. The joint stream enters the Euxine at Poti close to the large Palaiostom lake or lagoon with which it was formerly connected. Although now cut off both from the sea and the Rion, its partly marine fauna shows that this now fresh-water lake at one time communicated with the sea, thus forming the "Old Mouth" of the Rion, as still indicated by its Greek name.[1] Formerly navigable for nearly 100 miles, the Rion has now scarcely 2 feet at low water, and even during the floods from January to

[1] From παλαιός, old, and στόμα, mouth.

June is ascended by small craft only for 30 miles from its mouth.

Both the Rion and the Ingur, whose basins comprise the ancient Colchis of the Greeks,[1] flow through romantic upland regions. The magnificent gorges of the Ingur, with their steep granite walls, are 800 to 1000 feet high in some places, and often clothed with a luxuriant sub-tropical vegetation. Their upland valleys are accessible from the Poti-Tiflis railway running up the Rion valley to Kutais in Imeria (Imeritia).

South of the Great Caucasus the Kura (Cyrus) and Aras (Araxes), by far the largest rivers in Caucasia, now flow through one mouth to the Caspian. Yet they belong mainly to two distinct basins; for although both rise in the Armenian highlands, the Aras remains an Armenian river for most of its course, whereas the Kura soon emerges in the Georgian plains and receives all its large affluents from the Great Caucasus. Hence the Aras is historically the Armenian, the Kura the Georgian river pre-eminently, and even within the historic period both reached the Caspian through independent mouths.[2] The mingling of their waters throughout their lower course is comparatively recent.

The Kura has its source in the Kizil-Gyaduk, 10,340 feet above the sea, whence it flows along the east base of the Arsiani and Ajara ranges north-eastwards to about the 42nd parallel, where it receives several small feeders from the east slopes of the Mesk water-parting. It now trends eastward to Mtzkhet and Tiflis, whence it pursues a south-easterly course through a continuously broadening

[1] It is remarkable that the Pasis-mta, or "Pasis Peak," source of the Rion, still preserves the old name *Phasis* (read *P'hasis*), by which the Rion was known to the ancients.

[2] Strabo tells us expressly that in his time the Kura and Aras entered the sea through separate mouths.

valley to the Caspian below the Apsheron peninsula. Above the plains of Tiflis it is almost a mountain torrent, rushing through a succession of wild gorges and rapids, in one of which it descends nearly 750 feet in a distance of 15 miles. Soon after receiving the Yora and Alazan, its great tributaries from the north, it becomes navigable for vessels drawing four feet for a distance of 450 miles through the Karabagh and Mugan steppes to its mouth. Its lower course is one of the most productive fishing grounds in the world, teeming as it does with enormous quantities of sturgeon and white fish.

The discharge of the united Kura and Aras rises from nearly 7000 cubic feet per second in winter to 25,000 in summer. Much of this water might easily be applied to irrigating the now arid but formerly productive Mugan and Karabagh steppes. Like most of the other Caucasian rivers, the Kura is continually encroaching on the sea, its delta having advanced over 50 square miles between the years 1830 and 1860.

The Aras has even a longer and a much more winding course than the Kura. Rising south of Erzerum, at the foot of the Bingol-dagh, it flows for some miles through Turkish territory north-east to the recently-advanced Russian frontier. Here it turns eastwards to the Erivan plain north of Ararat, whence it sweeps in a semicircle mostly between the Russian and Persian empires round to its confluence with the Kura.

Of the three great closed basins of the Armenian highlands, Lake Gok-cha is the smallest. Yet it fills a vast triangular cavity 540 square miles in extent, at an elevation of 6400 feet above the sea on the plateau, nearly midway between the Aras and Kura valleys. It has an extreme depth of 250 feet, and when swollen by the melting snows from the surrounding hills discharges its surplus waters through the Zanga towards the Aras in the Erivan district. On an islet at its north-west extremity lies the historical Convent of Sevan in one of the most desolate spots on the globe. At the opposite end the horizon is bounded by the huge volcanic mass of the extinct Ala-Polarim volcano, whose lava-streams descend in two channels to the edge of the lake.

4. *Natural and Political Divisions: Cis-Caucasia—The Northern Steppes and Slopes of the Caucasus; Trans-Caucasia—Colchis; Georgia; Russian Armenia.*

The old historical divisions of Caucasia—Georgia, Lesghistan, Imeria, Mingrelia, Kabardia, Abkhasia, Circassia—have been completely swept away under the new order of things. They were based almost exclusively on ethnical considerations, and in some instances, as in the case of the Circassians and Abkhasians, the very races themselves have all but disappeared which gave rise to these distinctions. No regard has even been had for the great historic nations of this region, and the ancient kingdoms of Georgia and Armenia are now in official language replaced by the Russian Governments of Tiflis and Erivan.

Our survey must therefore follow the great natural divisions, which are in the north the steppe-lands stretching from the Manich depression to the foot of the hills, and the zone of fertile and inhabited uplands between the steppe and the alpine regions of the main range; in the south the Rion-Kura and the Aras basins.

Travellers approaching the Great Caucasus from the north, after crossing the Manich depression are not for a long time sensible of any marked change of scene. The boundless level steppes of Southern Russia are still continued without any perceptible break south-eastwards far into the Ponto-Caspian isthmus. From the borders of the

Government of Saratov and the Don Kossak domain right away to the advanced spurs of the Great Caucasus the land remains almost perfectly flat, broken only here and there by a few low hills or ridges.

Yet this extensive region, comprising fully half of all Caucasia, is divided into two natural sections, each with its special physical features. First come the lowlands stretching southwards to the Kuban, Malka, and Terek rivers, and following the line of their course seawards. This is a true steppe land, interrupted only here and there by a few deep furrows. It is marked by an almost total absence of timber beyond a few small plantations in the neighbourhood of Stavropol. There is also a great lack of moisture, most of the streams running quite dry in summer, while the lakes in the northern districts are mostly brackish. Here the great evil is the want of water for irrigation purposes. The black loamy soil is naturally highly productive, yielding heavy crops of cereals and rich pasturage whenever the rainfall is sufficiently abundant. But the summers are mostly rainless, the long droughts and great heat reducing the steppe vegetation to a fine dust, which forms dense clouds sometimes covering the whole horizon, and wafted to great distances by the winds.

The second section stretches along the foot of the main range between the Euxine and the Caspian for a distance of over 450 miles, but contracting in some places to a width of from 20 to 30 miles. Here the luxuriant growth of grasses and the genial climate remind the traveller of the Mississippi prairies. The steppe now soon merges in a boundless park-land, bordered southwards by the mighty central range with its frowning granite and basalt crags and glittering glaciers, northwards by broad rivers, east and west by two seas. Yet even here agriculture has been but slightly developed, and apart from a few patches of cultivated land, the whole region is still virgin soil. It produces when tilled magnificent crops, and the grass grows to a height of 5 or 6 feet.

At either extremity of the range are the two remarkable naphtha-producing peninsulas of Taman and Apsheron. Of its other natural products very little is yet known, although the presence of silver, lead, and copper has long been placed beyond doubt. Many of the hillmen cast their bullets from the lead or copper they pick up on the surface, and near Elbruz there is an abundance of common pyrites, which they utilise in the manufacture of their gunpowder. Granites, magnificent green and red porphyries, various-coloured marbles, and rock-crystal exist in great quantities, while extremely copious mineral waters of every description and coal combined with the inexhaustible naphtha springs promise to prove a future source of permanent wealth to the country.

Trans-Caucasia, comprising the Rion-Kura and Aras basins, possesses even greater economic importance than the northern division. Here is the great commercial route between the two seas leading from Poti through Tiflis to Baku. Here is the historic domain of the Georgian nation, by far the most important of all the Caucasian races, and Tiflis the former capital of the Georgian States has naturally been selected as the political centre of the new Russian Government of Caucasia. Here all the streams are perennial, while the Kura-Aras alone discharges a far greater volume than that of all the northern rivers combined. Here are some of the most productive fisheries and one of the finest wine-growing countries in the world. Here amongst other mineral treasures is Mount Kulpi, a prodigious mass of rock-salt in the Upper Aras valley, the salt-mines of which are in some places over 200 feet thick. Although almost continuously worked since prehistoric times, as shown by the implements frequently picked up dating from the stone age, these mines show no sign of exhaustion, and the Armenians have a tradition that Noah drew his supplies from this source. The present average yield is about 16,000 tons yearly, although the workings are carried on in the most primitive fashion.

The Rion basin, the Colchis of the ancients, has been famous from the remotest times for its surprising fertility and resources of every kind. The legendary Argonautic expedition was fabled to have been fitted out by the Greeks to recover from this region the golden fleece, emblem of boundless wealth. It is completely enclosed by an amphitheatre of hills sweeping round from Sukhum-Kaleh to Batum, and now crossed by the Poti-Tiflis railway at the Suram Pass in the Mesk range some 3000 feet above the sea.

This pass leads directly down to the ancient kingdom of Georgia, comprising the greater part of the Kura basin. The depression through which this river flows may be regarded as a dried-up fiord or inlet of the Caspian, which formerly penetrated between the Great Caucasus and the Armenian highlands across the southern portion of the Ponto-Caspian isthmus westwards to the Mesk and Anjara ranges. The lower section of this basin, comprising the Mugan and Karabagh steppes, is now mostly waste land. But the traces of ancient canals, and the ruins of many villages, caravansarais, and even towns, show that it was once highly cultivated and thickly inhabited.[1] It is now visited only by the Tatar nomads, in spring, when the rainfall produces herbage.

[1] Some of these canals were nearly 100 miles long, and on one of them stood the great city of Bilgan, destroyed by Jenghis Khan. When Timur restored the canal Bilgan rose again from its ruins and continued to

Higher up rice was formerly cultivated along the banks of the Kura above the Alazan confluence. But the raids of the neighbouring Lesghian marauders caused the irrigation works to be abandoned, and during the present century the Karayazi steppe between the Kura and Yora has reverted to a state of nature. An attempt, however, has now been made by the construction of the "Mary Canal" to bring this tract once more under cultivation. The recently-executed surveys also show that over 5,000,000 acres in the Lower Kura valley might again be easily rendered productive. Much of the soil consists of a rich black loam, and many of the old canals might be restored and extended at a moderate outlay.

At the same time the whole of this region is notoriously malarious, and farther north the Baku coast district is still subject to violent earthquakes. The centre of the seismic action seems to be the town of Shemakha, which was nearly destroyed in the seventeenth century, and again suffered so severely in 1859 that the administration has since been transferred to Baku.

The Aras basin, comprising that portion of Armenia which is now included in Russian territory, differs in no respect from the Armenian highlands politically belonging to Turkey and Persia. Here also the most salient characteristics are the long ranges intersecting the plateau in every direction, and dividing it into a number of upland arid steppes, sparsely peopled and almost treeless. The hills falling abruptly towards the Rion basin are here and there thinly wooded; but the Traletes and their spurs overlooking the Kura valley often present nothing but bare rocky surfaces for miles together.

Immediately north-west of Alexandrapol rises the

flourish till the close of the seventeenth century. But the subsequent wars with the Daghestan hillmen again caused the works to be abandoned, and Bilgan again disappeared.

extensive Chaldyr plateau. North of it flows the Upper Kura, here intersecting the Armenian frontier hills, and

CONVENT OF ECHMIADZIN.

forming a natural approach to the frontier district of Ardahan between Kars and Batum. The whole of this district also consists of a treeless tableland about

4500 feet above the sea, enclosed north and south by almost inaccessible hills. Southwards stretches the rugged plateau of Kars, east of which the Arpa-chai valley leads down to the Upper Aras, which here flows through the plains of Erivan between Ararat and Ala-goz. This is almost the only comparatively low and well-watered level tract in the whole of the Armenian highlands, and here is, so to say, the focus of the Armenian nation, where are centred all its most hallowed associations. The highway approaching it from the north-west is fringed with the ruins of Ani, Vardzia, and other ancient cities, recalling the former greatness of the land.

But a far more sacred spot is the venerable Convent of Echmiadzin at the southern foot of Ala-goz, residence of the Armenian "Katholicos," who rules with a plenitude of spiritual jurisdiction over the two millions of Gregorian Christians scattered over the continent from the Bosphorus to the Ganges.

5. *Climate: Rainfall.*

No other region of the same extent presents a greater diversity of climate than Caucasia. This is due partly to its peculiar position between two inland seas, at the southern verge of the Russian steppes and at the north-western edge of the Iranian plateau, partly to the extreme deviations in the general relief of the land ranging from the low-lying Mingrelian plains to the Elbruz Peak nearly 19,000 feet above the sea.

Being exposed to the northern winds sweeping over the Russian steppes, Cis-Caucasia is both drier and colder than the southern slopes. Hence many of the rivers here run out during the summer heats before reaching the coast, and are ice-bound in winter: whereas in Trans-

Caucasia all the streams are perennial, and frozen only in exceptionally hard seasons. For analogous reasons the western section, receiving the moist and relatively warm atmospheric currents from the Euxine, enjoys a far higher winter temperature and greater abundance of moisture than the eastern slopes facing the Caspian and arid Turkoman deserts. Here the contrast between the rainless and sultry Mugan and Karabagh steppes of the Lower Kura and the moist and moderately hot Rion basin is very striking. In general, the rainfall is three times heavier on the western slopes than on the Central Caucasus, and from eight to ten times more copious than on the east side of the Daghestan ranges. So little moisture is brought from the Caspian that at times no rain falls for six months together in the Lower Kura basin.

Although the extremes of heat and cold are much greater in the Caucasus than in the Alps and Pyrenees, the mean annual temperature is much the same in all these highland regions. Thus, while Caucasia and Switzerland have a common mean, the temperature varies in the latter about 18°, in the former as much as 25° or 26°, between winter and summer. Hence in its extremes the Caucasian climate resembles the Asiatic, in its general mean the European, so that the region is a land of transition in its climatic as well as in other features.

6. *Flora and Fauna.*

This transitional character is especially conspicuous in its vegetable and animal kingdoms. In some respects Caucasia seems to be a land of dispersion, where certain vegetable species, such as the peach, apricot, cherry, and other stone-fruit trees became differentiated, and thence distributed east and west over the two continents. Plants of this sort are found in such variety and abundance on both sides of the main range, but especially in the Rion basin, that they may be regarded as the typical vegetable order of this region.

The southern limits of Trans-Caucasia, lying under the same parallels as Central Italy, are the natural home of the laurel, orange, citron, vine, and mulberry. The vine arrives at great perfection, especially in the Georgian province of Kakhetia, which is famous for its fiery vintages. The plant has in recent years suffered from the ravages of the oidium; but large quantities are still produced of a very full-bodied wine, which is now largely used for improving the flavour of inferior sorts. The vine, like the stone-fruits, is probably indigenous in Caucasia.

Heavy crops of rice, maize, wheat, and other cereals of excellent quality are raised on all the lowland tracts on both sides of the Great Caucasus, wherever water can be obtained in sufficient abundance. But the abandonment of the old irrigation works, especially in the Middle and Lower Kura valley, and the increasing dryness on the north side, have reduced to barren wastes extensive districts where these crops were formerly widely cultivated. The fruits of Southern Russia, such as the pear, plum, cherry, and walnut, flourish on the northern slopes and along the banks of the Terek, Kuma, and Kuban.

In the profound and precipitous mountain gorges of the central range, where a solitary sunbeam seldom penetrates, not a blade of grass will grow. But emerging from these abysses, we sometimes fancy ourselves transported to the Alpine valleys of Switzerland, with their luxuriant pastures, rich woodlands, and foaming mountain torrents. Here the forest zone stretches along both sides of the Great Caucasus for a distance of 500 miles, with a breadth varying from 10 to 20 miles. On the heights grow the maple, lime, ash, fir, pine, beech, and larch; farther down the oak, chestnut, several species of poplar, the plantain, box, and walnut. In the valleys thrive most southern fruits, as well as the loveliest flowering shrubs; and in the more favoured spots the cotton and olive. Conspicuous amongst the flowering shrubs is the *Azalea pontica,* one of the glories of the vegetable world, rivalling the Himalayan rhododendron in the richness, variety, and splendour of its blossom. It reaches an elevation of 6000 feet, where the deep-red autumn tints of its foliage offer a surprising contrast to the sombre green of the surrounding conifers. A species of tea grows wild on the southern slopes of the Mingrelian highlands, and on the Caspian seaboard the Tatars raise crops of madder and saffron. In the hot moisture-charged atmosphere of Abkhasia and Mingrelia the vegetation is marvellously luxuriant; but man and nature alike are here still in a wild state. Wheat and rice no doubt flourish in the valleys, but the natives themselves grow nothing but millet, some barley, and maize. As a rule, all these cereals reach a higher elevation than in the Alps. Barley is cultivated by the Osses in the central ranges up to 8000 feet, while wheat thrives at 6500, and maize at 3000, in all the sheltered southern valleys.

The Russian botanist, N. Alboff, confirms the remarks of previous observers on the general resemblance of the West Caucasian flora to that of the tertiary period. The flowering plants present a great variety of forms, no less than thirteen species of evergreens, six conifers, and eighty-six deciduous trees and shrubs occurring in the Abkhasian forests alone. Here climbing plants prevail at the lower levels, and a striking feature is the extremely wide vertical spread of certain forest growths, most of which have a range of from 3000 to 4000 feet. The beech, walnut, *Ilex aquifolium,* azalea, rhododendron and several others extend from sea-level to about 7000 feet, which is the highest limit of arborescent vegetation in this region. At this great altitude they are found associated with the birch, mountain-ash, *Quercus pontica, Rubus Idæus, Rhanus alpina, Rosa mollis,* and other hardy alpine shrubs. In the Abkhasian forests about twenty-five species of ferns have been collected (*Memoirs of the St. Petersburg Society of Naturalists,* xxiii. 1893).

The flora of Russian Armenia, a land apparently of diminished moisture, is incomparably poorer than that of Caucasia proper. The few forest tracts consist chiefly of oaks, beeches, aspens, and especially poplars. But here the characteristic plant is the curious nölbönd, a magnificent species of elm, with enormous leafy branches, through which the solar rays never penetrate. This highly-ornamental tree is absolutely unknown beyond the limits of the Aras basin. Of cultivated plants the chief, besides cereals, including rice, are the apricot, cotton, sesame, and the vine, which in some places yields a highly-flavoured wine, somewhat like Madeira.

The Caucasian fauna, like that of similar highland regions elsewhere, is far less varied than its flora. The tiger sometimes ventures across the Persian frontier, and the leopard and hyena are also met in the Lower Kura and Aras basins. The lowland thickets and sedgy river-banks are the favoured haunts of the wild boar, and the Abkhasian and Mingrelian forests are still infested by the panther, wolf, lynx, and bear. The Caucasian bear, not a very formidable species, reaches no higher than about 5000 feet, above which a few herds of the bison or wisant, wrongly identified with the aurochs, still linger in the upland forests on the slopes of Elbruz. Still higher up the chamois and *túr,* a species of ibex, frequent the alpine valleys along the central range. A wider range is enjoyed by the marten, blue fox, squirrel, hare, fish-otter, and some other wild animals of smaller size. On the whole, game is abundant, but is found chiefly in the low-lying and unhealthy wooded tracts along the northern slopes of the Caucasus (Clive Phillips-Wolley).

Pre-eminent amongst the domestic animals are the horned cattle of the Ingur and Rion basins. Here there are two fine breeds, one small and active, the other of magnificent size and symmetrical proportions, sprung originally from Ukranian stock. The horses, mules, asses, goats, and other domestic animals of this region are all alike noted for their fine proportions and good qualities.

In general the Great Caucasus may be regarded as a parting line between the European and Asiatic vegetable and animal kingdoms, the Cis-Caucasian flora and fauna being more allied to the western, those of Trans-Caucasia to the eastern continent.

7. *Inhabitants: Varied Ethnical and Linguistic Elements —Tabulated Scheme of the Caucasian Aborigines—The Georgians, Mingrelians, Imerians, Circassians, Abkhasians, Chechenzes, Lesghians, Osses; Non-Caucasian Intruding Races.*

Caucasia is inhabited by a highland population, comprising a multiplicity of distinct ethnical elements elsewhere almost without a parallel. The most varied tribes, speaking fundamentally distinct languages, here dwell in the closest proximity, hemmed in on the north by the Russian Slavs, southwards by the Armenian, Kurdish, and Persian Iranians. But in comparatively recent times all these, besides the Tatars and other alien races, have penetrated into the Kura, Terek, Kuma, and Kuban basins.

The popular view is, that we have in the Caucasus

the remnants or fragments of the peoples who have from time to time been driven into these recesses from the surrounding lands, or who have passed through these highlands during the ceaseless flow of prehistoric and subsequent migration from Asia to Europe.

But this view was combated by Professor Virchow at the Archæological Congress held at Tiflis in the autumn of 1881. Here it will suffice to remark that the Caucasus could not have been a highway of migration when the ice-fields descended much lower than at present, and that the numerous Caucasian languages, with the single exception of the Ossetian, have no kind of affinity with those elsewhere current.

Partly on geographical, but mainly on linguistic grounds, all the indigenous Caucasian races are here grouped in four great divisions as under:—

I. Southern Division.

(*Kartveli Stock.*)

Georgian	East of Mesk range to Tiflis district	1,150,000
Imerian, Rachan	Imeria (Imeritia)	
Mingrelian, Gurian, Lechgum	Mingrelia	
Laz	Lazistan	
Svan	Upper Ingur and Tskhenis valleys	
Pshav, Khevsur	Sources of Alazan and Yora	

II. Western Division.

Cherkess	Ubych, Shapsuch, Dshiget	Left bank Kuban	188,000
Abkhasian		Coast of Euxine, N. of Ingur River	
Kabard		N. and E. of Elbruz	

III. Eastern Division.

Chechenz	Ingush, Galgai, Kist, Tush, Karabulak	Right bank Upper and Middle Terek	170,000

Lesghian	Avar, Kazi-Kumyksh, Andi, Dargo, Dido, Duodez, Ude, Kubachi, Kurini	Daghestan	540,000

IV. Central Division.

Oss or Ossetian	Both slopes of Great Caucasus about Kazbek	127,000

The Georgians, Mingrelians, and Imerians.

None of the Caucasian people except the Georgians possess any historic importance. Direct descendants of the old Iberians,[1] they still form the bulk of the population in the Governments of Tiflis and Kutais, and although essentially a lowland race, bear a marked resemblance to the Imerians and other highland members of the Kartvelian (Kartalinian) family. The term Grusia or Georgia[2] does not occur till mediæval times, when it came into use after the north-eastern division of Kakhetia became detached from the old Kartvelian kingdom. Since the annexation to Russia this old national name, traditionally traced to a Kartlos, grandson of Noah, has again come into favour.

The several branches of the Kartvelian stock afford a striking illustration of the often-repeated remark, that the less favoured the land the more industrious and intelligent are its inhabitants. The magnificent race occupying Lower Mingrelia, who would be amply rewarded by the labour bestowed on their fertile soil, are a hopelessly

[1] The present type and even the head-dress are absolutely identical with those of the statuettes found in the numerous graves dating from classic times scattered over the land.

[2] The name *Georgia*, of which *Grusia* is merely the Russian form, has been wrongly referred to the national saint, George, by whom they were supposed to have been converted from Paganism to Christianity.

indolent and poverty-stricken people. Allowance should doubtless be made for the enervating climate ; but everywhere the southern are far outstripped in energy and enterprise by the northern tribes.

The best conditions of existence are found at about an elevation of 4000 feet above sea-level. Here the vine still flourishes, sericulture is possible, maize and millet yield good returns, and wheat prevails on heavy soils. Here industry meets with a fair reward. The natives, although not grouped together in large centres of population, are still found in more compact masses than in the Mingrelian lowlands, where the agricultural villages are replaced by solitary farmsteads. Here rice, cotton, and sub-tropical fruits might easily be cultivated ; but nothing is done, and the people remain poor. Much more prosperous are their northern kinsmen, who are compelled by the less favourable conditions to work for their living, and are not free from anxiety for the winter season. Their dwellings are also more substantially built, their cattle require to be housed, the vine must be carefully dressed and pruned, the silkworm needs constant attention, the wooded slopes must be cleared, the ground requires hoe and spade to provide sufficient for the sustenance of the family. Hence the Imerian is a better agriculturist than the lowland Mingrelian, and the higher we ascend the more industrious become the kindred tribes.

Although speaking one of the harshest languages in Caucasia, where a surprisingly harsh phonetic system is the rule, the Georgian race is distinguished by a passionate love of song and music. In the home, in the tavern, of which they are unfortunately constant visitors, in the market-place, at all their feasts and social gatherings, the Kartvelians are perpetually singing or shouting to the accompaniment of their tambourines, their balalaïkas, and other stringed instruments. Even their daily occupations and their field operations are relieved by a concert of voices, whose cadence is adapted to the movement of their various pursuits.

The Circassians and Abkhasians.

Although their domain has been largely encroached upon by Tatars from the east, Armenians from the south, and Slavs from the north, the Georgians still constitute the most compact and homogeneous nationality in Caucasia. But more typical representatives of the Caucasian races are, or rather were, the Cherkesses or Circassians, formerly the most powerful and warlike of all the Western nations. Their domain seems to have at one time extended round the Euxine seaboard, as far as the Strait of Kerch. But they were for centuries confined by the advancing Little Russians to the left or southern bank of the Kuban, and now since their final reduction in 1864, after a heroic resistance maintained for generations, nearly all their lands have been occupied by the Great Russians. A few scattered groups still cling to their ancient homes along

A CIRCASSIAN.

the course of the Kuban and its affluents; but the great bulk of the nation withdrew after the conquest into Turkish territory, and isolated Cherkess communities are now found dispersed over Armenia, Asia Minor, Syria, and the Balkan peninsula. Here they have acquired an unenviable notoriety for lawless and turbulent habits. But the national character should rather be studied in the mountain fastnesses, where the race was moulded.

A similar fate has overtaken the kindred and neighbouring Abkhasians, of whom 20,000 migrated to Turkey after the late Russian war. Their territory is now reduced to a narrow tract on the coast of the Black Sea, north of the Ingur basin, where they are hemmed in by the Kartvelians, Tatars, and Great Russians. The Tatars, who are here isolated, separate the Abkhasians from their remote kinsmen, the Kabardians of the Central Caucasus.

The eastern division, whose most representative members are the Chechenzes and Lesghian Avars, have also been encroached upon, especially by the Nogai and Kumik Tatars, who have long been settled on the Caspian coast, south of the Terek, and more recently by the Russians, who have penetrated into the Chechenz territory as far south as Vladikavkaz.

The Chechenzes, Lesghians, and Osses.

The Southern (Kartvelian) and Western divisions, amid much physical diversity, are at least characterised by linguistic unity. For the original identity of all the numerous Kartvelian dialects on the one hand, and on the other of the Cherkess, Abkhasian, and Kabardian, is an accepted conclusion of comparative philology. But the utmost ingenuity of specialists, who have devoted a lifetime to the study, has hitherto failed to introduce much order into the Babel of tongues still current amongst the innumerable tribes of the Eastern division. Here there are at least five stock languages, probably more, which can be affiliated neither to each other nor to any other known forms of speech.

Even the Chechenzes, by far the most important nation in the Eastern Caucasus, are split up into some twenty different groups, each with a distinct language. They occupy the whole of West Daghestan, between the Osses and the Avars, and long maintained a hopeless struggle under Khazi-Molla and Shamyl (Samuel) against the Russians. Even more fanatical Muhammadans than the Western Cherkesses, they fought with the dauntless valour inspired by religious enthusiasm and a passionate love of freedom. After their reduction in 1859 large numbers migrated to Turkish Armenia, where most of them perished of want and hardships of every description.

Rivalling the Cherkesses in valour and physical beauty, the Chechenzes surpass them in generosity and self-respect. Their love of finery amidst the squalor of their wretched highland villages is very remarkable. Men and women dressed in rich flowing garments, worn with admirable grace, are often found residing in damp and gloomy underground hovels, or in huts formed of interwoven branches, or of huge stones thrown loosely together.

The Osses or Ossetians, who call themselves Iron, constitute the fourth division of the Caucasian races. But while the other three may be regarded as indigenous, the Osses are certainly intruders of Aryan stock. They occupy the most central part of the Great Caucasus along both slopes of the Kazbek, where they are conterminous with various tribes of all the other divisions—Kartvelians on the south-west and south-east; Chechenzes on the north-east; Kabards on the north-west. Of fair complexion, robust, of a somewhat heavy and sluggish temperament, and lacking the graceful carriage of the other highlanders, these Ossetians seem to resemble the Germans more than any other branch of the Aryan family. Yet their language belongs to the Iranian group, and the national name of Iron has been accepted as an argument in favour of their Persian origin.

Non-Caucasian Intruding Races.

Besides the already-mentioned Russian Slavs, Nogais, and other Tatars, the more important intruding peoples are the Armenians, who have advanced from the Aras to the Kura basin; the Kurds, some of whom have penetrated north to the Rion basin; the Greeks, numerous in the district west of Tiflis; the Tats and Talyshes of the Baku district, akin to the Iranian Tajiks; lastly, a German colony from Wurtemberg settled in a few isolated communities in the Kura basin east of Tiflis.

There is also a very old Jewish element in several parts of Caucasia, but nearly everywhere assimilated in speech and habits to the surrounding peoples. A number of places known by the name of Jut-Kend, or "Jewish Town," are now occupied by communities claiming Tatar descent, and the Jews of the Baku district have adopted the Persian garb and speech. These latter are said to have arrived from Persia during the time of the Sassanides, and are by some writers supposed to be descended from the Israelites who were removed to Persia after the first destruction of the Temple by Salmanazar. This view seems confirmed by the family names still current amongst them, which belong to the period of the Judges, and which have been elsewhere obsolete for over 2000 years.

The prevailing religions in Caucasia are Christianity and Muhammadanism, with almost everywhere a substratum of the old pagan superstitions. Thus the

Khevsurs, who belong to an extremely interesting group of tribes clustered round Mount Borbalo, about the sources of the Aragva, Yora, and Alazan, have developed a form of Christianity of a somewhat peculiar type. Having been followers of the Prophet before they were Christians, and heathens before their conversion to Islam, they keep the Friday with their Moslem neighbours, and the Sunday with the Georgians, intermingling the worship of trees and of the spirits of earth and air with more orthodox rites. Their chief deity seems to be the God of War, and they also do homage to the Mother of the Earth, the Archangel of Property, the Angel of the Oak, and many other lesser gods and angels. Yet they are very proud of their Christianity.

8. *Topography.*

In Western Caucasia there are no towns of any size. Yekaterinodar, capital of the Kuban province above the delta of that river, is an important agricultural centre, much frequented during the autumn fairs, when produce to the amount of over £400,000 is usually disposed of. Yeisk on the Sea of Azov is a thriving seaport and fishing station, and although founded so recently as 1848, has already become the largest place on the whole Caucasian seaboard. Taman, which gives its name to the peninsula, is a mere village, and Sukhum-Kaleh on the Abkhasian coast, possessing a safe and deep harbour, is noted for its dolphin-fishery.

Stavropol and Vladikavkaz.

Stavropol, capital of the government of like name, lies on the verge of the steppe about 2000 feet above the sea, and at some distance to the east of the south-eastern line of railway, whose present terminus is Vladikavkaz.

SUKHUM-KALEH.

This place, which is the capital of the Terek territory, lies almost in the heart of Central Caucasia, at the northern entrance of the Dariel Gorge, through which the great military road leads south to Tiflis. Although since the pacification of Caucasia it has lost its former strategic importance, Vladikavkaz has continued to flourish as a commercial emporium, and its commanding central position, 2300 feet above the sea, midway between the two seas, marks it out as the future capital of Cis-Caucasia. In the whole of this region its only rival is Tiflis, over which it possesses the great advantage of a genial and healthy climate. Piatigorsk, about equidistant from Stavropol and Vladikavkaz, is the chief centre of the Caucasian watering-places, which are nowhere surpassed for variety, copiousness, and health-giving properties. But although supplied with grand hotels, promenades, pleasure-grounds, and other attractions, it has hitherto failed to tempt many visitors from the West.

Derbent and Baku.

On the Caspian coast, Derbent, in lat. 42°, occupies a peculiar position between the spurs of the Daghestan hills and the sea, completely guarding the narrow defile on the great historic route along this coast. It covers a long, narrow strip, enclosed by walls running from the citadel of Narin-Kaleh eastwards to the sea. The line of fortified works is continued over the hills for some distance westwards, and is traditionally supposed to have formerly stretched right across the isthmus from the Caspian to the Euxine. But a much more important place is Baku at the neck of the Apsheron peninsula, centre of the most productive naphtha district in Asia. This trade, which employs quite a little fleet of coasting vessels and steamers on the Caspian, yielded 4,490,000 tons of raw naphtha in 1892. Close to Baku hundreds of wells have been sunk, and the industry has now extended to

DERBENT.

Northern Caucasus, where nearly 27,000 tons were produced in 1890. At Baku is a famous shrine of the Persian fire-worshippers, which is directly fed with inflammable gases from the subterraneous fires.

In the Rion-Kura depression the chief places are the port of Poti, Kutais, and Tiflis, all connected by the line of railway which runs through Tiflis to Baku.

Poti and Tiflis.

In spite of its exposed and shallow roadstead and its pestilential climate, Poti had rapidly progressed before the

TIFLIS.

last Russo-Turkish War. It still retains some importance as a coast station on the railway which has its present terminus at Batum. But since the acquisition of Batum in 1878 its shipping has been greatly reduced.

By far the most important city in Caucasia is the capital, Tiflis, which lies on the left bank of the Kura, a little south of Mtzkhet, the ancient capital of the Georgian kingdom. It is a half-European, half-Asiatic town, consisting of a Russian quarter with some ambitious buildings in the modern style, a clean and picturesque German suburb, and a Persian district with a decidedly Eastern appearance. The most prominent feature of the place is the fine open "Golavinsky Prospect," of which any provincial town in Europe might be proud. Less inviting is the quarter containing the Armenian bazaar, although historically interesting as recalling the time when Tiflis still acknowledged the authority of the Shah. The old castle of the Georgian princes, a reminiscence of a still earlier period, now lies in ruins on a hill rising precipitously above a wild romantic stream. Tiflis has been described as a city of contrasts. Cairo alone presents a similar mingling of Oriental poetry and decay with some of the humble types of European society.

Kars, Batum, Erivan, and Alexandrapol.

South of the Rion basin Russia acquired in 1878 as the prizes of victory, besides the frontier town of Ardahan, the much-coveted seaport of Batum and the formidable stronghold of Kars. A straight line drawn from Batum to Erivan, capital of Russian Armenia, will pass through Ardahan, leaving Kars a little to the right. Regarding the great natural strength and strategic importance of Kars there never could be any question. But the descriptions current in Europe of Batum vary somewhat. It lies some 30 miles south-west of Poti, and its harbour, formed by the delta of the Chorukh advancing westwards, is completely sheltered, about 60 feet deep, and capacious enough to accommodate twelve large vessels. But this harbour is being gradually encroached upon by the very delta to which it owes its

MOSQUE IN THE BAZAAR AT ERIVAN.

existence, a danger, however, which may easily be remedied by giving the river a better scour, or connecting it by canal directly with the port. When these works are carried out, Batum, already connected by rail with Poti, must become the commercial emporium as well as the naval station of Trans-Caucasia. It was declared a free port by the Berlin Congress of 1878, but this has not prevented the Russians from converting it into a second Sevastopol.

Erivan, capital of Russian Armenia, stands on the Zanga, which flows intermittently from Lake Gok-cha, and is here diverted into innumerable little irrigation canals before reaching the Aras. It occupies an important strategic position at the entrance to the route leading over the Gok-cha plateau to the Kura basin and Tiflis. But the climate, with its sudden changes of temperature, malaria, and dust-storms, is one of the worst in Caucasia. Hence Erivan has always remained a small place, and has already been outstripped by the comparatively new fortified town of Alexandrapol, now the largest place in the Aras basin. This stronghold was founded in 1837 on the then Turkish frontier. The possession of these places not only renders the Russian position impregnable in the Aras basin, but gives that power complete command of the head-waters of the Euphrates. The road through Ardahan to Batum is extremely rugged; but the works now in progress will soon render it an easy military route from the Black Sea to the Turkish frontier.

Shusha and Nakhichevan.

East of Erivan the only places in the Aras basin calling for mention are Shusha and the ancient town of Nakhichevan. Shusha lies in the heart of the plateau near the water-parting between the Aras and Kura, some miles above their confluence. It is the largest place in the Yelizavetpol Government; but standing 3500 feet above the sea on an exposed terrace, its climate is excessively severe in winter and correspondingly hot in summer. Nakhichevan lies close to the left bank of the Aras near the Persian frontier. It enjoys the distinction in Armenian tradition of being the oldest city in the world,[1] founded in fact by Noah himself after planting the vine on the slopes of Ararat. But it is now chiefly inhabited by Tatars, while its finest monuments, including the old palace gateway and the "Tower of the Khans," date from the Persian epoch.

9. *Highways of Communication.*

All the routes from Southern Russia through Rostov, Astrakhan, and other points, converge at Vladikavkaz on the north side of the Central Caucasus. From this place the great military and commercial highway leads through the magnificent Dariel Gorge and right under the Kazbek down to Tiflis on the south side of the main range. Here the roads again diverge in all directions—west through Gori, over the Suram Pass of the Mesk range, down the Rion valley, and through Kutais to Poti; south-east down the Kura valley through Yelizavetpol to Baku on the one hand, and on the other to Lenkoran on the Caspian near the Persian frontier; south direct to Alexandrapol and Kars, and over the Gok-cha plateau to Erivan, Echmiadzin, and Ararat. From Poti the road skirts the coast to Batum, whence the military route leads over the Lazistan and Arsiani ranges through Ardahan to Kars, down the Arpa-chai valley through Ani to Erivan, and down the Aras valley to Nakhichevan and Julfa on the Persian frontier. From Kars and Julfa two short roads, crossing the Turkish and Persian frontiers,

[1] Nakhichevan, of which the classic form was Naxuana, is explained to mean "The First Abode," *i.e.* of man after the deluge.

strike the great trade route from Trebizond to Tabriz at Erzerum and Khoi.

The old and difficult coast road from the Taman peninsula to Poti, followed by trade and warlike expeditions from the time of Mithridates down to the Middle Ages, and which had been abandoned, will be reopened for traffic. The coast route from Baku through Derbent to the Terek and Kuma basins is little used, although easier than the Black Sea road.

The Caucasian railway system comprises two trunk lines, the Rostov-Vladikavkaz in the north, and the Trans-Caucasus line in the south. The northern line forms a south-eastern extension of the Russian system, running from Rostov at the head of the Azov Sea through the Kossak stanitzas of Yekaterinovskaya and Kavkaskaya to Georgiyevsk for Piatigorsk, and so on across the Upper Terek to Vladikavkaz, its present terminus. The southern line runs from Batum through Poti and Kutais over the Suram Pass down to Mtzkhet and Tiflis, and thence to Baku and Surachany.

Several projects have been proposed for connecting both sides of the Caucasus. Of these the most practicable seems to be a line from Vladikavkaz along the military route through the Dariel Pass to Tiflis. Another will ere long be carried round the east coast through Derbent to Baku facing the Mikhailevsk terminus of the new Turkestan line. This will give direct railway communication from the heart of the empire to within "a measurable distance" of Herat, interrupted only at the narrowest part of the Caspian.

10. *Administration: Results of Russian Rule—Armenian Politics—Administrative Divisions.*

The brief period during which Caucasia has formed an integral part of the Russian Empire has already produced great social and ethnical changes. A succession of able and energetic administrators have gradually succeeded in stamping out the last spark of independence amongst the highland tribes, and this mountainous region no longer forms a weak point in the colossal empire.

This result has been brought about by very simple means—a steady but determined pursuance of purely practical ends. The administration has everywhere respected the local customs and usages, restricting itself to the maintenance of the preservation of social order. After the heroic defenders of their highland fastnesses were sufficiently reduced by military operations and by wholesale expulsion, efficacious military steps were taken to place the supreme authority beyond the reach of attack from the survivors, while interfering as little as possible with their local affairs. The several tribes were allowed to retain their primitive usages and social institutions, the judicial functions were carried out more in a paternal than in a bureaucratic spirit, and the administration was, wherever possible, entrusted to natives.

The authorities have from the first entirely abstained from interfering with the peculiar religious views of the people, so that it is scarcely correct to assert that the Russian rule has mainly resulted in the extinction of national life.

It is remarkable, in this connection, that the opposition of the Armenians was long directed rather against the Russians than the Turks. But since the acquisition of Erivan and Kars Russia has begun to exercise a growing influence over the Armenian people. In the neighbourhood of Erivan is the Convent of Echmiadzin, the residence of the Armenian Katholicos or Patriarch, whose spiritual authority is absolute over all orthodox Armenians wherever settled.

But the rising generation of Armenians has shown

itself less submissive, in secular matters, to the authority of its spiritual guides. A political party has been formed in Constantinople, inspired by the modern revolutionary spirit, which has undertaken to quicken the slumbering sentiment of nationality, and direct the efforts of the people towards independence. Since the late political changes the Armenians have also made their voice to be heard, determined not to be overlooked in the midst of the innovations which promise to fundamentally modify the social condition of the Eastern races.

Meanwhile Russian Armenia forms a simple division of Caucasia, which constitutes a single administrative government under a viceroy or lieutenant-general responsible only to the Czar. To this government was at first attached the Trans-Caspian territory, which in 1889 was constituted a separate administration. Since then the whole of Caucasia comprises fourteen separate administrations, variously named governments, provinces, territories, or circles, as in the subjoined tabulated scheme of the several administrations, with their areas, populations, and chief towns. The populations are estimated for 1893, and the areas are reduced from the data published by the Statistical Commission of the Caucasus for the Caucasian section of the Russian Geographical Society.

11. *Statistics.*

Areas and Populations.

	Area in sq. miles.	Estimated Pop. 1893.
Cis-Caucasia	89,700	2,860,000
Daghestan	11,300	610,000
Trans-Caucasia	80,800	4,400,000
	181,800	7,870,000

Cis-Caucasia.

Provinces.	Area in sq. miles.	Pop.	Chief Towns.	Pop.
Stavropol	27,500	672,000	Stavropol	35,600
			Piatigorsk	15,000
			Belaglina	12,000
			Praskoveya	9,000
			Alexandrovskaya	8,000
Terek	23,700	780,000	Vladikavkaz	44,000
			Kizlar	10,000
			Grozniy	9,000
			Mozdok	9,000
Kuban	37,500	1,408,000	Yekaterinodar	66,000
			Yeisk	34,000
			Maikop	24,500
			Temruk	13,000
			Novo Petrovskaya	9,000
Total Cis-Caucasia	88,700	2,860,000		

Trans-Caucasia

Provinces.	Area in sq. miles.	Pop.	Chief Towns.	Pop.
Baku	15,095	777,556	Baku	103,000
Daghestan	11,332	609,330	Derbent	16,000
Elizabethpol	16,721	793,969	Elizabethpol	20,000
Erivan	10,075	670,400	Alexandrapol	24,000
Kars	7,308	212,000	Batum	20,000
Kutais	13,968	998,620	Kutais	23,000
Tiflis	15,306	875,181	Tiflis	146,000
Zakataly	1,541	74,449	...	...
Total Trans-Caucasia	91,346	5,011,505		
Total Caucasia	180,046	7,871,505		

Inhabitants of Caucasia grouped according to Races and Religions.

Stock	Races	Religions	Number
Kartvelian Stock	Georgians, Imerians, Mingrelians	Christians, Greek rite	1,150,000
	Svans, Pshavs, Khevsurs	Nominal Christians	
	Lazis	Sunnis	40,000
West Caucasian Stock	Cherkesses, Abkhasians	Sunnis	188,000
	Kabards	Christians	
East Caucasian Stock	Chechenzes, Lesghians[1]	Sunnis	610,000
Semite Stock	Jews	Jewish	51,000

[1] All the Lesghians are Sunnis except the Dido tribe of the Upper Koisu valley, who are said to be "Devil-worshippers."

Stock	People	Religion	Number
Slav Stock	Great Russians Little Russians Bulgarians Bohemians	Orthodox Christians and Dissenters	2,100,000
Iranian Stock	Osses	Nominal Christians	128,000
	Armenians	Gregorian Christians	804,000
	Kurds	Sunnis	97,000
	Tats	Shiahs	125,000
	Talishes	Shiahs	88,000
	Greeks	Christians, Greek rite	57,000
	Germans	Lutherans	24,000
Mongolo-Tatar Stock	Tatars Nogais Turkomans Kumiks Kirghiz [1] Turks	Sunnis mostly; some Tatars are Shiahs	1,550,000
	Kalmuks [1]	Buddhists	

[1] The Kirghiz and Kalmuks occupy a large area in the Lower Volga District, whence some of them reach southwards beyond the Manich depression along the right bank of the Kalaus and as far as the left bank of the Kuma.

CHAPTER III

RUSSIAN TURKESTAN

1. *Boundaries—Extent—Area—Nomenclature.*

HERE we enter a region in which geographical, ethnographic, and political conditions are still far from being reconciled. Although somewhat simplified by the recent progress of Russian arms in West Central Asia, the very nomenclature is still in a confused state, and few even of the leading authorities are of accord as to the exact meaning of such common expressions as Turkestan or Central Asia. The Russians themselves often designate as Central Asia the second great administrative division of their Asiatic possessions, which is mainly comprised within the Aralo-Caspian depression. But this expression is misguiding in a geographical sense. To the portion of this division directly administered by the Governor-General, whose headquarters are at Tashkent, they give the still more questionable name of Eastern Turkestan —the true Eastern Turkestan, if there be any, lying beyond his jurisdiction in the Chinese province of Kashgaria. The confusion of nomenclature is increased by the distribution of the land, portions of which are attached either to the European governments of Orenburg and Perm, or to the administration of Turkestan, while the south-eastern section beyond the Upper Oxus belongs politically to Afghanistan, and thus forms part of the

British political system. Much of the confusion may be avoided by the plan adopted in this work of applying to the whole of Turkestan the three general expressions Russian, Chinese, and Afghan Turkestan, answering to the three present political divisions of this Central Asian region. In consequence of its historical importance, the term *Kashgaria,* from the ancient city of Kashgar, is retained as an alternative name for Chinese Turkestan.

This vast region is mostly comprised within the limits of the Aralo-Caspian basin, stretching west and east between the Caspian Sea and the Central Asiatic highlands, limited southwards by the scarp of the Iranian plateau, merging northwards in the west Siberian steppe. The south-eastern boundary line along the Afghan and Indian frontiers was roughly laid down in 1885, and finally determined by the Anglo-Russian Commission of 1895 (see vol. ii. ch. i.). The south-western boundary along the Persian frontier has been greatly extended by the reduction of the Akhal Tekke Turkomans in 1881. The rectified frontier, as determined by the agreement with Persia in 1882, now runs from the south-east corner of the Caspian up the Atrak valley to Chat at the junction of the Sumbar, thence eastwards along the water-parting to a point south-east of Askabad.

Russian Turkestan is bordered on the west by the Caspian, the Ural river and mountains, on the east by the Pamir plateau, the Tian-Shan and Ala-tau ranges separating it from the Chinese Empire, northwards by the low ridge crossing the Kirghiz steppes about the 51st parallel, and forming the water-parting between the Aralo-Caspian and Ob basins. But here again the administrative overrides the geographic division, for a large portion of West Siberia beyond this natural boundary is now attached to the Turkestan Government.

Including this tract, which is alone about 400,000 square miles in extent, Russian Turkestan has an extreme length from the Caspian to Lake Issik-kul of 1400 miles west and east, with a breadth of nearly 1000 north and south, a total area of about 1,550,000 square miles, and a population of 5,620,000.

2. *Relief of the Land: The Pamir—The Kizil-art and Alai Ranges—The Tian-shan and the Ala-tau Highlands—The Mugojar Hills—The Turkestan Depression—The Dried-up Central Asiatic Mediterranean—The Turkestan Deserts.*

Russian Turkestan is commonly supposed everywhere to consist of vast low-lying sandy or saline plains. But nothing could be more opposed to the actual conditions, for the relief of the land here presents absolutely greater contrasts than are elsewhere found on the surface of the globe. The misconception is due to the failure to distinguish between the Aralo-Caspian depression and the Aralo-Caspian basin. The basin—that is, the whole area of drainage—consists of about even parts highlands and lowlands; and while the lowlands fall in the Caspian Sea as much as 85 feet below sea-level, the highlands in the culminating points of the Tian-shan and Great Pamir rise to over 26,000 feet above sea-level. In no other region are such vast differences of relative level to the surface of the sea brought into such close juxtaposition.

The highlands, which lie mainly in the east, consist substantially of the Pamir and Tian-shan systems, with the Alai and other sections, all converging westwards between the Tarim and Aralo-Caspian depressions.

The Pamir.

The Great Pamir or Bam-i-Dunya—that is, the "Roof of the World," as it has been not inaptly termed—forms the nucleus of the whole Central Asiatic highland system. Here converge the Hindu-Kush and Himalayas from the south-west and south-east, the Kuen-lun from the east, the Tian-shan from the north-east, while the plateau itself merges westwards in the snowy highlands and icefields about the sources of the Zarafshan, between the Oxus and Jaxartes valleys.

Since 1838, when Wood first penetrated into the Pamir from the Afghan side, and discovered Lake Victoria, source of a main branch of the Oxus, this region has been approached from all quarters and traversed in every direction by explorers, naturalists, sportsmen, and even by military expeditions. Its more salient features were first revealed by the Forsyth mission to Kashgaria, Hayward and Shaw's excursions, and the journey of Kostenko, who in 1876 crossed the Trans-Alai range by the Kizil-art Pass (14,260 feet), and was thus the first to enter the Pamir from the north. And now all important details have been filled in by the surveys of Grombchevsky, Bogdanovich, and other Russians, undertaken primarily for political purposes (1889-92), and by the journeys of Mr. and Mrs. Littledale, who traversed the whole region from north to south in 1890; of M. Henri Dauvergne (1889), who entered from the north-east, and followed a route which enabled him to determine with some certainty the farthest sources both of the Yarkand and Oxus rivers; of Dr. G. Capus (1887), to whom we are indebted for a careful study of the climatic and ethnological relations; lastly, of Captain Younghusband, who, during the years 1889-91, explored all the central and eastern districts, approaching the plateau both from the glacial heights of the Mustagh and Aghil ranges and from the relatively low levels both of Yarkand (3923 feet) and Kashgar (4040).

A fairly accurate idea has thus been gained of the general configuration and physical structure of the Pamir, which may be described as a north-western continuation of the Tibetan tableland, projecting at a mean elevation of about 12,000 feet from the Mustagh and Hindu-Kush heights northwards between Chinese and Russian Turkestan to the Trans-Alai and Mustagh-tau ranges, that is, the western extension of the Tian-shan system. The plateau, which has an extreme length of 280 miles by 120 to 150 in breadth, with an area of about 30,000 square miles, is thus encircled on three sides by snowy ranges, whose culminating points are in the north the Kizil-agyn (21,700 feet) and the Kaufmann (23,000) peaks; in the extreme south-west Mounts Lunkha (22,500), Sad Ishtrag (24,000), and Nushau (24,600); in the east Mount Charkum (22,500), and the Tagharma (25,800) and Mustagh-ata (26,000 ?) peaks. Within these encircling heights the surface is intersected in various directions, but mainly from east to west, by snowy ridges from 14,000 to 16,000 or 17,000 feet high, enclosing, not deep mountain gorges, but somewhat flat and broad plains or valleys, which deepen rapidly westwards, and which are traversed by the ramifying headwaters of the Oxus descending to Bokhara and the Aralo-Caspian depression. Some idea of the general incline is afforded by the Murgh-ab branch, which falls from 12,300 feet a little below the Ak-baital confluence to 6550 feet at the Panj confluence, while the Panj itself falls from 11,340 feet at Sarhad to 8560 at Ishkashem, where it trends sharply round from west to north.

These characteristic river valleys take the general

name of *pamir*, a term which has been explained to mean "desert," but which really means an upland plain, valley, or plateau. Thus the whole region, which may in a general way be called "The Pamir," consists in fact of a considerable number of "pamirs," nearly all watered by the Oxus affluents, often studded with lakes, and separated from each other by lofty snow-clad ridges, which are crossed here and there by rugged mountain passes. Such are, going northwards, the Tagh-dum-bash Pamir, which, however, lies so far to the east that it is comprised within the Yarkand basin; the Little Pamir, which communicates over the Andemin Pass (15,500 feet) with the Great Pamir, from which the Khargosh Pass (14,550) leads down to the Sarez Pamir, watered by the Alichur River and the large lake Yashil-kul, which discharges west to the Panj above Wamar.

All these elevated plains appear to have been originally profound river gorges, which have been gradually filled up to their present levels by the detritus from the enclosing ridges accumulating too rapidly to be carried away by the running waters. At least such is the theory put forward by Younghusband, who remarks that "in the Hindu-Kush and Himalayan region the valleys as a rule are deep, narrow, and shut in; but on the Roof of the World they seem to have been choked up by the *débris* falling from the mountains on either side, which appeared to me to be older than those farther south, and to have been longer exposed to the wearing process. The valleys have thus been filled up faster than the rainfall has been able to wash them out, and so their bottoms are sometimes as much as four or five miles broad, almost level, and of considerable height above the sea. The Tagh-dum-bash Pamir runs as low as 10,300 feet, but on the other hand, at its upper extremity, the height is over 15,000 feet, and the other Pamirs vary from 12,000 or 13,000 to 14,000 feet above sea-level. That is, the bottoms of these Pamir valleys are level with the higher summits of the Alps."

The general slope is from the west side upwards to the horse-shoe rampart overlooking Kashgaria, to which Hayward and Shaw have extended the name of Kizil-art, and which the Chinese call Kung-ling, or "Onion Mountains," from the garlic growing on their flanks. Here is the culminating point of the whole region, which between the Mustagh and the Tian-shan appears to maintain a mean altitude of about 20,000 feet, or 16,000 feet above the plains of Kashgaria. But at a distance of 120 miles due west of Yarkand the system develops a prodigious mountain mass, the highest summit of which has since the time of the Forsyth mission figured on the maps as the Tagharma Peak, with an altitude of 25,300 feet, this being the estimate made by Captain Trotter of that mission. M. Bogdanovich, however, a member of the Pevtsof expedition, who carefully surveyed this alpine district in 1889, found that the proper name of the mountain group is Mustagh-ata, "Father of the Ice Mount," while Tagharma, *i.e.* "corn-mill," is the name of the river valley at its base, and has reference to the numerous corn-mills which have been set up by the Sarikoli natives in the arable tracts. On the other hand, Younghusband, who sighted the Mustagh from the Little Kara-kul Lake on his journey from Kashgar to the Pamir in 1891, resolves Trotter's Tagharma into two peaks, the Mustagh-ata (over 26,000 feet?), and another, for which on his map he retains the name of Tagharma, giving it a height of 25,800 feet. He distinctly saw two crests about 20 miles apart, appearing to differ little in height, and both remarkable for their great massiveness. "They are not mere peaks, but great masses of mountain, looking from the lake as if they were bulged out from the neighbouring plain. The appearance of these two great mountain masses rising in stately grandeur on either side of a beautiful lake of clear blue water is, as may be well imagined, a truly magnificent spectacle" (*Geo. Proc.*, 1892, p. 230).

The Tagharma valley is described by Bogdanovich as mainly a broad pasture land, visited in spring by thousands of Kirghiz nomads from the neighbouring mountains. The Kara-tash Pass, by which the Mustagh-ata was crossed, stands at an altitude of 16,500 feet, and is reported to be the most difficult in the eastern borderland between the Pamirs and Kashgaria. The descent towards the Little Kara-kul Lake, itself 12,790 feet high, is extremely rugged, the track for over 13 miles clinging to the side of the mountain along huge slopes of detritus. From the lake M. Bogdanovich passed round the west foot of the Mustagh-ata by the wild Iki-bel-su River gorge and over the Ulugh-Rabat Pass down to the Tagharma River, thus crossing the divide between Russian and Chinese Turkestan at this point. From the Tagharma, one of the farthest head-streams of the Yarkand-darya (Tarim), the explorer struck eastwards over the Kok-moinak (15,000 feet) and Chichiklik Passes, crossing the south side of the Mustagh-ata group. Here all the valleys have the aspect of old glacier streams, with alluvial deposits in their lower parts similar to those of the Kara-tash-su below the last Kirghiz camping-ground. In the very wildest part of these gorges slightly ferruginous hot springs were met, with temperatures ranging from 127° to 135° F., all rising from thick masses of conglomerate rock overlying granite and crystalline schists. But in the Mustagh-ata and its subsidiary ranges the prevailing formation is gneiss, with a general north-west and south-east trend. Comparing this with the Pamir ranges running east and west and north-east and south-west, M. Bogdanovich infers that the Mustagh-ata system is the older, and that it was not greatly modified by the later upheaval of the Pamir ranges. Beyond Chichiklik the route lay over the Torut Pass (13,000) eastwards to Yarkand, where M. Bogdanovich joined the Pevtsof expedition.

The whole of the Pamir region is destitute of trees or shrubs, and even the grass grows only in isolated patches along the banks of the streams and lakes. Here, however, it affords some of the very finest pasture in the world to the flocks of the Kara-Kirghiz nomads, who visit the Pamir during the summer season. According to M. Ximenes, who visited the Pamirs in 1892, these nomads comprise three distinct tribal groups, with over 17,000 *kibitkas* (tents), or a total population of from 68,000 to 70,000. This estimate corresponds with that of Colonel Kostenko, made in 1880.

The hills consist of a soft stone, so that the passes are less abrupt and easier to cross, while the tracks are everywhere tolerably good. The streams also are seldom very rapid, and the soil consists either of grit or sandy loam, dotted over with salt or brackish tarns, which when dried up are covered with an incrustation of dazzling white magnesia.

Of these basins by far the largest is the Kara-kul, which stands at an altitude of 12,800 feet due south of the Kizil-art Pass in the northern section of the Pamir. When it was visited by Younghusband in the autumn of 1890, "a violent storm was raging, the whole lake was lashed into a mass of foam, and heavy snow-clouds were sent scudding along the frowning mountain sides; but above all rose the great snowy peak Mount Kaufmann, calm and serene amid the warring elements raging at its base." Kara-kul has no visible outlet, and is evidently at present a closed basin; but it formerly stood at a higher level, as shown by the water-marks distinctly visible on the lower slopes of the encircling hills. At no very remote epoch it probably sent its overflow eastwards to the Markhan-su, which joins the right bank of the Kizil-su (Upper Kashgar-darya) below the Utch-tash confluence.

The Kara-kul freezes hard in winter, the ice attaining such a thickness that, according to the Swedish traveller Sven Hedin, "St. Petersburg might be built on the western and Moscow on the eastern side of the lake." This traveller, who visited the Pamir early in 1894, found that in the eastern part of the basin the water is comparatively shallow; but in the western section the soundings revealed depths of from 800 to 825 feet. This result corresponds to the conformation of the surrounding plateau, which is flat on the east side, while the western shores of the lake are enclosed by huge mountain masses.

South-east of the Kara-kul lies the much smaller, but perhaps more interesting, basin of the Rang-kul, visited by Ney Elias and Younghusband, and identified by Sir Henry Rawlinson with the famous "Dragon Lake," the holiest spot in the whole Buddhist cosmogony, which is placed in this region by Hwang-Tsang, the Chinese pilgrim of the seventh century. The Dragon myth which is associated with this lake, and which is still credited by all the surrounding populations, is due to a peculiar light effect in the Chiragh-Tash ("Lamp Rock"), a cavernous cliff standing about 100 yards from the water's edge. All observers have noticed the beam of light visible from the entrance to the cave, which is popularly attributed to a large diamond set in the head of a dragon, guardian of vast treasures stored in the recesses of the rock. The phenomenon had been generally attributed to phosphorescence, until the true cause was discovered by Younghusband, who was the first to scale the cliff and solve the mystery. On gaining an entrance to the cave he found that "the light came neither from the eye of a dragon nor from any phosphorescent substance, but from the usual source of light—the sun. The cave, in fact, extended to the other side of the rock, thus forming a hole right through it. From below, however, one cannot see this, but only the roof of the cavern, which being covered with a lime deposit reflects a peculiar description of light."

The water of the lake is of a deep blue colour, and now saline, although in the time of Hwang-Tsang it appears to have been fresh. From this it might be inferred that, like the Kara-kul, the Rang-kul has subsided, and formerly stood at a level high enough to discharge perhaps to the Ak-baital tributary of the Ak-su branch of the Oxus.

The Tian-shan Orographic System.

It is difficult to assign a beginning or an end to the Tian-shan, or "Celestial Mountains," which separate the Tarim from the Issik-kul and Ili basins south and north, and which are generally supposed to extend thence eastwards to about 120 miles east of Hami (Khamil) in 95° E. longitude. They are even continued still farther east by M. Pevtsof, whereas the brothers Grum Grijimailo fix their eastern limit at 90° 50′ E. longitude—that is, considerably to the west of Hami. These naturalists, who in 1889-90 traversed the whole range from Kulja to Hami, and then crossed the Gobi to Su-Chau in Kansu (North-West China), transfer the Hami Mountains, that is, the easternmost section, from the Tian-shan to the Altai system, and show that there is a distinct break between this section and the Tian-shan

THE TIAN-SHAN MOUNTAINS.

proper. The gigantic Bogdo-ola (Topatar-aulich) group, rising far above the snow-line, falls abruptly westwards near Urumtsi, and is continued eastwards by an unbroken snow-capped chain skirting the north side of the Turfan district as far as 90° 50′ E. longitude. Here the "Nameless Range," as the Russian explorers designate this terminal section of the Tian-shan, suddenly merges in a plateau studded with isolated cones and falling in steep escarpments on the north and south sides for a distance of 10 miles. The plateau is followed by rolling ground with rounded hills for a further distance of 40 miles, where it abuts on the Barkul Mountains, which have a north-westerly trend, and which unite with the Hami range a little east of the town of Hami. This range runs in the same direction, and under the name of the Mechin-ola encloses the Barkul depression, while according to native report it crosses the Gobi south-eastwards to Su-Chau. Thus the Hami-Barkul Mountains, which at their junction are dominated by several snowy peaks, are clearly separated by a break of 50 miles from the Tian-shan; but whether they are to be assigned to the Altai, or regarded as an independent system, cannot be decided without further and more accurate surveys of this little-known region.

Crossing the Nameless Range by the Builuk and Ulan-ussu Passes near its eastern extremity, MM. Grum Grijimailo struck southwards in the direction of Lob-nor, by a route far to the east of that followed by Prjevalsky. Here they found that the Tarim-Hami wilderness, so far from being a moderately elevated plain, as hitherto supposed, is a mountainous region throughout, being traversed by numerous apparently disconnected chains with a normal north-westerly trend. Such are, going southwards, the Tuz-tau, which divides the Turfan district into two distinct sections; the Chol-tau, and several parallel ridges collectively grouped under the name of the Tiuze-tau Mountains. Many of the ranges are over 200 miles long, varying in height from 6000 to 10,000 feet, and belonging in their geological structure rather to the Hami-Barkul than to the Tian-shan system.

Towards its western extremity the snowy Nameless Range culminates in the magnificent Bogdo-ola group, long known by name, but first visited by the Russian naturalists. This huge uplifted mass, which may be approached by a fair road from Urumtsi, presents an imposing sight from whatever side it is viewed. It is clothed from foot to summit by spruce forests and meadows; from the upper slopes numerous torrents rush down to the north, and at its base lies a magnificent alpine lake. For ages these highlands have been regarded as hallowed ground, and Bogdo-ola, that is, the "Lordly Mountain," is a kind of Mongolian Olympus, the Throne of God, who is supposed at times to descend from its snowy heights to the romantic lake at its foot.

Near the eastern extremity of the section of the Tian-shan running from Urumtsi westwards in the direction of Kulja, the two Russian explorers discovered the remarkable Doss-Megen-Ora group, which, although attaining an altitude of over 19,700 feet, had hitherto been unvisited, and apparently even unsighted or unheard of, by any previous modern traveller. Doss-Megen-Ora, that is, "Loftiest of Mountains," forms the nucleus of three distinct ranges: the chain separating Zungaria from the Ili basin, the Urumtsi Chain running eastwards, and the main range stretching south-westwards in the direction of Khan-Tengri. From its northern flanks descend numerous streams, some of which flow through wild gorges or cañons over 1000 feet deep north to the Manas oasis, west to the Kash (Ili), and south-east to the Yulduz. The Khusta, most copious of these rivers, is confined to a narrow rocky bed 10 to 11 feet wide, 1000 feet deep, and 26 miles long; beyond this remarkable chasm it breaks into seven deep branches, which ramify as irrigation canals through the Manas depression.

Beyond the Doss-Megen-Ora the whole system of the Tian-shan continually expands westwards, developing two or more lateral and parallel ridges, and in the extreme west ramifying into several distinct branches, which spread out like a fan far into the Turkestan lowlands. Of these branches the south-westernmost are the Alai and Trans-Alai, which stretch in parallel lines for 240 miles along the northern edge of the Pamir, down to the Turkestan plains. They are separated from the Tian-shan proper by the Kog-art[1] and Terek-davan[1] passes, but their diorite and granite formations show that they belong none the less to that system. The Alai, or Kichi-Alai, rises to over 18,000 feet in the east, and is crossed by several passes, of which the lowest is the Isfairan (12,000). From this point a view is afforded of the "Kaufmann" Peak, over 25,000 feet, one of the very highest, if not the highest, in the whole Tian-shan system.

The Alai and Trans-Alai merge westwards in the snowy plateau and ice-fields about the sources of the Zarafshan, whence the Shchurovsky and other enormous glaciers descend towards the surrounding upland valleys.

North of Ferghana, the most important western branch of the Tian-shan are the Alexander Mountains, which run at an elevation of 15,000 feet from the closed basin of Lake Issik-kul along the northern edge of the Narin (Upper Jaxartes) valley, and are continued by the Aksai,

[1] In the Tian-shan, *art*, *davan*, *bel*, and *kutal* are the general names for passes. The *art* is a high and dangerous gap, the *davan* a difficult rocky defile, the *bel* a low easy pass, the *kutal* a wide opening between the hills.

Talas-tau, and Kara-tau (6000 feet) north-westwards between the Middle Jaxartes and the Chui valleys. Here the culminating crests are the Hamish (15,550) in the Alexander chain, the Kara-bura (11,000) in the Talas-tau, and the Min-jilke (7000) in the Kara-tau.

West of Hami the eastern section of the Tian-shan soon attains an elevation of 8000 to 10,000 feet.

THE BUAM PASS NEAR KUTEMALDI, ON LAKE ISSIK-KUL.

Between Hami and Barkul it is crossed by the Kosheti-davan Pass, over 9000 feet. Farther on there occurs a profound gap or break of continuity, through which the historical route leads from the Gobi desert through Turfan north to Urumtsi and the Ili basin. West of this defile the central section, under the name of the Katun range, rises far above the snow-line, and attains an elevation of at least 16,000 feet. Here there are few passes, and some glaciers were discovered by Regel about the sources of the Kash, which flows from these highlands westwards to the Ili valley. At this point are developed as many as four parallel snowy ridges, enclosing the two extensive dried-up basins known as the Great and Little Yulduz, or "Stars," 7000 feet above sea-level.

THE YULDUZ VALLEY.

During his exploration of the Central Tian-shan in 1894, Roborofsky crossed by the Karagai-tasnin-davan Pass from the valley of the Kok-su, which flows to the Tekes affluent of the Ili, to the Chaidik-gol, which drains through the Yulduz valley to Lake Bagratch. A survey of the Great and Little Yulduz showed that these basins were at one time occupied by an alpine lake, whose waters have been discharged through a narrow rocky gorge at a level of nearly 5000 feet below that of the surrounding region.

From the Yulduz basin Roborofsky made his way by a comparatively easy track over the northern crests down to the Algo river, which also flows in a deep bed down to the Lukchun depression discovered by the brothers Grijimailo. The surface of the salt lake Bodshaite, by which this remarkable depression is flooded, was found to be more than 980 feet below the sea, a level considerably lower than had been supposed by its discoverers. The western and northern districts are tolerably fertile, and yield fairly good crops of corn, tobacco, cotton, fruits, and vegetables to the surrounding Turki populations.

An excursion was also made by Roborofsky's companion, Lieutenant Kozloff, over the Tian-shan range south of the Yulduz valley to the Bugur oasis on the route between Kuchar and Kurla. Here both the ascent and the descent were extremely steep, the wild rocky summit of the pass attaining an absolute elevation of no less than 14,000 feet above sea-level. Farther on the expedition penetrated as far as the Sa-chu oasis, from which three expeditions were made for a space of 300 miles along the northern and 175 miles along the southern slopes of the Nan-shan Mountains. The range was crossed a long way to the west of Sa-chu, and after a survey of the lacustrine district of north-west Tsaidam-khuntei the course of the River Su-lei-khe was followed for 200 miles from the mountains to its entrance into Lake Khalachi.

South of the Great Yulduz basin the main southern section runs under diverse names westwards to Lake Issik-kul. Here the chief sections are the Muz-art-tau, crossed by the historical Muz-art Pass (12,000), and the Khan-tengri (24,000), the most imposing and dominant

SANDHILLS AT KURLA.

mass in the whole Tian-shan system. Although exceeded in height by the Trans-Alai peaks, the Khan-tengri contains far more numerous glaciers, ice-fields, and snowy crests. Some of the glaciers in the little-known highlands west of the Muz-art rival the Aletsh of the Valais, and from this pass the main range runs for over 60 miles westwards at a mean altitude of nearly 17,000 feet. Here all the peaks are higher than Mont Blanc by over 3000 feet, and yet are dominated southwards by the magnificent Khan-tengri or Kara-gol-bas.

West of the Yulduz basin the extensive dried-up marine basin of the Tian-shan-pelu rises westwards towards the plateau occupied by Lake Sairam, which is skirted north and south by the Zungarian Ala-tau and the Boro-Horo chains. The Boro-Horo range, which extends from the Doss-Megen-Ora group westwards to the meridian of Kulja, is described by the brothers Grum Grijimailo as a lofty, continuous wall-like chain forming the divide between the Ili basin and the Zungarian waters. At the Doss-Megen-Ora end the range attains a great elevation, and is crossed only by the Ulan-ussu Pass, and even this is practicable only in summer. Here large masses of snow lie on the summits, and patches on the principal spurs. But farther west the northern slopes are too steep to allow any water to collect in great rivers. Hence all the streams, such as the Kijtyk and Jirgalty, are mountain torrents still flowing in deep rocky chasms like those on the flanks of Doss-Megen-Ora.

Facing the Boro-Horo is the imposing Nan-shan (Temurlik) range, which rises abruptly above the south side of the Kulja plains, between the rivers Tekes and Kegen, east and west. North of the Narin valley the main range takes the name of the Ala-tau Terskei—that is, the "Shady Ala-tau"—to distinguish it from the Ala-tau Kungei, or "Sunny Ala-tau," which skirts the opposite side of Lake Issik-kul.[1] It culminates with the Ugus-bas (17,750 feet), and is crossed by the Barskaun Pass

THE SOURCE OF THE KOPA, AFFLUENT OF THE ILI.

(12,000 feet), near the source of the Narin. The Terskei

[1] As much confusion is caused by the numerous "Ala-tau" or "mottled" ranges in this region, it may be well to explain that the Zungarian Ala-tau runs north of the Ili valley, over against the Tarbagatai chain, the Trans-Ili and Kungei Ala-tau lie south of the Ili and north of Lake Issik-kul, where they are separated by the valleys of the Great Kebin and Chilik Rivers; the Ala-tau Terskei runs south of Issik-kul, between it and the right bank of the Narin (Upper Sir) River; while a fifth Ala-tau running west from Issik-kul has now been renamed the Alexander range by the Russians.

Ala-tau is continued westwards by parallel ridges, which enclose Lake Son-kul, and effect a junction north-westwards with the Alexander chain.

Thus is completed the vast system of the Tian-shan, which is about 1500 miles long east and west, with an average width of nearly 250 miles, and a total area of 400,000 square miles, or rather more than that of all the European highland systems taken together.

The long-suspected presence of still active volcanoes in the Tian-shan, and especially towards the Kulja frontier, seems to be now settled in a negative sense. General Kolpakovsky, Governor of Semirechinsk, in the autumn of 1881 explored Mount Bia-shan, twelve miles north-east of the city of Kulja, in the Ailak highlands, and discovered that the fires which have been burning there from time immemorial are not volcanic, but proceed from ignited coal-beds. The caves in the side of the mountain emit smoke and sulphurous gases. In 1889 the brothers Grum Grijimailo also visited what they call "a burning coalfield" south of Jirgalty (Shi-ho) in the Boro-Horo range. Here they noticed smoke issuing from fissures in the ground, the edges of which were encrusted with crystals of sulphur; funnel-shaped cavities were even observed emitting jets of steam; the ground was also so heated that walking was difficult and even dangerous. "The fire is visible from the River Ebteh and from Shi-ho, whence it has a striking appearance." They also mention the occurrence of naphtha springs near Urumtsi, and of sal-ammoniac beds in Central Zungaria. But no traveller speaks of igneous cones, and still less of true craters, even in a quiescent state, and it may now be presumed that such have no existence.

The Mugojar and Tarbagatài Hills.

From the Mugojar ridge, running from the Urals to the head of the River Emba, the low hills which form the water-parting between the Ob and Aralo-Caspian basins may be traced across the Kirghiz steppes about the 48th parallel eastwards to Lake Balkhash. From the exploration of the Mugojar range by a party of Russian savants in 1889, it appears that these hills undoubtedly form a continuation of the Ural Mountains, although now separated from them by a depression six miles wide. The highest point is Mount Airuk (1970 feet), while the water-parting falls at one or two points to 220 feet. But this water-parting is continued eastwards along the north side of Lake Balkhash by the Denghiz-tau, or "Sea Range," which near Sergiopol merges in the Tarbagatai highlands. From this point the Tarbagatai — that is, "Marmot Mountains" — stretch in two main sections under the 47th parallel between Lakes Zaisan and Ala-kul eastwards to the Upper Irtish valley. Above the left bank of this river rise the snowy Tas-tau (9850 feet), and Muz-tau or Sauru (11,320 feet), culminating point of the whole system. West of the Muz-tau the range is crossed by the Khabar-assu Pass, which, notwithstanding its great height (7628 feet), has always been more frequented than any other, especially by traders between Siberia and the Ili basin. The Tarbagatai has a mean altitude of 6000 feet, but few of its crests rise above the snow-line, which here falls to about 9000 feet.

The Turkestan Depression.

The Turkestan lowlands, which stretch from the Caspian and Ural River to the foot of the Central Asiatic highlands, possess no bold natural limits towards the north, where they merge imperceptibly with the West Siberian steppe. Southwards they are limited by the western continuation of the Hindu-Kush as far as the Hari-rud valley, and beyond that point by the North Khorasan highlands as far as the Caspian. By far the greater part of this region is occupied by the Aralo-Caspian depression, which is the most extensive on the globe; for it properly includes the plains of south-east Europe, which drain through the Volga to the common basin of the Caspian. In this great inland sea the depression reaches its lowest level of 85 feet below the Mediterranean. Eastwards the Caspian is separated by the extensive Ust Urt plateau from the Aral Sea, which is the next largest reservoir in Asia.

The Chink, or eastern edge of the Ust Urt plateau, is 500 feet above sea-level along the west coast of the Aral Sea, but falls to 210 farther south towards the Sary Kamish lakes. This is probably little more than the mean elevation of the Turkestan lowlands, which are now known to have at one time formed part of a vast inland sea communicating through the Manich depression across the Ponto-Caspian isthmus with the Euxine, and through the Ob basin with the Arctic Ocean. This inland sea is supposed to have escaped through the bursting of the Bosphorus, an occurrence which some writers have connected with the legendary deluge of Deucalion about 1530 B.C. But however this be, were the Bosphorus again to be closed to a height of 220 feet, the former condition of things would again be gradually brought about. The Euxine, Caspian, and Aral, but not the Balkhash, would form a continuous sheet of water, draining through the Aralo-Caspian and Ob water-parting (220 feet at its lowest point) to the Arctic Ocean.

North of the Aral Sea the plains and steppes form a vast lacustrine region, dotted over with lakes or tarns with no outflow, and fed by intermittent steppe rivers. Some of these, such as the Chui and Sari-su, formerly reached the Aral Sea through the Jaxartes, while others found their way either directly or through the Emba to the Caspian. Lakes Balkhash and Kara-kul, however, which are fed by the Ili and Talas respectively, cannot have communicated through the Chui with the Aral Sea, at least so recently as quaternary times. But Balkhash is known to have formerly stretched much farther east than at present, forming a continual sheet of water with the Sassik-kul and the numerous other lakes strewn over the so-called "Zungarian Strait," which is supposed to have connected the Turkestan and Mongolian mediterraneans through the depression between the Ala-tau and Tarbagatai ranges. At the time when these inland seas were thus connected, their elevation must have been far more than 500 feet, for the present level of Balkhash is 514 feet. Consequently the inland waters may have at that time communicated over the Bosphorus barrier with the Atlantic, as well as over the present Aralo-Caspian and Ob water-parting with the Arctic. When, through the process of desiccation continually going on in Asia, the mediterraneans were reduced to about 220 feet, the northern and southern outflows were probably arrested, and then the pressure of such a prodigious body of water would help to account for the bursting of the Bosphorus and gradual draining of the Han Hai, or "Western Sea," as the Asiatic Mediterranean is called in the Chinese Chronicles.

The Turkestan lowlands proper are known by various names, such as the Kara Kum or "Black Sands," north of the Aral Sea; the Kizil Kum or "Red Sands," between the Oxus and Jaxartes; the Ak Kum or "White Sands," between the Alexander range and the Chui River; and the Khwarezm or Turkoman desert, between the Oxus and

Caspian. But all alike present the same monotonous and desolate aspect, in which a dull brown is the prevailing colour. "The gloom of the West Turkestan steppes, which first impresses one so forcibly in the Karakum deserts north of Aral, seems surpassed by the sadness of Kizil-kum near the south-east corner of the lake." [1]

3. *Hydrography: The Rivers Oxus, Zarafshan, Murgh-ab, and Sir-darya—The Aral Sea—Lakes Balkhash and Issik-kul.*

All the streams of the Aralo-Caspian basin have their natural outlet in the Caspian Sea, the lowest part of the Turkestan depression. But none of them now reach that outlet except the River Emba flowing from the Mugojar hills south-westwards to the north coast, which it reaches after a course of 250 miles through the Kirghiz steppe. All the rest either run dry in the sands like the Tajand and Murgh-ab from the Iranian plateau, the Zarafshan, Chui, and Talas, from the Tian-shan highlands, the Sari from the Ob water-parting, or else, like the Oxus and Jaxartes, are at present absorbed in Lake Aral. But as Aral stands about 160 feet above the Mediterranean, consequently 245 above the Caspian, its proper outflow should also be to the Caspian, with which, in fact, it formerly communicated. Hence the Oxus and Jaxartes must be regarded as rivers also arrested on their course to their natural outlet in the Caspian.

The Oxus Basin.

The Oxus—the Jihun of the Arabs and Amu-darya of the Persians—is the Vak-shu of Hindu writers, a term which has been wrongly derived from the Turki *Ak-su,*

[1] Major Herbert Wood's *Shores of Lake Aral*, p. 337.

"White Water," to which it bears only an accidental resemblance. Yule, Curzon, and others have shown that it is from *Wakhsh*—that is, the Surkhab of Karategin and East Bokhara. The Oxus collects most of the Pamir drainage through its two main head-streams, the southern Ab-i-Panj and the northern Murgh-ab, which converge at

THE SOURCE OF THE OXUS.

Kala Wamar in Roshan, thus enclosing between them the Little Pamir, the Great Pamir, the Shignan valley, and the northern section of Wakhan. The Ab-i-Panj, the longer and more copious of the two forks, is formed by the junction above Kala-i-Panj of the Pamir River descending from the Gaz-kul or Victoria Lake (13,980 feet), discovered in 1838 by Wood, and the Wakhan-su, whose farthest head-stream appears to be the river which joins its left bank at Bozai-Gumbaz, and which, according to Littledale, is called Varjer. This Varjer, flowing from the east down the Pamir-i-Wakhan valley, is evidently the same stream which was found by Henri Dauvergne issuing from three great glaciers near the northern approach to the Kilik Pass over the Hindu-Kush (14,200 feet) in lat. 37° 10′ N., long. 75° E., and which he followed down to Sarhad (11,340 feet) on the north side of the Baroghil Pass. As the glaciers terminate just west of the Wakhijrui (Wakhiji) Pass (15,600), which forms the narrow divide towards the Yarkand basin, it is obvious that here is the farthest source of the Oxus, which Dauvergne claims to have discovered on his adventuresome journey from Kashmir round by the Pamirs to Gilgit in 1889. The spot was again visited by Captain Younghusband in 1891, when he ascended the Pamir-i-Wakhan from Bozai-Gumbaz to the valley facing the Kilik Pass, whence an excursion was made to two of the most interesting water-partings in the whole of Asia, one between the Indian Ocean, the Tarim and Aral basins, the other between the Oxus on the west and the Tarim on the east. "If any point can be called the heart of Central Asia, I should think this must be it. Here on the Oxus side of the watershed are vast snowfields and glaciers, and among these, with three of its sides formed of cliffs of ice—the terminal walls of glaciers—we found a small lake, about three-quarters of a mile in width, out of which flowed the stream which joins the Panj branch of the Oxus at Bozai-Gumbaz" (Younghusband). In September 1894 the district was again visited by the Hon. G. Curzon, who confirms M. Dauvergne's views in every respect, except as regards the "three great glaciers." According to this traveller the Wakh-jir, the true name of Littledale's "Varjer," rises at the foot of a single glacier, which is undoubtedly the farthest source of the Oxus.

The northern branch of the Oxus has its source in another Lake Gaz-kul,[1] which, according to Littledale, the Kirghiz call Chak-mak-Kul, and which stands at a level of 13,850 feet near the head of the Little Pamir Valley at the south foot of the Andemin Pass (15,500 feet). From this basin the main stream of the Oxus, as some authorities still hold this branch to be, flows under the name of Ak-su ("White Water"), first east down the Little Pamir valley, then north and north-west round the Great Pamir to its junction with the Ak-baital from the north, where it takes the name of Murgh-ab or Bartang for the rest of its course westwards to the Panj confluence at Kala Wamar. Formerly the Murgh-ab was supposed to receive through the Ak-baital the overflow of the Great Kara-kul, by far the largest lake on the Pamir. But this lake is at present certainly a closed basin, although it may have discharged to some of the surrounding rivers at a time when it stood at a considerably higher level than it now does (12,800 feet).

Below the confluence at Wamar, the Oxus, here flowing at a level of 6550 feet, turns westwards between Afghanistan and Bokhara in the direction of Balkh; but before reaching that place it gradually trends round to the north-west, and retains this direction for the rest of its course to the south coast of the Aral Sea. During its upper course it receives the Surkh-ab (Kizil-su,[2]

[1] That is, "Goose Lake," a term applied to several basins in this region, no doubt because frequented during the season by flocks of Brahmin geese.

[2] *Surkh-ab* is simply the Persian translation of the Turki *Kizil-su* = "Red River." Both forms are common throughout Central Asia, and often give rise to misunderstandings. Thus the Kizil-su here in question has been confused with another Kizil-su which rises in the same district, that is, the north slope of the Kizil-Art, but which flows in the opposite direction eastwards to Kashgar, where it takes the name of the Kashgar-darya.

Waksh) from the Trans-Alai and Karateghin Mountains, besides numerous other feeders on both its banks from the Hindu-Kush and Bokhara highlands, but none lower down for a distance of 700 miles, or about half of its entire length. At Kilif its bed is narrowed to 350 yards by the advanced spurs of the Hissar hills, but in the plains it broadens to a mean breadth of 800 yards with a depth of 20 feet and a velocity of over 5 miles during the floods which last from May to October. At Pitnyak, about the head of the Khivan irrigation works, the discharge is about 125,000 cubic feet per second, which is nearly equal to that of the Nile. But fully half of this volume is absorbed in supplying the irrigation canals of the Khivan Oasis, by which over 4000 square miles of marvellously fertile alluvial land are kept under cultivation. The sedimentary matter yearly brought down exceeds 16,000,000 tons, some of which maintains the productiveness of the soil, while much of it forms shifting banks and bars in the Taldik and Yani-su, the two main channels through which the Oxus enters the Aral Sea.

These channels present the usual outlines of a delta. But the triangular space thus formed is not a true delta, as it consists, not of alluvial deposits, but of much older formations, through which the river has cut its way to the Aral. Separate little deltas, however, have been developed at the Taldik and Yani-su mouths, where vessels drawing over four feet are already excluded by the accumulating deposits. Nevertheless the Oxus has been regularly navigated by small craft since 1875, when the Petrovsky steamer, drawing about three feet six inches, forced its way up the Yani-su and so-called Kuvan-Jerma, or "New Cut," to Nukus, at the head of the false delta.

The tendency of the Oxus, like that of the great

Siberian rivers, is to press continually on its right or east bank. The consequence of this tendency, which is due to the rotation of the earth round its axis from west to east, is that the stream has been gradually deflected from the Kungrad channel, navigable in the seventeenth century, but now dried up, eastwards to the Taldik channel, now slowly disappearing, and thence to the Yani-su, or "New River," which thus at present receives the main discharge into the Aral. But in former times a far greater deflection took place. For it is now ascertained beyond doubt that no less than twice during the historic period the Oxus has oscillated between the Caspian and Aral Seas. In the time of Strabo it was a sort of eastern continuation of the Kura water-highway, affording a continuous trade route from Georgia across the Caspian and the Khwarezm desert, under the 39th parallel, to Charjui, and so on to Baktria (Balkh) under the Hindu-Kush. Its course across the desert in this direction seems to be still indicated by the Igdy and other wells dotted over the plains in a line with its former bed, which reached the Caspian not through the Atrak, as was at one time supposed, but most probably directly through the depression between the Great and Little Balkan hills.

Later on Edrisi and other early Arab and Turki writers find the Oxus flowing, as at present, to the Aral. But in the fourteenth century it had been again diverted to the Caspian; this time presumably through a fresh bed, known as the Uzboi. This bed is supposed to have run from near Nukus westwards to the Sary-Kamish steppe lakes, and thence southwards to the Igdy wells, and so on along the original bed between the Balkans to the Caspian, close to Mikhailovsk. Guédroich, however, who carefully surveyed the Oxus delta and the Sary-Kamish lakes in 1880, found no indications that the Oxus ever flowed through the Sary-Kamish basin. In any case it seems evident, in the light of recent geological surveys by Mushketof and Konshin, that the Uzboi is a very old marine channel, the last in fact through which the Aral communicated with the Caspian after the two basins became separated by the upheaval of the Ust Urt plateau and the Balkan ranges (*Ausland*, 1893). Konshin shows that the whole of the Kara-Kum desert originally formed part of this Aralo-Caspian basin, and that the series of "ungusses," or dried-up lakes or depressions, such as the Charjui-ungus and Kelif-usboi intersecting the desert from north-west to south-east, are in fact an ancient shore-line of the great inland sea, and not, as formerly supposed, an old bed of the Oxus.

The hopes till recently entertained by the Russians of restoring the Oxus to the Uzboi channel have now been abandoned for the apparently more feasible project of connecting the Aral and Caspian through the bed of the steppe river Chagan, round the northern edge of the Ust Urt plateau. The Chagan is only 65 miles from the Aral, and probably at a somewhat lower level, while it seems to have formerly reached the Caspian at the bay of Chuche-bas, through the deep Arys valley. The Aral being 243 feet above the Caspian, if its waters can be brought to the Chagan the connection will be established.

On the other hand, all these ambitious engineering projects are confronted with the fact that Turkestan, like most of Central Asia, comes within the area of desiccation that has been in progress since the remotest times. The Zarafshan and Murgh-ab, the two great former tributaries of the Middle Oxus on its right and left banks, are now both absorbed partly in the sands, partly in irrigation rills before reaching it. The Zarafshan, or "Gold Distributor," rises in the Alai range, at the foot of a stupendous glacier (9000 feet), which is still 30 miles long, but which Mushketof recently traced for 33 miles below its present limits. In this magnificent ice-world the Zarafshan is fed by no less than thirteen secondary glaciers, beyond which it receives the outflow of the romantic Lake Iskander, at an elevation of 7000 feet above the sea. Thence it flows westwards down to the Bokhara plains, being first tapped by the Russians at Samarkand, and then diverted into a thousand irrigating canals by the Bokhariots. Thus 60 miles before reaching the Oxus, the "Gold Distributor" is completely used up in bringing nearly 2,250,000 acres under cultivation.

The Murgh-ab (or "Water-Fowl River"), which flows from the Garjistan Mountains in North Afghanistan, also runs dry in the sands, after supplying the numerous irrigating rills of the Merv Oasis.

The Sir-darya Basin.

The Jaxartes — Sihun of the Arabs, Sir-darya (or "Head-River") of the Persians, twin sister of the Oxus— has its source in the very heart of the Tian-shan. Here the Narin, as its upper course is called, has its chief head-stream at the foot of the Petrov glacier in the Ak-shiirak hills (12,000 feet), whence it flows at first westwards through the former Khanate of Khokand, and present Russian province of Ferghana. Beyond Khojend it turns abruptly north-westwards, thenceforth running parallel with the Oxus to the north-east end of Lake Aral, which it reaches after a course of about 2000 miles, or 1150 from the confluence of the Narin and Kara-darya, where it takes the name of Sir-darya. Through the unexplored Kapchegai defile the Narin falls nearly 3300 feet in a space of 46 miles, and passes through two other romantic gorges before reaching the Ferghana plains.

At Khojend the Swedish explorer, Sven Hedin (1893-95), found the Sir to be 430 feet wide and nearly 13 feet deep, with a velocity of $2\frac{1}{2}$ feet per second, and a volume of nearly 12,900 cubic feet per second in winter. The summer rise at this point is nearly 15 feet, and the fall in some places 3 feet per mile. But below the gorge near Khojend it descends not more than 8 or 9 inches per mile for the rest of its course of 990 miles to Lake Aral. At Kasalinsk it is 1066 feet broad and 8 feet deep at the beginning of winter, with a discharge of nearly 20,000 feet. Much of this discharge is used especially for irrigating the productive rice-fields of the plains, so that the Sir-darya has been called the "Nile of Turkestan." It also overflows its banks, extensive marshes occurring between Khokand and Marghilan, while temporary morasses and lagoons are formed lower down, especially on the right bank and east of Peroffsky. In summer much water is absorbed in the sands and by evaporation, so that despite the heavy floods the volume reaching the Aral Sea at that season is probably not greater than in winter. Between Ferghana and Khojend the river never freezes, but in the lower reaches, as at Kasalinsk, it is often frozen from December to April.

Like the Oxus, the Sir has frequently shifted its lower course. But it can never have reached the Caspian directly, as has been asserted, but only through the Oxus, with which it formerly communicated through the Yani-darya. This channel branches off from the main stream 7 miles below Peroffsky, but although occasionally flushed during the floods, it now never gets beyond Lake Kukcha-denghiz, 180 miles from the Sir, and 60 from the Oxus delta. Below Peroffsky the main stream enters the Aral through one large branch and several small channels, forming a shifting marshy delta, haunted by an astonishing number of wolves, wild boars, and deer.

With a bar covered sometimes by scarcely three feet of water, the Sir cannot be regarded as a navigable river, although the light craft of the Russian flotilla have contrived to reach Peroffsky and return to the Aral without getting embedded in the shifting sandbanks.

The Aral Sea.

The Aral—that is, "The Inland Lake"—has a present area of perhaps 26,000 square miles. But it was formerly far more extensive, the water-marks on the Chink and many other indications showing that its level was at least 200 feet higher than at present. Yet it still stands 243 feet above the Caspian, and consequently 158 above the Mediterranean. It can scarcely be shown that in historic times the Aral has ever been deprived of both of its great influents, on which its existence entirely depends. It is abundantly evident that but for them the lake would disappear in a few years. Even as it is, a slow process of desiccation is steadily going on, by which its size has in a short time been reduced by 1400 square miles.[1]

Under the Chink the Aral is about 225 feet deep; but it shoals continually eastwards, until it becomes merely a flooded swamp along its east and south-east coasts. Hence the mean can scarcely be more than 40 feet, which would give a volume only eleven times greater than Lake Geneva, while exceeding that basin 116 times in area.

Although the Turkestan depression was at one time covered by a vast marine basin, the Aral Sea cannot be regarded, like the Caspian, as a relic of that period. In

[1] It is remarkable that neither the Greeks nor Marco Polo make any mention of the Aral.

its brackish waters, no doubt, both a fresh and salt water fauna exist; but the former greatly predominates, while the latter is also common to the Caspian. Seals have been spoken of by Pallas and others, which might point to an independent connection with the Arctic Ocean, at least if of a different species from those of the Caspian. But later observers have made it evident that none of these cetacea are found in the Aral basin, which must on the whole be regarded as a sort of intermittent steppe reservoir, oscillating between the conditions of a lake and a mere morass, according to the vagaries of its great feeders, the Oxus and Sir.

Recent surveys, and especially the researches of Colonel Koslowski in 1890-91, have shown that the deep narrow Aibu-ghir inlet at the south-west corner of the Aral Sea has no existence. But in this depression Koslowski has discovered a fresh-water lake quite distinct both from the Aral and the Sary-Kamish basins. Unlike the latter, which are mostly dry, it is permanently flooded, being fed by a fresh-water stream from the north-east, which, although not in direct communication with the Oxus delta, drains the marshes formed by the overflow of that river. This newly discovered lake appears to be part of the former great Aralian basin, which has become isolated by the general process of desiccation going on in the whole of the region. The absence of salt in solution may be due either to the formation of salt-beds under the sands, or to some great overflow of the Oxus filling the basin with fresh water, driving the salt water back to the Aral Sea, and at the same time throwing up a bar with the alluvia brought down by the floods, and thus preventing the return of the salt water from the Aral Sea.

Lakes Balkhash and Issik-kul.

A similar statement might almost apply to Lake Balkhash, if regard be exclusively paid to the fact that this basin also has in recent times diminished enormously in size. But other considerations show that it could at no time have been exhausted, so that it may be regarded as a true remnant of an old lacustrine basin. It is evident from the study of the shells in these depressions made by Mr. W. Bateson in 1886-87 that the Quaternary Aralo-Caspian Sea cannot have extended eastwards so far as to include Lake Balkhash.

Even within the historic period this lake, which is at present 514 feet above the sea, spread out westwards to three or four times its actual area, while stretching nearly 250 miles towards the east, where it absorbed the now isolated Sassik, Jalanash, and Ala Lakes, south of the Tarbagatai range. Even still its area, which is somewhat fluctuating along its low-lying south coast, cannot be less than 8500 square miles, with a total length of 330 miles and a circumference of 880 miles. But the depth nowhere exceeds 56 feet, so that it has only twice the volume of Lake Geneva, which it surpasses in extent thirty-six times. Its waters, which are very slightly brackish (M. Krasnof says " nearly sweet "), abound in fish, although generally frozen over from December to April. Besides the Ili, it receives several affluents on its south coast, which is over 450 miles long. Such a quantity of alluvia is brought down by these streams from the Zungarian Ala-tau that the lake seems to be slowly filling in. The surveys of MM. Krasnof and Ignatief (1886-87) show that the Ili delta is shifting eastwards. Till lately the main branch reached the lake near its western extremity; but this and all the other western channels have now become mere pools of standing water, while most of the discharge has been diverted to the eastern arms of the delta.

Of the upland lakes by far the largest is Issik-kul, which lies at an elevation of 5300 feet in the heart of the Tian-shan between the Ala-tau Kungei and the Ala-tau Terskei. But notwithstanding its great altitude this lake belongs still to the Aralo-Caspian basin; for it drains intermittently through the Kutemaldi to the Chui, which formerly reached the Sir at some point below Peroffsky. Like the steppe lakes, the Issik-kul has greatly diminished in size, as is evident from the water-marks fully 200 feet above its present level. Yet it has still an area of about 2300 square miles. Its blue and somewhat brackish waters abound in fish, and are overlooked from the east by the Khan-tangri, the giant of the Tian-shan.

4. *Natural and Political Divisions: Uralsk — The Daman-i-koh—Khiva—Bokhara—Ferghana—The Kirghiz Steppes—Semirechinsk.*

As already remarked, the Russian administrative divisions are not based upon ethnical and physical considerations. Thus the province of Uralsk, east of the Ural River, includes portions of the districts of Uralsk, Gurief, and Kalmikov, west of that river, while the Nikolayevsk district of the adjoining province of Turgai lies within the limits of the Ob basin. On the other hand, the Sari-Suisk district of the Akmolinsk Province, West Siberia, encroaches across the Ob water-parting on North Turkestan.

The eastern section of Turkestan between the Aral Sea and the Chinese frontier is divided politically between Russian territory and the still nominally independent Khanates of Bokhara and Khiva. The Russian portion, collectively called "Eastern Turkestan" in official documents, comprises several natural divisions, which to some extent coincide with the administrative provinces and circles. Thus the province of Semirechinsk, or "Land of the Seven Streams," lies mainly between the Zungarian and Trans-Ili Ala-tau and Lake Balkhash; the province of Ferghana, the former Khanate of Khokand, comprises the upland valley of the Narin (Sir); while the circles of Zarafshan and Amu Daria embrace the tracts watered by those rivers mainly between Bokhara and Khiva.

Uralsk includes all the land between the Ural and the Obshchy Syrt, and between the Caspian and Aral north of the Ust Urt plateau. Here the West Siberian grassy steppes are continued southwards to within 100 miles of the Caspian, the intervening space being occupied by saline wastes. These are succeeded by argillaceous desert tracts, which extend eastwards between the grassy steppe and the north coast of the Aral Sea to the Kara-kum sands. The same argillaceous formation prevails throughout the Ust Urt plateau, and along the east coast of the Caspian southwards to the old bed of the Oxus.

But the Daman-i-koh,[1] or northern skirt of the Khorasan range, is fringed by an almost continuous strip of fertile tracts or rich pasture-lands stretching between the hills and the desert from the Little Balkans to the Tajand valley. Here is the domain of the Akhal Tekke Turkomans, who were reduced in 1881 by the Russians. This district, which is dominated by the Kerawul (5000 feet), the Kuh-Giffan (7770), the Kuh-Bughun (8000), and other lofty crests of the Kuren and Kopet ranges, now forms by far the most important division of the Russian Trans-Caspian territory. It is now traversed in

[1] Daman-i-koh is sometimes used by English writers in the sense of a mountain range. But it simply means the skirt of any mountain range, and is especially applicable to such ranges as slope somewhat rapidly down to the plains.

its entire length by the Trans-Caspian railway. Abundant fuel for the locomotives on this line is supplied by the rich naphtha wells on the coast of the Caspian and the islands close to the Mikhailovsk terminus. A horse tramway now also runs from the Bala-Ishem station for 40 miles to the centre of the oil-producing districts.

Khiva itself properly belongs to the same sandy formation, and is only rendered fertile by the alluvia of the Oxus combined with a fine system of irrigation developed on an extensive scale. Hence the difficulty of assigning any definite limits to the khanate, which is almost everywhere encircled by desert and steppe lands, except on the north and east, where it is bounded by the Aral Sea and Oxus. Even here the Lower Oxus flows for some distance through wastes of shifting sands, saline marshes, or morasses overgrown with sedge and reeds.

In the centre of this inhospitable region lies the Oasis of Khiva, which lies mainly between the towns of Pitnyak and Kungrad near the now dried-up bay of Aibughir at the south-west end of the Aral Sea. But the settled population is mainly grouped along the left bank of the Oxus as far north as Khojeili over against the advanced Russian fort of Nukus at the head of the false delta. Here there is a rapid flow, by which the intricate system of irrigation works is greatly facilitated. From Pitnyak the land is seen to be covered by a complete network of canals fed entirely by the Oxus. To guard against the danger of excessive floods the Khivans have constructed a dam or level all along the left bank of the river, and through the pipes laid across this dam the flow of water to the irrigating rills is regulated. Works have also been erected to raise the water to the higher grounds; yet notwithstanding the skill and labour spent on these works, they have failed to render productive much more than one-third of the whole deltaic region, which has a total area of some 5500 square miles.

The Oxus is also the main stream of the southern Khanate of Bokhara, which, like Khiva, is still suffered to enjoy a certain show of political independence. It has, however, been deprived of the eastern and more important section of the country, which now forms the Russian circle of the Zarafshan, and includes the city of Samarkand.

Northwards, also, the two khanates are separated by the intervening Russian territory, which here reaches the Oxus across the Batkak and Khalata sands. Between the Russian station of Petro-Alexandrovsk on the Oxus, and Ak-kamysh farther up, the country, though thinly peopled, is well cultivated. But from that point the road crosses a sandy steppe at a considerable elevation above the river, but here and there falling to the level of the stream, where it forms little projecting oases covered with "jidda" and other steppe plants called "tugai" by the natives.

Farther south the land is being continually encroached upon by the shifting sands of the steppe. From the fortress of Ustin to Kara-kul there stretches a sandy waste for a distance of 15 miles, which is dotted over with the ruins of abandoned buildings and other silent witnesses of better days.[1] Even within the last thirty years several flourishing towns in this region have been swallowed up in the sands, drifting like ceaseless billows from the north. Year after year the sand-storms encroach upon the last cultivated plots, and the Kara-kul district between Bokhara and the Oxus already presents a hopelessly desolate aspect. Besides the diminution of

[1] The vast field of ruins stretching from Shahri Gulgula through Shahri Saman to Termes on the Oxus was for the first time explored by M. Bonvalet during the summer of 1881.

moisture, the more immediate cause of this widespread ruin is the increasing absorption of the Zarafshan waters for irrigation purposes in the Samarkand district.

Between Bokhara and Samarkand the land is in some places admirably cultivated. The circle of Kermin in the Miankal valley even surpasses the district of Bokhara, and the country becomes more flourishing as we approach the Zarafshan district. Along the broad valley of this river there stretches an almost uninterrupted chain of townships between the capital and Katti-kurgan. But this flourishing tract is hemmed in by the Kizil-kum sands advancing southwards to the Zarafshan, and scarcely arrested by the isolated ridges of argillaceous schists and igneous rocks here rising to a height of about 1000 feet. Vambery describes this region as a boundless sandy ocean.

The Narin, or Upper Sir, separated from the Zarafshan valley by the Alai ranges, flows through the heart of the Khanate of Khokand, now the Russian province of Ferghana. This upland region is entirely enclosed on three sides by the lofty mountain barriers of the Western Tian-shan, and is open only towards the west, where the Sir escapes down to the Turkestan lowlands. Here the western spurs of the Tian-shan fall in a series of terraces to the plains, and send down numerous mountain torrents, which, however, are mostly absorbed in irrigation works before reaching the Sir. To this abundant supply of water Ferghana is indebted for an exuberant fertility nowhere surpassed in Central Asia. But lower down nearly all trace of vegetation disappears, and the vast region stretching from Ferghana to Lake Aral, between the Sir and Oxus, is mostly occupied by the Kizil-kum sands. Nevertheless there are still several fertile tracts in the strip of land stretching from the right bank of the Sir to the Ak-sai and Kara-tau, the last spurs of the Tian-shan projecting north-westwards between the Kizil-kum and Ak-kum deserts. Here are even several considerable towns, including Tashkent, the present centre of Russian authority in Turkestan.

The region about the Lower Sir everywhere presents the appearance of a land that had once been under water. The saline and clayey soil towards the Aral Sea has been rendered arable by the skilful irrigation works of its former and present rulers. But under the cloudless summer skies the land beyond the reach of the irrigating channels has the aspect of a wilderness strewn with salt, and producing nothing beyond a few prickly plants. The reedy morasses about the delta are infested by dense clouds of mosquitoes, the plague of the Russian sailors navigating these waters.

North of the Aral and Caspian the sands merge everywhere in the grassy Kirghiz steppes, which, with a mean elevation of about 300 feet, consist mainly of vast rolling or gently sloping tracts. The steppe is, however, occasionally intersected by broad and deep furrows. But scarcely a tree or shrub is anywhere visible, the whole region presenting the aspect of a boundless sea whose rolling billows have suddenly become solidified. The only relief to this monotonous picture are the Mugojar hills, a continuation of the Ural range, nowhere reaching 2000 feet in height. The hilly portion of the steppe consists everywhere of feldspar and porphyry, with which are often associated lead, copper, silver, and occasionally even gold. Between Aral and Lake Balkhash the steppe is in many places strewn with lakes and tarns, often strung together like pearls on a string, but everywhere showing the same tendency as the Aral and Balkhash themselves to disappear. All have long been closed basins, nor do the Sari, Chui, or any other of the steppe rivers now ever reach either the Aral or the Sir.

Beyond the water-parting of the Chui and Ili begins

the Semirechinski-Krai, or "Land of the Seven Streams," consisting of the steppe which here stretches at a mean elevation of about 1000 feet between Balkhash and the Zungarian Ala-tau north and south, and between the Lower Ili and Lepsa rivers west and east. Although thus partly severed from the Central Asiatic highlands, this region is connected with them through the deep valley of the Ili leading to Kulja, recently restored by the Russians to China.

The seven streams whence the land derives its name are the Lepsa with the Baskan, the Ak-su with the Sarkan, the Biyen and the Karatal with the Kok-su. The Lepsa, Ili, and Karatal alone reach Balkhash throughout the year, all the others either losing themselves in the sands or discharging their waters into the lake only during the floods. They all rise in the snowy Ala-tau, and flow through fertile upland valleys before reaching the open plains. These plains are dotted over with brackish lagoons, and their vegetation resembles that of the Aralo-Caspian depression, while the cultivated and well-watered upland tracts recall the lowlands of West Siberia and the East of Europe.

5. *Climate: The "Fever Wind."*

Notwithstanding the extraordinary difference in the relief of the land, the climate of the Aralo-Caspian basin is everywhere characterised by a remarkable uniformity. It is distinctly continental in its main features, intense heat being followed by equally intense cold, while great dryness prevails over the whole area. This general uniformity is largely due to the low elevation of the Aralo-Caspian and Ob water-parting, which offers no obstacle to the full play of the northern winds from the Arctic Ocean across the Turkestan depression. Thus the difference in elevation between this depression and the Tian-shan and Pamir uplands is to a great extent neutralised in winter, when the glass falls to 30° or 40° below freezing-point in the Kizil-kum sands 300 feet above the sea, as well as on the Pamir plateau, standing at an altitude of 14,000 feet. The chief difference between the uplands and lowlands is the excessive rarefaction of the atmosphere on the great tablelands, due to their enormous elevation above sea-level. A greater quantity of moisture also falls on the highlands, the rain-bearing clouds from the south-west being first arrested by the Pamir and Tian-shan uplands. But even here the rainfall is much slighter than in the European highlands; while at times whole years pass without any rain falling on the Kara-kum and Kizil-kum sands. Such are altogether the climatic conditions that most of the Aralo-Caspian basin would be uninhabitable but for the moisture deposited mostly in the form of snow on the elevated lands. From this reserve the great rivers are fed, which in their turn supply the irrigation works on which the Bokhara, Khiva, Merv, and other oases are dependent.

A special study of the Pamir climate was made in the year 1887 by M. Capus, who limits the summer season, characterised by absence of frosts, to at most three weeks in July; spring and autumn last each a little over two months, while winter prevails for the rest of the year. But even in this season very low temperatures are seldom continuous for any great length of time. On the other hand, the variations are excessive between day and night, sun and shade, and the rise and fall of the thermometer very rapid, often ranging from the freezing-point of mercury to over 32° F. within the twenty-four hours. Snow is very unevenly distributed, being determined by the degree of exposure to the winds, aspect of slopes, altitude, nature of the soil, and such-like conditions. In general the Pamirs proper have a far less heavy snowfall than the Altai, owing to the fact that most of the moisture is intercepted by the border ranges. Many parts are bare even in winter, when they still afford a little pasturage. The snows are also quickly melted, owing to the large amount of radiant heat absorbed by the generally dark soil. Hence also glaciers scarcely exist, while the drainage is evidently much less than formerly, so that the Pamirs would appear to be affected by the general process of desiccation going on throughout the zone of Central Asian depressions (*Bulletin*, Paris Geo. Soc., 1892).

Along the Lower Oxus, which is usually frozen for four or five weeks only, summer begins in April and lasts occasionally into November. The long hot and dry season is rendered still more oppressive by the dust-storms prevalent throughout the lowlands. But more dreaded still is the "tebbad," or "fever-wind," to which the Kizil-kum and other desert tracts are exposed. At its approach the camels of the caravan utter loud moaning sounds, and cowering to the ground stretch their long necks flat on the parched land, or seek to bury their heads in the hot sands. Behind them crouch the terrified guides and travellers, while over them the pestiferous blast sweeps with a dull soughing. The whole caravan is soon covered with a layer of hot sand, which falls like a shower of fiery sparks.

6. *Flora and Fauna: The Saxaul—Mosquitoes and Locusts—The Turkoman horse.*

Vegetation is represented in the wilderness chiefly by the saxaul, the jidda or wild olive, the poplar, and other hardy or prickly plants, which have invaded this domain from all quarters since the subsidence of the waters. As a rule, the flora of the Iranian tableland advancing northwards has prevailed over that of Siberia, and "it is interesting to observe how all those plants gradually adapt themselves to the changed conditions of soil and climate in the steppe. To resist the wind they acquire a more pliant stem, or present a smaller surface to its fury by dropping their foliage. To diminish the evaporation their bark becomes a veritable carapace, and their pith becomes mingled with saline substances. They clothe themselves with hairs and thorns, distilling gums and oils, whereby the evaporation is still further reduced. Thus are able to flourish far from running waters such plants as the saxaul, which, though perfectly leafless, produces both flowers and fruits. So close is its grain that it sinks in water, and emits sparks when struck with the axe." [1]

In the oases the cultivated plants, both cereals and fruits, are noted especially for their great abundance and excellent quality. In Khiva wheat will yield sixty, rice seventy, and the *jugara* as much as three hundredfold. This latter grain takes the place of oats, and its stalk that of hay for horses and cattle. Barley, lentils, and pease are also cultivated, besides cotton, hemp, *kunshut* (an oil-yielding fruit), madder, flax, and tobacco. The lack of grazing grounds is here obviated by the lucerne clover, which is mown three times and yields excellent fodder. But the special glory of this oasis are its fruits, which are remarkable for their fine flavour. Here flourish choice apples, pears, plums, apricots, peaches, the grape, the pomegranate, and, above all, the melon. Of trees, the poplar, naruan, and elm are grown for their timber, and the mulberry for the silkworm.

The mulberry, flax, and maize form the staple pro-

[1] E. Reclus, vi. p. 196.

duce of Ferghana, which also grows fine crops of wheat, rice, sorghum, cotton, and tobacco. In Semirechinsk fertile and well-watered upland valleys are succeeded by grassy steppes stretching away to the low-lying swampy shores of Balkhash. Higher up a splendid wooded zone clothes the slopes of the Zungarian Ala-tau, between 4500 and 8500 feet. In the forests of the Central Tian-shan the prevailing trees are the mountain ash and the spruce (*Picea Schrenkiana*), which are now supplanting the apple and apricot. The spruce "attains a height of 70 to 80 feet, with a thickness of stem 2, 3, and often 4 feet in diameter. It grows very much in the sugar-loaf shape, its thick branches hardly projecting from the general mass, so that the whole tree has the appearance of having been cropped by a barber" (Morgan's *Prjevalsky's Lob-nor*, p. 40). The spruce grows as far as 8000 feet and upwards above sea-level. These varied advantages, combined with a healthy climate, have attracted numerous Russian settlers to Semirechinsk, which has become a chief centre of Slav culture in Central Asia. Even though sparsely wooded, the elevated lands, with their abundance of water, supply at least some of the necessary essentials for the development of social culture. The low, flat steppes, with their arid wastes, here, as elsewhere, arrest the progress of civilisation. These waterless and treeless tracts admit of nothing but a nomad existence.

In the whole of Russian Turkestan, about 162,000,000 acres in extent, it is calculated that not more than 945,000 acres are under timber on the uplands, and less than 16,000,000 under bush and scrub on the steppes. As this is quite inadequate to supply the demand of the steadily increasing population, even for building purposes alone, and as the woodlands have suffered much of late years from reckless cutting, attempts are now being made to replant the land both on the mountains and the plains. Some success has attended these efforts, especially in the province of Samarkand. The saxaul, which formerly fringed both banks of the Sir-darya, has been almost exterminated, or, as the natives say, "has fled into the depth of the steppes." It is hoped that its restoration and the enlargement of the forest area may have the effect of arresting the deterioration of the climate, which is everywhere becoming drier. Both glaciers and rivers continue to lose volume; the lakes are shrinking, and the extremes of temperature become more marked, while the sands of the desert are steadily encroaching on the cultivated zones.

In the Turkestan lowlands a characteristic feature of animal life are the scorpions, lizards, snakes, and other reptiles, with which all the fissures in the ground are alive. But the special plague of these regions are the mosquitoes and locusts. "Mosquitoes are serious evils in many other parts of the world, and stories have been told of seamen driven to jump overboard, and so to commit suicide on this account, in the Rangoon River. But it may be doubted whether any more exquisite torture can be suffered than that inflicted by the mosquitoes of the Lower Amu" (Major H. Wood). The same traveller, speaking of the prodigious quantities of locusts which swept over the Khivan Oasis in July 1874, remarks that "one of such clouds was estimated to measure 15 miles in length by 2 miles in breadth, and to have a depth of half a mile. . . . It might be inferred that the Khivan Oasis must be exposed to great danger from this plague; but in practice the large number of small birds in the planted groves of the khanate seem to afford a sufficient safeguard against their ravages."

Large beasts of prey, such as the wild boar, tiger, ounce, wolf, haunt the thickets of the swampy Oxus and

Aral deltas. But the open desert is frequented only by swift gregarious animals, like the wild ass and gazelle.

The camel, horse, sheep, and cattle are everywhere the prevailing domestic animals. In Russian Turkestan camels are not so numerous as might be supposed, scarcely amounting to 400,000, as compared with over 1,600,000 horses, 1,160,000 cattle, 1,350,000 sheep. The Kirghiz horses are a hardy, active breed, which traverse distances of 40 to 50 miles at a stretch. But a far finer animal is the Turkoman horse, which possesses many of the best points of the Arab and English with some excellent qualities peculiar to itself. "Well do these noble animals deserve all the care that is lavished on them, for in courage, speed, and endurance combined they stand at the head of the equine race. It is probable that the race dates back, like our own thoroughbred, to the Arab. But the race is now distinct; and, besides being much larger, they far excel the Arabs both in speed and endurance. . . . In appearance they more nearly resemble the English racehorse than any other type, and average about the same height."[1]

7. *Inhabitants: Table of the Turkestan Races.—The Usbegs—Kara-Kalpaks—Kara-Kirghiz—Kirghiz-Kazaks—Turkomans—Tajiks—Sarts—Galchas—Russians.*

The Aralo-Caspian basin is commonly supposed to be the exclusive home of the Turki race, from whom this region takes the name of Turkestan, or "Land of the Turk." But the statement is true only of the unarable but still inhabitable grassy upland and lowland steppes, whose first occupants seem to have been the nomad tribes of Turki stock, by whom they are still mainly inhabited. On the other hand, the arable tracts, especially in Khiva,

[1] *Clouds in the East*, p. 214.

Bokhara, and Ferghana, have apparently from prehistoric times been the joint home of men of Turki and Iranian blood. Here an incessant intermingling of the two races has been going on for ages, resulting in a profound modification of both types, now represented by every shade intermediate between the two extremes. A third element, entirely distinct from the Turki, and allied to but not identical with the Iranian, is found in almost exclusive possession of the productive upland valleys of Ferghana, the Zarafshan, and the Oxus. To these highlanders Ch. de Ujfalvy has given the collective name of Galcha, and these Galchas, whose true position seems to be intermediate between the Iranic and Indic branches of the Aryan family, are obviously allied to the Wakhis, Badakhshis, Siah-Posh Kafirs, Chaganis, and other highland races holding the upland valleys on both sides of the Hindu-Kush.

To the primitive Galcha, Iranian, and Turki stocks are therefore reducible all the varieties of mankind from time immemorial in possession of the Aralo-Caspian basin, as in the subjoined scheme:—

I. TURKI STOCK.

Usbegs . .	Kungrad . . .	Bokhara, Ferghana, Khiva . . .	2,000,000
	Naiman		
	Kipchak		
	Jalair		
	Andijani . . .		
Kara-Kalpaks	Baymakli . . .	S. and S.E. shores Aral Sea mainly .	50,000
	Khandelki . . .		
	Achamayli . . .		
	Ingakli		
	Shaku		
	Ontoturuk . . .		
Kirghiz-Kazaks .	Great Horde (Ulu-Yuz) . . .	Steppes between Lake Balkhash and Lower Volga . .	450,000
	Middle Horde (Urta-Yuz) . . .		1;100,000
	Little Horde (Kachi-Yuz) . . .		1,000,000
	Inner Horde (Bukeyevskaya . .		200,000

Kara-Kirghiz (Buruts)	Right Section ("On")	Tian-shan and Pamir	300,000
	Left Section ("Sol")		100,000
Turkomans	Tekke Goklan Yomud Sarik Salor Kara Ali-Eli Ersari Chaudor	Ust Urt, Khwarezm, Daman-i-koh, left bank Middle Oxus	? 600,000

II. Iranic Stock.

Tajiks "Sarts" Persians	Khiva, Bokhara, Ferghana	? 1,000,000

III. Galcha Stock.

Maghians Kshtuts Falghars Machas Fangs Yagnaubs Karateghins	Ferghana, Zarafshan, and Karateghin highlands, Upper Oxus valleys	? 250,000

The Usbegs.

Of all the Turki peoples of Central Asia the Usbegs are by far the most civilised. The great majority in the two khanates have long abandoned the nomad life, and are now the chief agricultural element in Khiva and Bokhara. The term itself is rather political than ethnical, being the collective name of the numerous Mongolo-Tatar tribes, who became the dominant people in this region after the dissolution of Jenghiz Khan's empire, But although the ruling race, the Usbegs are intellectually inferior to the conquered Tajiks, with whom they now live in social harmony, and through alliances with whom they have become largely assimilated in appearance to the Iranian type. Many are settled in all the large towns, where they are partly occupied with trade, and furnish the principal contingent to the army. Besides the Osmanli, the Usbegs are the only Turki people who possess a written language and a literature. Sultan Baber, founder of the so-called Moghul Empire in India, who belonged to this race, composed his well-known Memoirs in the Jaghatai, which is still the standard literary language of all the Central Asiatic Turki peoples. Like the Tajiks, the Usbegs are all Sunnis.

AN USBEG MUSICIAN.

The Kara-Kalpaks—that is, "Black Caps"—formerly a widespread branch of the Turki family, now occupy a restricted area round the shores of Lake Aral between the mouths of the Oxus and the Yani-darya. They are a harmless, but feeble and somewhat sluggish race, evidently in process of extinction or absorption by their former oppressors, the Usbegs of Khiva.

The Kirghiz.

The Kara-Kirghiz, or "Black Kirghiz," and the Kazaks, or Kirghiz-Kazaks, represent respectively the highland and lowland nomad elements all along the northern and eastern border-lands of the Aralo-Caspian basin. The Kazaks, who are by far the more numerous division, have never accepted the name of Kirghiz, which has been imposed upon them by the Russians to distinguish them from their own Kossaks.[1] Hence it

A KIRGHIZ.

would be more correct to speak of a Kazak than of a Kirghiz steppe, the Kazaks being here the exclusive

[1] The unaccented *o* in Russian being pronounced like *a*, Kossak and Kazak have very much the same sound in the mouth of a Russian.

element, whereas the Kara-Kirghiz dwell in the Tian-shan highlands and Great Pamir. At the same time there is no substantial difference between the two peoples.

A KIRGHIZ TENT.

The Kara-Kirghiz, who are the "Buruts" of the Chinese and Kalmaks, live partly in Zungaria and Turkestan, partly in the Western Altai, in the hilly districts about the source of the Sir and its tributaries, in the Alexander range, in the highlands about Issik-kul and southwards to the sources of the Oxus on the Pamir. They speak an almost pure Turki dialect, and their two great sections, *On* or "Right" and *Sol* or "Left," are again subdivided into numerous tribes and septs. North of the Sir their grazing grounds are limited northwards by the Kazaks, while stretching southwards to the Hindu-Kush. Their camping-grounds in the Tian-shan are here and there overlapped by the lands of the warlike Galchas. The northern Kara-Kirghiz have no common bond of union, nor any kind of political organisation. Even the lesser tribes are often split up into independent hordes still living in a state of constant feud amongst themselves. Thus all their energies have been wasted in internecine strife, or else in chronic hostilities with the Kazaks. Hence in spite of their personal courage they have at various periods been easily subdued by the Chinese, or by the Khans of Khokand. In recent years all the right section and many of the left have accepted Russian supremacy. Some years ago a small number penetrated to the Sarikia pasturages on the Karatash River near Sanju, the southernmost point ever reached by the Kara-Kirghiz.

The Kazaks occupy a somewhat intermediate position between the Turki and Mongolian races, possessing many physical traits in common with the latter, while still speaking a pure Turki dialect. Their four hordes, all now subject to Russia, occupy a vast domain stretching from the Lower Volga to Zungaria, and from beyond the Aralo-Caspian and Ob water-parting southwards to the Aral Sea.

The Kara-Kirghiz is of a sullen, rude, and fierce temperament, but he is more straightforward and good-natured than the Kazak. Both are Muhammadans in little more than the name, without mollahs, mosques, or fanaticism, their whole religion being limited to a few simple rites, strongly tinged with the traditions of the old Shamanist cult. The Kirghiz cultivate more land than the Kazaks, but both are essentially stock-breeders, living mainly on the produce of their herds. Their chief drink is *kumiss,* fermented mare's milk, which is preserved in skins, and largely consumed throughout the spring, summer, and autumn. Kumiss is very wholesome, and a specific against all consumptive diseases.[1]

The monotony of nomad existence was formerly relieved by tribal warfare, and by the so-called "barantas" or marauding expeditions generally directed against the encampments of their neighbours. The attacks were usually made towards dawn, when man and dog alike, wearied with the night-watch, were buried in sleep.

The Kazaks have all been gradually induced to acknowledge the sovereignty of the Czar. The Russian Government has been the more anxious to effect their submission, that all the caravan tracks between the Caspian and Altai highlands traverse their domain. Towards the south-east a few of the Kara-Kirghiz tribes still maintain their independence beyond Lakes Balkhash and Zaisan, within the Chinese frontier. But some of these have of late years shown a tendency to migrate westwards into Russian territory.

The Turkomans.

The Turkoman nomads have now also tendered their submission to the "White Czar." Even those of the Merv Oasis, struck with the irresistible power of the

[1] In *Koumiss . . . and its Uses in the Treatment and Cure of Pulmonary Consumption* (Blackwood, 1881), Dr. G. L. Carrick fully establishes its claim to be regarded as a sovereign remedy for all affections of the chest and lungs.

Russian arms, and flattered by the courteous reception accorded in the spring of 1881 to their chief, Tikma Sirdar, by the St. Petersburg authorities, were at last induced to accept the proffered protectorate of Russia in the year 1882. The example of the Merv tribes was soon followed by those of Yuletan, 35 miles farther

A TURKOMAN.

south, and in 1885 those of Penjdeh, the last independent section of the race, were incorporated in the Russian empire.

Throughout the historic period the Turkomans or Turkmenians seem to have been a plundering nomad race, never at any time united under an organised political system. They are divided into "Khalks" or tribes, each comprising several "tayfe" or hordes, who are again grouped in a number of "tir" or septs. Of all the Khalks the most friendly and civilised are the Goklans, who have long been settled in the Persian province of Astrabad. Although recognising no general leader, except, perhaps, for a short time in great emergencies, the Turkomans do not live in a normal state of anarchy. Offences against their own unwritten code are even rarer amongst them than amongst other Muhammadan peoples. Everything is regulated by the all-powerful "dab" or "custom," religion exercising but a slight influence.

The various tribes formerly lived in mutual hostility to each other, showing little fear of the Persian Government, but great respect for the Russian power. To their individual tribes they remain true to the last, and even little children, five or six years old, know exactly the tayfe and tir to which they belong. The Turkomans are distinguished from other Asiatics by a bold, penetrating glance, developed by the dangers surrounding the "alamans," or marauding excursions, to which they have been addicted from the remotest times. These alamans were preconcerted affairs, in which every precaution was taken against failure. The attack usually took place about midnight or at sunrise, and was generally successful. The Persian caravans were constantly taken by surprise, all who showed any resistance being cut down, and the rest carried off into slavery. But since the predominance of Russia in Turkestan the Khivan and Bokhara slave-markets have been closed. This was the first blow given to the Turkoman power, which was almost reduced to national bankruptcy by the stoppage of a traffic on which its very existence depended. Then came the massacre of the Yomud Turkomans, followed in 1881 by the crushing defeat of the Akhal Tekkes at Geok-tepe.

The Tajiks and Slavs.

The Tajiks are to be distinguished both from the Sarts and Persians. They are the original Iranian element settled in all the arable lands throughout Turkestan from the remotest times. It is probable that the diffusion of the Iranian race in this region, regarded as the peculiar home of the Turki peoples, was brought about by the extension of the old Persian empire through Margiana (Merv) and Baktriana (Balkh) to Sogdiana (Bokhara). The Tajiks, direct descendants of those early Iranian settlers, differ in many respects from the Persians, who have in comparatively recent times settled in most of the Turkestan towns and oases. Many of them are descendants of those carried into captivity by the Turkomans, and sold in the Bokhara and Khivan slave-markets. They are everywhere the most industrious and intelligent section of the community, and are occupied chiefly in the cultivation of the land.

Sart is a term which has given much trouble to ethnologists, who have sought for a distinct race in a social distinction. The word seems to have originally meant "dealer" or "trader," and never had any ethnical value at all. It is applied to the settled in opposition to the nomad element in Turkestan, irrespective of race or nationality.

The Galchas, by De Ujfalvy at first described as "Tajik highlanders," but afterwards by him separated from that connection, present some obscure problems which still await solution. Most of them have been assimilated in speech to the Tajiks and Persians. But the Yagnaubs speak a distinct Aryan language, which will probably be found to be allied to the Wakhi and Siah-Posh of the Hindu-Kush.

The Russian Slavs, who are the latest intruders into

this region, threaten ultimately to absorb all the others, at least in the settled and cultivated districts. They already form a continuous cordon stretching from the Urals round the northern and north-eastern borders of Turkestan beyond Barnaul and Semipalatinsk, and have also founded agricultural settlements in Semirechinsk, the Issik-kul uplands, and Ferghana. In the Aralo-Caspian basin their political status and higher culture give them a preponderance out of all proportion with their numbers, and this preponderance must go on increasing according as the Russian authority becomes more consolidated.

8. *Topography: Askabad—Merv.*

A region consisting mainly of sandy wastes, grassy steppes, bleak upland plateaux and highlands, cannot contain many large centres of population. In an area of nearly a million square miles there are scarcely half a dozen towns with over 30,000 inhabitants, and these are all concentrated in the eastern districts of Russian Turkestan and Bokhara. Elsewhere, except in the Khivan Oasis, there are few places attaining to the dignity of a town. A few forts and military stations are scattered over the Russian circles of the Sir and Oxus; but the monotony of the Kirghiz steppes and Western Turkestan deserts is unrelieved by a solitary hamlet. But in the recently-organised Daman-i-koh of the Akhal Tekke Turkomans, a few places along the new line of railway have already acquired some importance. A few places on the Caspian possess a certain historical interest in connection with the events leading up to the conquest of the Turkoman nomads. Such are Chikislar, near the mouth of the Atrak; the military station of Krasnovodsk and Mikhailovsk, near the Caspian terminus of the Trans-

VIEW OF MERV.

Caspian railway. Beyond this point the chief stations are Kizil-Arvat, Bami, and Askabad, nearly midway between Krasnovodsk and Herat.

East of Askabad the sandy wastes beyond the Tajand valley are broken only by the Merv Oasis, where the famous historical city of Merv[1] had ceased to exist after its destruction in 1784 by the Amir of Bokhara. Before its occupation by the Russians in 1882 it was a mere collection of mud huts, and even as a strategical point had been replaced by the neighbouring fort of Kala Kaushid Khan, which was protected by the Murgh-ab River on two sides, and which was said to be spacious enough to contain 50,000 Turkoman tents. Since the Russian occupation a new city, well laid out, with broad shady streets and canals, has sprung up on both banks of the Murgh-ab, here crossed by a wooden bridge, and by a viaduct carrying the lines of the Trans-Caspian railway. This place cannot fail to prosper, and may one day rival in size and splendour the great historical cities that have successively flourished in the district before and since the destruction of the Græco-Bactrian *Antiochia Margiana*, overthrown by the Arabs in 666 A.D.

Khiva—Bokhara—Samarkand.

Khiva, capital of the khanate to which it gives its name, lies near the head of the irrigation works at some distance from the left bank of the Oxus. It is intersected by two artificial canals, and surrounded by a mud wall 4 miles in circumference and 10 feet high. The palace of the Khan in the interior of the city, besides the houses of the officials and some religious buildings, are all pro-

[1] *Merv* is the Persian form of the old Aryan Meru, whence the Turki *Mar*, and the Greek *Margiana*, *i.e.* the region watered by the river *Margus*, the present *Murgh-ab*.

VIEW OF KHIVA.

tected in the same way, the whole forming a sort of inner town and citadel with three gates, and defended by twenty guns. The outer town contains a large bazaar and the summer palace of the Khan; but the whole place has a population (1894) of scarcely 5000.

By far the largest place in Khiva is Urgenj, near the capital, a fortified town with a wall mounting several guns. But these and other little strongholds are completely overshadowed by the Russian fortress of Nukus, conveniently erected at the head of the delta on the Russian side of the river. From Nukus and Petro-Alexandrovsk, facing the capital, the whole khanate could be at any time occupied in four-and-twenty hours.

Near the point where the Zarafshan runs dry in the ever-encroaching sands, stands Bokhara, capital of the khanate of like name. But sufficient water still remains to supply the magnificent gardens, cotton, jugara, and other plantations, for which the surrounding district has long been famous. The great feature of the city is its well-stocked bazaar, whose vast size is a constant surprise to the stranger. Here all the shops and caravansarais are gorged with Russian and "Kabuli"—that is, English and Indian—wares. No less astonishing is the number of colleges, schools, mosques, graveyards, and "saints" of all orders. Yet Bokhara "the Noble" has fallen far below its former greatness, and in the 50 years between 1830 and 1880 its population had been reduced from 140,000 to 70,000, of whom about two-thirds were Tajiks. Since then its prospects have improved, thanks to the Trans-Caspian railway, and in 1894 the number of inhabitants was estimated at 100,000.

But under the native administration Bokhara can scarcely hope ever to recover its former prosperity. It suffers, and must continue to suffer, from the gradual loss of water from the Upper Zarafshan, which is being drawn off in ever-increasing quantities by the Russians for the irrigation works of Samarkand. This renowned metropolis of Timur lies near the left bank of the river due east of Bokhara, 2154 feet above the sea, on a western spur of the Alai range, which here merges gradually in the Turkestan lowlands. The plains terminate east of the city, which, notwithstanding the splendour of its

TOMB OF TIMUR, SAMARKAND.

ancient buildings, now differs little from other Central Asiatic towns. Here we have the same belt of blooming gardens and orchards encircling the same confused mass of narrow, gloomy streets, mud hovels, and crumbling walls, the whole pervaded by the same oppressive stillness, broken only in the vicinity of the great bazaar. Yet here are some of the grandest monuments of Islam, dating mostly from the time of Timur, and including

MARKET AT BOKHARA.

several magnificent colleges, and the Shah-Zindeh, the most sumptuous mosque in Central Asia. The old palace of the Emir has been converted into a hospital by the Russians, whose administrative and military offices occupy a large portion of the ancient citadel. The same cause that is hastening the doom of Bokhara is furthering the prosperity of Samarkand, the population of which has increased between 1834 and 1894 from 8000 to upwards of 40,000.

Tashkent—Khokand—Verniy.

It is probable that to Samarkand will ultimately be removed the headquarters of the administration in Russian Turkestan, which are at present centred in Tashkent.[1] Next to Tiflis, Tashkent is the largest city in Asiatic Russia, and its population, which rose from 86,000 in 1874 to 120,000 in 1885, already nearly equals that of the Georgian capital. But beyond its size it presents few points of interest except the Museum, where is now preserved the Surma-tash (" Black Stone"), an inscribed stone which stood originally on the banks of Lake Yashil-kul, on the Pamir. The inscription, which appears to be in the Uighur or East Turki dialect, has given rise to much controversy, but is now generally supposed to refer to a comparatively recent event—the occupation of the district in 1759 by the Chinese up to the point where the monument was erected, and the flight of the Khoja rulers of Kashgar to Badakhshan. In view of the pending rectification of frontiers in the Pamir region this stone may be found to possess some political importance. Like most of the large towns in the Sir valley, Tashkent lies at some distance from the main

[1] The natives always say *Tashkand*, but the Russian pronunciation *Tashkent* seems to have gradually established itself in the West.

stream, on the Chirchik, a small tributary flowing from the Aksaitagh, and in a healthy district 1400 feet above the sea. In proportion to its population, Tashkent covers a very large space, being nearly 8 miles long and 4 broad. The Russian quarter has already a population of over 5000; but the great bulk of the inhabitants are Tajiks, who also form the chief element in Khokand, Namangan, Andijan, Marghilan, and the other large towns in Ferghana.

Khokand, capital of the former khanate of like name, scarcely deserves the title of the "Delightful," which has been conferred on it; for goître is here so prevalent that the Russians were compelled to transfer the centre of administration to Tashkent, lower down the Sir valley. Yet its bazaar is still one of the best stocked in Central Asia, and does a considerable trade in local produce and European wares. On the right or opposite side of the Sir, and near the Narin confluence, lies Namangan, the next largest place in Ferghana. It occupies the centre of a rich oasis at some distance from the river, and is the chief mart for the flocks of the Kara-Kirghiz nomads from the surrounding upland steppes. In the neighbourhood are some rich naphtha wells and coal-beds.

Besides Nukus and Petro-Alexandrovsk at the northern and southern extremities of the Khivan Oasis, Charjui higher up the Oxus occupies a position of great importance at the point where the river is crossed by the caravan route from Bokhara to the Merv Oasis.

Verniy (Vernöe), the old Almati, although a Russian town only since 1867, has already acquired importance as the capital of Semirechinsk. It lies near the southern base of the Tans-Ilian Ala-tau, 2430 feet above the sea, nearly midway between Lake Issik-kul and the left bank of the Ili. Although the centre of the Russian agricultural settlements in this region, its trade is still mostly in the hands of the Chinese dealers from Kulja, towards which it is the most advanced Russian outpost. Verniy is the mart for the Russian copper ware, which is distributed from this point over Central Asia and Mongolia.

In the months of May and June 1887 the Verniy district was the theatre of a series of violent earthquakes, from the disastrous effects of which it has scarcely yet entirely recovered. In Verniy itself several hundred people were killed, and much damage was done for a distance of over thirty miles round about, especially by numerous landslips and deluges of mud rushing down the northern slopes of the Trans-Ilian Ala-tau, filling up the river gorges to a depth of over 70 feet, and flooding the fields and plains far and wide along the course of the Keskeden Almaty and other streams flowing northwards to the Ili. But the movements were felt even in the Kibin River valley, between the Trans-Ilian and the Kunge Ala-tau, as well as along the northern margin of Lake Issik-kul, where the shore for a distance of over half a mile about Chulpan-ata was submerged or subsided. But the absence of true volcanic phenomena, and the limited area of the disturbance, show that it was of a purely local character, due probably to hydro-chemical agencies. One notable incident occurred in a defile near the Lesser Almaty, where a house disappeared under a landslip, burying the wife and children of a forester. But after an interval of eight days all were found still alive, the mother having fed the little ones with the bread and samovar (tea-urn) which had been prepared just before the disaster. She took nothing herself, and sank from exhaustion soon after their rescue.

9. *Highways of Communication.*

From Orenburg, the present terminus of the Russian railway system towards Central Asia, the northern postal and trade route passes through Orsk and Turgai across the Kirghiz steppes to Verniy, and thence up the Ili valley across the Chinese frontier to Kulja. The former Kossak stanitzas, established to curb the Kirghiz nomads, have now become so many postal stations along this line, which will some day be replaced by the "Great Northern Asiatic Railway."

At Orsk the south-eastern postal route branches off across the Kirghiz steppe through Kara-Bulak and Irghiz to Kasalinsk, thence following the Sir valley through Peroffsky and Yasi to Tashkent. "The post-stations, which have been built along the route crossing these desolate regions, afford excellent accommodation for travellers, and wells have been dug along the whole distance, though it is true that the water in many of them is not of good quality" (Major Herbert Wood).

From Tashkent the great historical military and trade route leads by Chinaz across the Sir, through the Jilanuti defile, over the Kara-tau to Samarkand, and thence down the Zarafshan valley to Bokhara. West of the Jilanuti pass stands the so-called "Gate of Tamarlane," a pyramidal slaty rock covered with Persian inscriptions, and marking the site of many a fierce struggle for the possession of the Zarafshan and Sir valleys.

Two parallel routes run from Samarkand and Bokhara through Karshi and Koja Sali across the Oxus southwards to Balkh, while a third leads from Bokhara across the Oxus at Charjui to Merv, and up the Murghab to Herat. An alternative line runs from Merv by Sarakhs on the Persian frontier, up the Tajand valley to Herat.

From Khiva several tracks radiate across the Kwarezm desert southwards. But there appears to be only one recognised highway, which follows the right bank of the Oxus to Charjui for Bokhara. The desert tracks are:—1. The Orta Yolu, nearly by the Usboi, or old bed of the Oxus, between the Great and Little Balkans, to the south-east corner of the Caspian; 2. The Tekke Yolu, west of and parallel to the previous, to Kizil-Arvat and the Atrak valley; 3. The Hazaresp, through the Dara-gez district to Kushan for Mashhad; 4. Direct to Merv for Sarakhs, Herat, and Mashhad.

From Ferghana to Kashgaria the route leads through Osh to Gulcha (4140 feet) and Sufi Kurgan (40° N. lat., 73° 30′ E. long.), whence two roads run over the passes of Terek (12,500 feet) and Shart (13,000), which again unite at the outpost of Irkeshtam. Two other roads also run from Sufi Kurgan over the passes of Archat (11,500) and Taldyk (11,800), which unite on the Alai, and lead thence through the Khizil - art gorge and passes (14,000) over the Trans-Alai to the Pamir. The Pamir itself is crossed in all directions by easy tracks, some of which would present no difficulties to the passage of large armies and artillery. The Alai passes, formerly supposed to have been closed from September till late in spring, were shown by Severtzof in 1877 to be free of snow at heights of 13,000 feet till the end of October.

Since the reduction of the Akhal Tekke Turkomans in 1881 the locomotive has penetrated into Central Asia. The Trans-Caspian line, built chiefly for military purposes, runs from Uzun-ada, 16 miles from Mikhailovsk, the original starting-point, on the south-east side of the Caspian, along the Daman-i-koh south-eastwards by Kizil - Arvat, Askabad, and Lutfabad to a point near Kalat-i-Nadir, where it turns abruptly north-east to Merv. From this central station it is continued in the same direction to Charjui, where it crosses the Oxus on a somewhat frail viaduct 2000 yards long, and runs thence through Bokhara to its present terminus at Samarkand. It is a single line of 5-feet gauge begun in 1880 and completed in 1888 at a cost of 46,120,000 roubles, the total length being about 900 miles. But it is liable to be flooded or blocked by sand, despite the saxauls planted on both sides as barriers against the drifting dunes. There appears to be little regular passenger traffic beyond one mixed daily train and three mail trains a week running each way at the rate of fifteen miles an hour. Nevertheless the line was "officially" stated to cover its expenses in 1890, and in any case its strategical and commercial importance is too obvious to call for comment.

Arrangements were made in 1894 for the immediate extension of the Trans-Caspian railway by two sections, one to run from Samarkand through Ferghana eastwards to Andijan on the Sir-darya, the other from Andijan north-westwards to Tashkent. From this extension great strategic and economic advantages are expected to result, the chief being the consolidation of Russian power in Turkestan, and the direct connection of the manufacturing districts of European Russia with the Central Asian cotton-growing regions. The preliminary surveys of the projected routes are practically completed, and both lines will probably be finished before the close of the century.

10. *Administration: Resources—Products—Trade.*

The administration of Russian Turkestan is of a purely military character. The Governor-General, or Yarin-padishah—that is, "Half-King," as the natives call him—has his headquarters in Tashkent. Appointed by the Czar, to the Czar alone he is responsible for the exercise

TRANS-CASPIAN RAILWAY AT UZUN-ADA

of the supreme civil and military functions centred in his person. He even enjoys the privilege of entering into diplomatic relations with the neighbouring States, an arrangement by which negotiations entered into in Central Asia may be confirmed or revoked by the Emperor according to circumstances. His jurisdiction embraces the Siberian provinces of Turgai, Akmolinsk, Semipalatinsk, and Semirechinsk; the extreme west and south comprise the Trans-Caspian territory formerly attached to the Government of Caucasia.

The governors of the various provinces and circles are appointed by the Minister of War, and assisted by provincial councils chosen by the Governor-General. These governors are directly responsible for the revenue and maintenance of order in their several districts. All religions are tolerated, and the tribal usages of the Kirghiz and other nomads respected as far as consistent with the general interests of the State. The towns appoint their own magistrates, who, however, may be removed at the pleasure of the Governor-General.

Public instruction does not seem to have been as yet undertaken by the new masters of the land. The only education that receives any encouragement is the harmless reading of the Koran, as taught in the Medresseh or colleges attached to the mosques. In the whole of Turkestan there are scarcely 8000 Muhammadan children receiving regular instruction.

The chief source of revenue is the land-tax, and the chief source of expenditure the army, which averages about 40,000 men. As all the supplies have to be brought from Russia, this item alone absorbs the whole of the revenue, so that there is a normal deficit amounting in some years to £1,500,000 or over £2,000,000.

Of the products of this region perhaps the most important next to live stock is cotton, of which there are two varieties. It has already become indispensable to the Russian manufacturers, and Vambery declares that it is of better quality than the Indian, Persian, or Egyptian, if not quite equal to the American. The best description is grown in Khiva, which is the chief area of the cotton cultivation in Central Asia. With the view of increasing the yield and still further improving the quality, several American varieties have been introduced, with the result that the exports rose from 120,000 lbs. in 1880 to 80,000,000 in 1890, valued at £1,500,000. Sericulture, originally introduced by the Chinese, has also been long established in Turkestan, and especially in the eastern districts. Important articles of export are, further, the black lambs' wool, known in Europe as "Astrakhan"; the Turkoman horses now supplying splendid remounts for the Russian cavalry; wool, hides, dyes, cereals, and fruits.

The local retail trade has of late years been greatly developed, and the remotest nomad hamlets are now supplied with all kinds of wares from the bazaars of the central marts. Hence the importation of Russian and English goods has largely increased, somewhat to the detriment of the native industries. The Russians have the advantage over their English rivals in being first in the field, and more carefully studying the tastes of the Eastern nations. Their goods are also exempt from the heavy duties imposed by the Russian Government on English and Indian wares. The rapid organisation of the Trans-Caspian Territory has enabled them to send their manufactures direct to the Mashhad and Herat bazaars.

Samarkand is now (1895) the chief centre for Russian trade in Central Asia, nearly 100,000 cwt. of manufactured goods being yearly brought to this place chiefly from Baku by the Trans-Caspian railway. Over half that quantity is taken by Bokhara, which has become Russia's best customer in Turkestan. Askabad is also an important centre of distribution, and in 1893 nearly 1000 bales of textiles, mainly from the Moscow manufacturing region, were exported to Khorasan, and as far east as Herat and Kabul. Altogether about 18,000 bales of Russian and Polish textiles are annually forwarded through Baku to all parts of Central Asia.

In the Russian *Official Messenger* for 1893 a summary is given of the results of the ten years' occupation of the Trans-Caspian Territory, which is now organised in five administrative districts—Manghishlak, Krasnovodsk, Askabad, Tejen, and Merv—with a total area of 220,450 square miles, and a population of 297,000, of whom 280,000 are natives (chiefly Tekke Turkomans) and the rest immigrants. The Turkomans are partly nomad, partly settled, their *auls*, or villages, having each their fields, gardens, and irrigating rills. Some live in permanent dwellings, while others camp with the herds, and they form socially two classes, the wealthy *charvà*, owners of sheep and camels, and the poor *chomur*, who have no live-stock, and remain in the villages. They are neither very religious, having but few mollahs, nor very industrious, nearly all their forts and canals having been built by their Persian slaves, while the housework falls to the women.

The *lex talionis* is still observed, but capital punishment and mutilation for stealing, and similar offences, have been abolished since the appointment of Russian judges. Raiding is also nearly a thing of the past, and encouragement is given instead to agriculture and irrigation. Loans without interest are made to the auls from a special irrigation fund to help in repairing the old wells and canals and digging new ones. Wheat and barley, the chief crops, yield from 12 to 20 fold, and lucerne, rice, millet, cotton, and melons also give good returns. In 1891 the sources of the Ghermal River were cleared out and the water led into artificial canals, thus increasing the daily supply by about 100,000,000 gallons. The irrigation of the Merv, Tolstan, and Penjdeh oases has also been greatly improved, and the new Russian military stations on the Kushk River have been connected by carriage-roads with Merv and Sarakhs.

Various crafts (leather, silver, arms, etc.) are carried on, and much salt is now extracted, 10,000 tons having been exported to Persia alone in 1889. Great stores of naphtha are found in the Cheleken Peninsula; asphalt and ozokerit in the Balkan Mountains; sulphur in the Askabad and Krasnovodsk districts; lignite and coal in the Manghishlak Peninsula. The imports and exports already amount to nearly £1,000,000, and in 1891 over 72,000 tons of goods were forwarded from Central Asia to Russia by the Trans-Caspian railway.

11. *Statistics.*

Areas and Populations.

	Provinces.	Area in sq. miles.	Population (1889-94).
Kirghiz Steppe	Akmolinsk	229,609	500,180
	Semipalatinsk	184,631	576,578
	Turgai	176,219	364,660
	Uralsk	139,168	559,552
	Lake Aral	26,166	...
Turkestan	Samarkand	26,627	680,135
	Ferghana	35,654	775,600
	Semirechinsk	152,280	971,878
	Sir-Darya	194,853	1,214,300
Trans-Caspiana		214,237	276,709
Caspian Sea		169,381	...
Total Turkestan and Administrative Dependencies		1,548,825	5,619,592

Inhabitants of Russian Turkestan classed according to Races and Religions.

Usbegs	Sunnis
Kara-Kalpaks	Sunnis
Kara-Kirghiz	Sunnis
Kirghiz-Kazaks	Sunnis
Turkomans	Sunnis
Tajiks and Persians	Sunnis and Shiahs.
Galchas	Sunnis and Fire-worshippers.
Slavs	Christians.
Kurumas	Sunnis.
Turuks	Sunnis.
Mazang, settled Gipsies	Sunnis mostly.
Luli, nomad Gipsies	Pagans.

Chief Towns in Aralo-Caspian Basin.

Tashkent	121,000	Andijan	33,000
Bokhara	100,000	Marghilan	26,000
Khokand	54,000	Urgenj	22,000
Samarkand	40,000	Shehr-i-Sebs	20,000
Khojent	35,000	Karshi	20,000
Namangan	33,000	Osh	20,000

Agricultural Returns, Russian Turkestan.

Provinces.	Under Crops. Acres.	Pasture. Acres.	Waste. Acres.	Total. Acres.
Semirechinsk	2,356,000	50,000,000	50,000,000	102,356,000
Sir-darya	984,000	50,000,000	68,512,000	119,496,000
Ferghana	1,650,000	8,250,000	8,525,000	18,425,000
Zarafshan	626,000	3,625,000	2,497,000	6,784,000
Amu-darya	126,000	3,625,000	19,949,000	23,690,000

Live Stock, Russian Turkestan.

Provinces.	Camels.	Horses.	Cattle.	Sheep.
Semirechinsk	97,412	892,007	523,200	6,296,000
Sir-darya	242,130	395,563	293,550	3,183,000
Zarafshan	1,248	51,991	84,463	283,000
Ferghana	38,294	213,760	220,717	1,260,000
Amu-darya	11,267	48,000	38,070	329,000
Trans-Caspian Territory (1890)	115,320	83,890	52,255	1,818,615
	505,671	1,685,201	1,212,255	13,169,615

Army—Peace footing 40,000; war footing, 80,000.

Revenue (mean)	£450,000
Expenditure (mean)	£1,400,000
Deficit (mean)	£900,000
Cotton crop (mean)	6,000 tons.
Silk (Bokhara)	2,500,000 lbs.
Live Stock	£18,000,000
Wool exported to Russia (mean)	£90,000

Distances.

	Miles.		Miles.
Chikishlar to Kizil-Arvat	200	Bokhara to Balkh	290
Ashurada to Kizil-Arvat	221	Samarkand to Balkh	220
Mikhailovsk to Kizil-Arvat, by rail	146½	Bokhara to Tashkent	320
		Samarkand to Tashkent	180
Kizil-Arvat to Sarakhs	320	Tashkent to Khokand	120
Sarakhs to Herat	210	Orsk to Fort Karabulak	120
Herat to Merv Oasis	240	Orsk to Irgiz	240
Merv to Sarakhs	70	Orsk to Kasalinsk	500
Merv to Charjui	140	Uzun-ada by Trans-Caspian Railway to Samarkand	890
Charjui to Bokhara	90		
Bokhara to Samarkand	140		

Caravans.

	Days.		Days.
Orenburg to Tashkent	50 to 60	Bokhara to Samarkand	6
Namangan to Semipalatinsk	40	Samarkand to Khokand	6
Bokhara to Herat	25 to 30	Khokand to Ush	4

CHAPTER IV

SIBERIA

1. *Boundaries—Extent—Area.*

A REASON analogous to that which awards the Caucasus to Asia gives the Urals to Europe. For the west Asiatic Mediterranean, which was formerly connected through the Ponto-Caspian Strait with the Euxine, also communicated with the Arctic Ocean over the low ridge forming the present water-parting between the Ob and Aralo-Caspian basins. This ridge transversely crosses the deep furrow stretching northwards along the Tobol valley to the Ob, through which the inter-continental strait flowed between the inland marine basin and the Arctic. Hence the Urals were at that time entirely cut off from the Asiatic continent, of which they form the present north-western boundary.

From this point Siberia stretches uninterruptedly eastwards across 130 degrees of the meridian to the Pacific Ocean. Its northern boundary is formed by the Arctic Ocean, whence it extends across 30 degrees of latitude southwards to China and Turkestan, forming its southern frontiers. But these southern frontiers are in many places extremely vague, and at some points purely conventional. Towards Turkestan the natural line follows the Aralo-Caspian and Ob water-parting, between 48° and 51° N. lat., which at its lowest elevation rises a few feet only above the surrounding Kirghiz steppe. Farther east the frontier towards China generally follows the line of the Altai from the Irtish valley to the Upper Amur valley, running thence along the course of that river to the confluence of the Usuri. Here it is deflected southwards along the Usuri valley, and beyond Lake Kenka to the Sea of Japan, about 43° N. lat. below Victoria or Peter the Great Bay. At this point the Russian territory thus impinges on the north-east frontier of Korea, and shuts off Chinese Manchuria from the Pacific seaboard.

South of the Upper Irtish valley the Russo-Chinese frontier line still remains to be definitely fixed. Here the political boundary running north and south has to be drawn *across* the highlands and depressions, which mostly run east and west. The line from the south-western extremity of the Altai across the Irtish valley at Lake Zaisan, and thence over the Tarbagatai range down to the Emil (Churtu) valley, seems never to have been definitely settled, and its settlement has recently been again postponed to some future period. But the line which thence follows the crest of the Zungarian Ala-tau down to and across the Ili valley to the Trans-Ilian Ala-tau was at least temporarily determined by the treaty ratified on 19th August 1881, in virtue of which Russia restored to China the province of Kulja, held by the Czar's troops during the troubles in the neighbouring districts. This line is drawn from the Boro-khoro hills along the Khorgos River down to the right bank of the Ili and thence across the valley to the Tengri-khan, culminating point of the Tian-shan. It thus leaves all the broad upper portion of the Ili valley—that is, Kulja proper—to China, Russia merely reserving a strip of land.

Lying mainly between 46°-78° N. lat., and 60°-190° E. long., and thus occupying the whole of North Asia,

Siberia stretches from Orsk for over 4200 miles north-eastwards to Cape Vostochni on Bering Strait, and from Cape Severo (Chelyuskin) for about 2000 miles southwards to the Tarbagatai range, with a total area estimated at over 4,830,000 square miles, and a population (1894) of 4,540,000, or rather less than one to the square mile.

2. *Relief of the Land: The Altai, Sayan, Ergik-Targak, Yablonovoi, Stanovoi, Sikhota-alin, and Kamchatka Ranges.*

A region of such vast extent is naturally of very diversified configuration. Thus, while the south-western portion is exclusively a lowland country, considerable highland tracts are comprised in the southern and eastern sections. These highlands, often comprehensively spoken of as the Altai system, begin properly north of Lake Zaisan and the Upper Irtish valley, by which their westernmost extremity is clearly separated from the Tarbagatai range. On this account the Tarbagatai, although usually included in the Siberian mountain systems, has here been regarded rather as the northernmost extension of the Tian-shan. Its true position is that of a water-parting between the Arctic and the Central Asiatic closed basins. For it sends down streams northwards to the Irtish, flowing to the Frozen Ocean through the Ob, southwards to Lakes Ala and Sassik, which formerly communicated westwards with Lake Balkhash, eastwards with the Ebi-nor and the Mongolian Mediterranean.

From the Irtish valley the Altai, or "Gold Mountains," stretch mainly north-eastwards through the Sayan range to the Daurian Alps, and thence beyond the Baikal basin under diverse names, such as the Yablonovoi and Stanovoi, to the volcanic masses filling the greater part of Kamchatka, and gradually falling through the Chukchi domain towards the north-easternmost extremity of the continent at East Cape. But it will be seen that the system is by no means continuous, being not only broken up into distinct sections by the deep gorges of the Upper Yenisei and Selenga Rivers, but merging round the Sea of Okhotsk in a moderately elevated plateau, where high ranges are figured on most of our maps.

Even the western section—that is, the Altai proper—is not so much a distinct mountain range as an aggregate of more or less detached chains running in various directions between the upper Irtish and Yenisei valleys. South of these valleys the main direction is rather west and east, or north-west and south-east, but north of them the normal direction is north and south, while the whole system inclines towards the north-east. The portion to which the term Altai is more specially applied, and which scarcely comprises more than one-fourth of the whole western section, stretches from the River Bukhtarma, an affluent of the Irtish on its right bank, and from the Smeinogorsk, or "Snake" mountain, north-eastwards to the romantic Lake Altyn (Teletzkoie) and to the Chulishman River, joining the lake from the east. Here the Altai is crossed by the much-frequented Suok Pass leading from Siberia to Mongolia. But this eastern limit of the chain is somewhat conventional, for east of the pass the system is continued by the Sayan range with no perceptible interruption to the Upper Yenisei valley.

Perhaps the most important of the Mongolian offshoots is the Hurka range, the position of which in relation to the Altai system was first determined by Captain Younghusband, who traversed it in its entire length of about 220 miles during his Central Asian journey of 1887. It crosses the Gobi desert between 102°-106°

E. longitude in an oblique direction from south-east to north-west, culminating towards the north in peaks apparently about 8000 feet high. Here it comes to an end in a barren sandy depression about 80 miles long,

TWO-HUMPED SIBERIAN CAMEL.

where a series of isolated hills "form connecting links between the Hurka range and the Altai Mountains." The range itself, which presents a bare, sterile aspect throughout, is flanked southwards at a mean distance of

30 miles by a similar but somewhat lower ridge, the intervening depression being traversed by the main route which leads from Kwei-hua-Cheng on the Shansi frontier north-westwards to Man-chin-tol. This route was followed for the first time by Younghusband, who crossed Prjevalsky's itinerary at Borston, near the south end of the Hurka chain. It was towards the north of this region that he first heard of the wild camels (two-humped), of which Mr. Littledale later procured the specimens now in the Natural History Museum, London. Here also he picked up several heads of the argali (*Ovis poli*), and saw "considerable numbers of wild asses, which appeared to be perfectly similar to the Kyang of Ladak and Tibet; and wild horses, too, the *Equus Prjevalskii*—roaming about these great open plains" (*Geo. Proc.*, 1888, p. 485).

The whole range has a mean altitude of perhaps 5000 feet, with numerous crests from 6000 to 10,000 feet, culminating in the Bieluka, or "White" mountain, whose twin peaks rise to 11,100 feet. The term "Great Altai," commonly applied to the little-known chains penetrating across the Chinese frontier into Mongolia, belongs rather to the Bieluka chain, which encloses the Kobdo plateau on the west, and several peaks of which rise above the snow-line. Hence Russian explorers now designate as the "Little Altai" the "Great Altai" of most geographers.

The western section — that is, the Kolyman or Russian Altai — abounds in ores, and encloses the romantic little Lake Kolyman, whose rugged granite banks are here and there clothed with fine timber. Elsewhere the numerous and rapid streams, the varied forms and colours of the hills, impart great variety to the scenery of the Altai. Between the detached chains there everywhere stretch extensive upland plains covered with snow or morasses, and intersected here and there by low rocky ridges or granite masses.

The southern spurs also consist largely of granites with crystalline schists and a hornblende porphyry, presenting fantastic, bare, and rugged outlines. Here lies the famous mining region of the Altai, which forms part of the Imperial domain, and has altogether an area of perhaps 200,000 square miles. The works at Serianovsk yield gold, silver, copper, lead, and tin. In the Smeinogorsk district the matrix of the metalliferous ores is augite porphyry, varied with schists and huge masses of auriferous quartz. The mines are worked exclusively after the German method, but water and horse power have not yet been supplemented by steam.

Beyond the wooded Sayan section the system is continued across the Bei-kem, or Western Yenisei valley, by the Ergik-Targak and other ridges rising here and there above the snow-line, and crossed by passes over 7000 feet high leading from Siberia to Mongolia. The Ergik-Targak on the Chinese frontier has an altitude of at least 10,000 feet, and is connected with the Baikal uplands by snowy masses which have been only recently explored. Conspicuous amongst them is the Munku-sardik, or "Silver Mount," covered with ice-fields, and first ascended by Radde in 1859. This pyramidal mass forms an important water-parting between the great western and eastern branches of the Yenisei, and in the neighbourhood are the vast deposits of graphite discovered about thirty years ago by Alibert.

The geological explorations undertaken in 1893 in connection with the Siberian railway have also revealed the presence of brown coal of fair quality in several districts in the southern parts of the Yeniseisk Government. The Kushun beds, within 10 miles of the projected railway station of Esaulova, contain about 1,000,000 tons of workable lignite, and other rich deposits occur at Antropova on the Chulin river, and at Kubekova on the Yenisei, 13 miles from Krasnoyarsk.

Beyond the Baikal region the plateau prevails over the strictly highland formation. Here the "Great Dividē" between the Lena and Amur, or rather between the Arctic and Pacific Oceans, consists not of a distinct mountain range, as laid down on the maps, but of a vast tableland, contracting gradually north-eastwards towards the Chukchi peninsula, and intersected by a number of moderately-elevated ridges running mostly parallel, or at slightly-diverging angles. Hence Russian geographers now propose to substitute the expression *Stanovoi Vȯdorazdyel*, or "Main Parting Line," for *Stanovoi Khrebet*, or "Main Dorsal Range," hitherto applied to this upland system. The Russo-Chinese frontier line was doubtless laid down by diplomatists along the crest of the Stanovoi; but from the first this frontier line was purely fictitious, and has been altogether dispensed with since the Russians have established themselves along the left bank of the Amur.

The plateau has a total length of about 2400 miles between Transbaikalia and Bering Strait, and the loftiest and best-defined range in the whole system is the Yablonovoi, or "Apple" range, running south of Baikal, near the Chinese frontier, and culminating southwards with the massive Sokhondo or Chokhondo (8370 feet). The upper crests are composed of granitic and palæozoic rocks, which nowhere reach the line of perpetual snow. The range is easily crossed by the road from Lake Baikal to Khita. East of the Yablonovoi stretches the Daurian steppe, which has been compared to a fragment of the Gobi desert transplanted to Russian territory. It was formerly crossed by a wall attributed to Jenghis Khan, and is separated by extensive pine forests and the River

Onon from the Nyerchinsk steppe, which extends thence eastwards to the Argun valley.

North of the Amur the Yablonovoi section of the Stanovoi runs between the Rivers Aldan and Żeya, at a mean altitude of 7000 feet, beyond which the Aldan or Jugjur ridge falls to little over 3000 feet. Yet here the formations are most varied, comprising granites, porphyry, gneiss, underlying palæozoic, and even jurassic rocks, followed by coal-measures towards Verkhoyansk, and basalts and trachytes near the Sea of Okhotsk. At this point the water-parting is deflected so far to the east that the head-streams of the Aldan have their source within a short distance of the Pacific, whence the gold, silver, lead, and iron ores of the neighbouring Aldan hills might easily be procured. A little farther north the "Captain" rises west of Okhotsk to a height of 4360 feet, and although falling short of the snow-line, it overlooks deep valleys filled with masses of perpetual snow and ice.

In 1891 the Verkhoyansk and Stanovoi uplands were visited by an exploring party under M. Cherski, which passed from Yakutsk over the Aldan and up the Chandyga River into the heart of the mountains, which, although of an alpine character, nowhere rose above the snow-line. Farther on the route bent sharply round south-eastwards to the Dyba tributary of the Tyra; and thence down to the Omekon, which the natives regard as the true head-stream of the Indigirka. Here were crossed the sources of the Kunku and Kente, the Uchugei-urach being tributary to the latter, and not to the Omekon, as shown on the maps. Then the Verkhoyansk range was crossed at the point where it branches off from the Stanovoi, the former running nearly due north, while the latter has an easterly trend. Between the Indigirka and Kolima rivers the Stanovoi system ramifies into three main ridges: the Tas-Kystabyt, forming the crest of the chain nearest to the

Indigirka; the Ulachan-chistai, a "great woodless region," between the Nera and Moma systems, more to the north-west; lastly, the divide between the Indigirka and Kolima, running north-west, and corresponding to the southern section of the range figuring on the maps as the Tomus-chaya. This chain decreases in height south-eastward in the direction of the Kolima, gradually merging through a plateau in the tundra, in which is situated the little settlement of Verkhnoye-Kolymsk, the farthest point reached by the expedition.

The whole region consists of silurian and triassic folds, and appears to be continued seawards by the New Siberian island of Kotelnyi, which is of the same geological structure. In general the valleys are broad and gently sloping, and show no trace of terraces, these having been obliterated by the shifting streams, which, instead of deepening the bottom-lands, tend to level them up by filling them with detritus. The process seems to be somewhat analogous to that by which Younghusband supposes the Pamir valleys to have been formed (p. 92).

Farther south lay the field of M. Joseph Martin's researches during the years 1882-86. This explorer's attention was first directed to the auriferous districts of Vitim and Olekma in the Upper Lena basin, where the prevailing formation is slate with iron pyrites and quartz reefs. The auriferous deposits, to which mining operations are at present confined, appear from the fossil remains to be of the same age as those of the Amur. But they occur at a much greater depth, and the gold is extracted in the form of spangles and nuggets from pits sunk in some places to a depth of 170 feet.

From Olekma, M. Martin extended his explorations to a great part of the little-known region between the Upper Lena and the Sea of Japan. The route, which lay generally in the direction of the south-west, crossed in succession seven or eight bare and rocky ridges ranging in altitude from 1300 to 2700 feet. Beyond these rugged heights the explorer entered a more inviting region, in which was situated the deep Lake Nichatka, which was sounded to a depth of 490 feet. This basin, which receives contributions from the surrounding ice-capped heights, sends its overflow eastwards to the Chara, affluent of the Olekma. South-west of the lake the Chara valley is separated from that of the Vitim by a narrow lofty ridge, which was crossed at an elevation of about 9000 feet. Farther on the track led down the Chara and over the water-parting to the Vitim valley, and thence to a lucustrine plateau with a whole group of flooded depressions, amongst which was the large Lake Amidisse. This plateau is continued by a series of tablelands abounding in iron, copper, coal, lead, and other minerals; and on a marshy plateau commanding the valley of the Kalar River were discovered two other basins at an altitude of about 3000 feet, to one of which the explorer gave the name of Lake Martin. From this point the expedition reached the foot of the Stanovoi Mountains, which here develop rounded crests overgrown with birch and larch forests, between which rise several peaks from 4500 to 5000 feet high, destitute of vegetation and snow-clad during the long winter season.

After crossing the Stanovoi Mountains, which in this region ramify into a number of nearly parallel ridges, the explorer entered the valley of Amazar, an upper affluent of the Amur, and followed its course down to the main stream some 30 miles from Albazine. The following year (1884) M. Martin revisited Albazine, and from that point, following the track along the southern slopes of the mountains, reached the upper valley of the Zeya, tributary of the Amur. In this part of its course the Zeya traverses a wild mountainous region covered with magnificent forests, and abounding in romantic scenery. The contrast is everywhere very striking between the northern and southern slopes of the Stanovoi water-parting, those draining to the Lena being bleak and arid, while the streams descending to the left bank of the Amur flow through a region covered with magnificent woodlands.

East of the Stanovoi proper a wooded range, variously known as the "Little Khingan," the Bureya, or Dauss-alin, runs at a mean height of 2500 feet from the Amur north-east to the south coast of the Sea of Okhotsk, and culminates with the Lagar-aul, 3450 feet. Still farther south and east the Primorsk Province is traversed in its entire length by the Sikhota-alin, or so-called "Manchurian Mountains," which really consist of an extensive plateau intersected by innumerable ridges, with a mean elevation of scarcely 3000 feet. But notwithstanding their low elevation, these ridges are of very difficult access, so that but few passes lead from the Usuri valley across the plateau to the coast. Communication, however, is effected southwards through the depression of Lake Kenka, whence an easy pass leads down to the Suifun coast stream. Although commonly supposed to be of igneous origin, the Sikhota-alin seems to be mainly a sandstone formation. It culminates with Mount Golaya, 5550 feet.

The Stanovoi water-parting is still continued north-eastwards to the Bering Strait by the low straggling eminences traversing the Chukchi peninsula, and separating the head-streams of the Kolima flowing to the Arctic from those of the Anadyr flowing to the Bering Sea. Here the continental system, nowhere more than 5000 feet high, falls to about 2000 as it approaches the coast.

But farther south Kamchatka is occupied by a totally different formation, belonging in its igneous character rather to the oceanic than to the continental system. The peninsula is traversed in its entire length by a chain of lofty burning mountains, fourteen still active volcanoes rising close to the east coast, amongst which is the Klyuchevskaya Sopka (16,000 feet), the highest active volcano in Asia. This igneous system is one of the grandest instances of a connected series in the world; yet it forms merely a link in the endless chain which stretches from Alaska through the Aleutian Islands, Kamchatka, the Kuriles, and Japan, to the Philippines and the Eastern Archipelago. The Kuriles are thoroughly igneous, and contain from eight to ten still active volcanoes.

3. *Hydrography: The Ob, Yenisei, Lena, Yana, Indigirka, Anadyr, and Amur Rivers — Lakes Baikal and Kenka.*

Siberia presents the most extensive, but economically perhaps the least serviceable water system of any country in the old world. The land has a general inclination towards the north, so that all the great rivers flowing from the southern highlands pursue a normal and nearly parallel northerly course to the Arctic Ocean. But most of the large tributaries flow rather north-west and north-east to the left and right banks of the main streams, thus affording an almost uninterrupted water highway from the Urals to the Pacific, as well as from the Southern highlands to the Arctic. From the River Ural to Yakutsk, a distance of 6000 miles, this magnificent waterway is broken only by two short portages between the Ob and Yenisei, and between the Yenisei and Lena respectively. The whole country is in this way covered with a network of rivers, affording altogether some 30,000 miles of navigable waters. Unfortunately all these rivers are ice-bound for the greater part of the year, while the estuaries are open only for about ten or twelve weeks during the warm season. Even the Amur, which is the only great river draining from the southern watershed to the warmer Pacific seaboard, is blocked for six months at a time. Hence, notwithstanding the repeated efforts that have been made, especially since Nordenskjöld's successful expedition round the north-east passage, to open up a trade with Siberia through these arteries, it is not probable that the markets of the world will be affected by the agricultural produce from Northern Asia, at least until the completion of the Siberian railway now in progress.

All the countless streams from the Urals and Southern highlands are collected and discharged into the Arctic mainly through seven independent channels, which, going eastwards, are the Ob, Yenisei, Khatanga, Olenek, Lena, Indigirka, and Kolima. In the same way those flowing towards the Pacific are grouped in two systems only, those of the Anadyr and Amur.

The Ob Basin.

Although not the longest of the rivers, the Ob drains the greatest extent of country, and with its tributaries affords the longest stretch of navigable water highways. The basin, merging eastwards almost imperceptibly with that of the Yenisei, bordering westwards and southwards on those of the Volga, Ural, and Aralo-Caspian, and penetrating south-eastwards far into the Mongolian plateau, has a total area of over 1,400,000 square miles. At Troitsk, where all the great affluents are gathered into one channel, the main stream, still nearly 700 miles from its estuary, has a width of no less than three miles.

The great head-streams are the Tobol, with the Tavda

from the Urals, the Ishim from the Aralo-Caspian water-parting, the Irtish from the Kobdo plateau, and the Ob with the Katun, Biya, and Tom and Ket, from the northern slopes of the Altai. The Tobol and Ishim are collected on its left bank by the Irtish, which ought to be regarded as the true upper course of the main stream; for the Urungu, its farthest head-stream, has its source on the Kobdo plateau, south of the Russian Altai, whence it flows first to Lake Ulungur, or Kizil-Bash. From this lake there seems to be an intermittent surface discharge, and a perennial underground outflow to the Black Irtish, a torrent from the snowy upland valleys on the west side of the Chinese Altai. This connection of Ulungur with the Black Irtish is not shown on our maps, but there can be little doubt of its existence. At a certain point below the level of the lake the volume of the Irtish, without receiving any visible influent, is suddenly increased from about 640 to 1900 cubic feet per second. Whence this great access except from the neighbouring reservoir? Mattus-sovski, who visited the district in 1870, could detect no apparent connection, but he ascertained on the other hand that there are no intervening elevations between the lake and the river.

After receiving the Kaljir from Lake Marka, the Black Irtish, which even in Chinese territory is already a considerable stream, 500 feet broad and 10 feet deep, enters the east end of Lake Zaisan, a vast steppe lake on the Russo-Chinese frontier, 60 miles long and over 25 feet deep, with a mean area of 720 square miles. This lake, which abounds in fish, yielding over 1,500,000 lbs. annually, has already been visited by a steamer, and it is now proposed to establish regular steam communication between Tiumen and the Black Irtish, a total distance of about 1000 miles.

From the west end of Lake Zaisan the river emerges

as the White Irtish, and thenceforth pursues a somewhat winding north-westerly course between the Kirghiz and Baraba steppes, west and east, to its junction with the Ob, about 1900 miles from its farthest source. After its junction with the Bukhtarma it pierces the western spurs of the Altai through the wild Ust-Kamenogorsk defile, here falling from 1300 to 1150 feet above sea-level, and from this point to the Ob it receives probably over a thousand tributaries, of which by far the largest are the Ishim and Tobol, both on its left bank. But the large Lake Chany, as well as most others of the Baraba steppe, have become closed basins, no longer sending even intermittent discharges to the Irtish.

The Baraba steppe is skirted eastwards by the Ob, which is formed by the junction of the Katun and Buja from the Altai, and which is joined by the Irtish 300 miles below Tobolsk. With a fall of scarcely 300 feet from the advanced spurs of the Altai to its estuary, the Ob pursues an extremely sluggish course, expanding here and there into broad steppe lakes, and occasionally almost undecided whether to flow west to the Irtish, east to the Yenisei, or in an independent channel north to the Arctic. Varying in breadth from half a mile to two miles, it expands during the spring floods into a great inland sea, which, even above Tomsk, is so broad that the opposite banks are quite invisible. After receiving the Tom and Chulim, it is joined near Narim by the Ket, which in some respects, though not the largest, is its most important tributary. For the Ket is navigable for no less than 600 miles towards the Yenisei, with which it is now proposed to connect it by means of a canal $2\frac{1}{3}$ miles long from Lake Kosovskoie across the portage to the Kas flowing to the Yenisei below Yeniseisk.

Below the Irtish confluence the Lower Ob flows to its estuary beyond Obdorsk, in two separate channels, known as the Great and Little Ob, the latter of which is most available for up-stream traffic, the former for craft going seawards. Both branches, which are everywhere connected by innumerable intermediate channels and backwaters, enter the gulf in a joint stream about two miles wide, and from 40 to 90 feet deep. The gulf or fiord runs first east and then north for over 480 miles beyond the Arctic Circle. This great water highway, which is navigable throughout nearly the whole of its course of over 3400 miles, was thrown open to the trade of the world by the expedition of Dahl, who reached the Ob from the Kara Sea in 1877. With the tributaries, there is a total navigable highway of perhaps 9000 miles, which has hitherto been but slightly utilised by steamers.

The Yenisei Basin—Lake Baikal.

Next in importance to the Ob is the Yenisei basin, which occupies nearly the whole of Central Siberia between the Ob and the Lena water systems. Like the Ob and most other great Asiatic rivers, the Yenisei has its farthest sources on the central plateau behind the enclosing mountain barriers, through which it forces its way seawards. Of its two great branches, the Yenisei proper and the Selenga-Angara, the former is developed on Chinese territory, about 4000 feet above the sea, between the Sayan and the Tanu-ola mountains, by the junction of the Bei-kem and Khuà-kem, from the Ergik-Targak frontier range and the northern slopes of the Tanu-ola. Flowing first west to a point where it finds an outlet in the Western Sayan range, the united stream here enters Russian territory, through which it henceforth pursues a northerly course, mainly between the 88th and 94th meridians, to its estuary in the Arctic Ocean.

The eastern branch also rises within the Chinese

THE YENISEI AT KRASNOYARSK.

frontier, whence the Selenga, its farthest head-stream, flows from the north-eastern slopes of the Tanu-ola eastwards to a point where it receives the outflow of Lake Kosso (Kosso-gol) from the north. This romantic basin, fed by the glaciers of Munku-sardik at its northern extremity, stretches north and south for a distance of about 70 miles, with an area of about 1300 square miles. In its centre is the islet of Dalai-kui, which the Mongolian Buddhists hold in special reverence as the "Navel of the World."

Beyond this point the Selenga sweeps round to its junction with the Orkhon, which rises in the Gobi itself, near the ruins of Karakorum, the old Mongolian capital. The united stream flows thence between Kiakhta and the Khamardaban range, north and west to the east coast of the great highland Lake Baikal, by far the largest fresh-water reservoir in Asia. This lake, which fills two enormously deep fissures in the plateau at a present elevation of 1363 feet above the sea, seems to have formerly communicated directly with the Irkut valley. But its present outflow is through the Angara, which forces its way over a series of romantic gorges to the right bank of the Yenisei. The lower section of the Angara, where it trends west, takes the name of the Upper Tunguska in contradistinction to the "Stony" and "Lower" Tunguska, which join the main stream farther down.

Baikal, the Dalai-nor, or "Holy Sea," of the Mongolians, has a mean depth of 850 feet, sinking in some places to 4500, or considerably over 3000 below sea-level. Owing to this prodigious depth, its volume, with an area of scarcely 14,000 square miles, is more than double that of Lake Michigan, which has an area of 23,000 square miles, but a mean depth of only 300 feet. Recent soundings have revealed a rocky ridge about

3350 feet high, dividing the lake into two secondary but now united basins, at a point where there is a depth of scarcely more than 200 feet. These surveys also show that Baikal was formerly far more extensive than at pre-

THE ANGARA RIVER NEAR LAKE BAIKAL.

sent. Within comparatively recent times its level has fallen about 20 feet; but at some more remote epoch it was high enough to drain through the Irkut to the Yenisei through a channel distinct from that of the Angara. Its waters, which are remarkable for their great transparency, revealing objects at a depth of 40 to 50 feet, are frozen from December to May to a thickness of 4 or 5 feet. Yet such is the fury of the winter gales that its icy fetters are constantly broken, thus affording fresh supplies of air to the salmon, sturgeon, and other fishes with which the lake abounds. Amongst its fauna is a species of seal, in appearance exactly resembling the Spitzbergen *Phoca fœtida.*

In summer the communication along the shores of the lake, which is 360 miles long, with a mean breadth of 35 miles, is kept open by a steamer, affording an opportunity of visiting the lovely island of Olkhon, famous for its alpine roses. The north-west coast contains some very grand scenery, the rocky granite masses being here in many places clothed with larch and pine forests from their summits to the water's edge. The shores also abound in hot springs, which are associated with still active underground agencies and frequent earthquakes. Near the hot springs is a flourishing Russian settlement, where rye, barley, and potatoes are successfully cultivated. During the long winter months the peasantry occupy their time in pursuing the sable, squirrel, and other fur-bearing animals, which, however, here, as elsewhere in Siberia, are rapidly disappearing.

M. Chersky's explorations (1881) show that the rocks on the west side of the lake belong to three different epochs — pre-Silurian (Laurentian ?), Silurian, and Jurassic. This naturalist also confirms the view that Baikal forms, as above stated, two distinct longitudinal cavities, connected together by a central ridge.

The Angara, which has a discharge of at least 105,000 cubic feet per second, flows for 40 miles below its junction with the Oka through a series of rapids between sheer rocky walls, rising in some places 600 feet above the surface. Here the average fall is about two feet in the mile; but the steamers now plying on these waters pass safely over the rapids. As the Selenga is also accessible to light craft as far as the Orkhon junction, there is an uninterrupted navigable highway of 2700 miles from Kiakhta through Lake Baikal and down the Angara and Yenisei to the Arctic Ocean.

Nor is the navigation entirely interrupted even by the rapids over which the Upper Yenisei descends from the Mongolian plateau through the Sayan range down to the Siberian plains. Even above the Angara junction the current is very gentle, though nowhere quite so sluggish as that of the Ob, the elevation at Krasnoyarsk being 530, or 200 feet more than that of the Ob under the same parallel. At Yeniseisk below the Angara confluence it is still 230 feet above sea-level. Here it has a mean width of 6000 feet, expanding in the floods to upwards of 4 miles, with a rise of about 40 feet.

Through the Lower Tunguska, which has a total length of 1620 miles, with a breadth of over half a mile at the confluence, the Yenisei approaches near Kirensk to within 14 miles of the Lena; and as the portage between the Ob and Yenisei is only $2\frac{1}{3}$ miles long, the navigable water highway from the Urals across Siberia nearly to the Pacific seaboard is only interrupted by two breaks of less than 17 miles altogether. This highway is also of far greater economic importance than those of the main streams flowing northward to the Frozen Ocean. Lying mainly between the 58th and 60th parallels, it is open for a far longer period of the year than those of the great arteries running for several degrees beyond the Arctic Circle towards the North Pole. The navigation of the Lower Yenisei for 300 miles above its estuary, where it expands to 30 or 40 miles during the floods, is rendered extremely dangerous by the northern gales, here sweeping with great fury over the tundra and against the current. During some seasons it is also blocked from the end of August till the first days of the following July, leaving scarcely six weeks of open navigation in the whole year. Yet since the discovery of the convenient harbour of Dicksonhavn in its estuary by Nordenskjöld, several trips have been made to the Yenisei by English and Scandinavian skippers, who have returned with cargoes of grain and other produce brought down from the Minusinsk steppe in the Upper Yenisei valley. The rafts and light craft engaged in this traffic are broken up for fuel or timber after discharging at Dicksonhavn.

The Yenisei drains a total area of about 1,180,000 square miles, and with its tributaries has a navigable waterway of not less than 5000 miles. The western branch, or Yenisei proper, is nearly 2600, the eastern (Selenga-Angara) 2950 miles from their farthest sources to the common estuary, which is separated from that of the Ob only by a comparatively small peninsula, 300 miles broad at its widest part. In General von Tillo's list of the eight longest rivers in the world the Yenisei-Selenga takes the fifth place, being exceeded by the Missouri (4194 miles), the Nile (4020), the Yangtze-kiang (3158), and the Amazons (3063).[1]

The Lena Basin.

Of all the great North Asiatic streams the Lena alone has its source on the seaward slope of the mountain range enclosing the central plateau. But the Upper Lena, which rises on the hills skirting the west coast of Lake Baikal, and which for some distance flows parallel with the Angara, seems to have formerly communicated with that river through a now dried-up depression in the low water-parting between the two basins. It is joined at Vitimskaya by the Vitim, which flows from the east side of Lake Baikal round the elevated Vitim plateau

[1] Petermann's *Mitteilungen*, 1887. The list is completed by the Amur (2920), Congo (2883), and Mackenzie (2868). But the estimates for some of these rivers vary considerably, and the Mackenzie is certainly equalled if not surpassed by the Hoang-ho, Lena, Ob, Mekhong, and perhaps one or two others not included in Von Tillo's list.

north-westwards, and which, from its size and volume, might be regarded as the true upper course of the Lena. Below the confluence the main stream is deflected by the scarp of the Yakutsk tableland for hundreds of miles east-north-east to Yakutsk, where it again resumes its normal northerly course to its delta in the Arctic Ocean, over against the archipelago of New Siberia. Below Yakutsk the Lena is again nearly doubled in size by two great affluents, the Aldan from the Stanovoi uplands and the Viliui from the low water-parting between the Lower Yenisei and Lena basins. The vast basin of the Lena, draining a total area of 1,000,000 square miles, is thus enclosed by those of the Yenisei, Amur, and Indigirka, and presents a total water highway of perhaps 6000 miles, open, however, for only five or six weeks in the summer.

Expanding at the Aldan junction to 12 miles from bank to bank, the Lower Lena again contracts to 3 or 4 miles as it approaches its delta, which is nearly 9000 square miles in extent. Here the navigable channel is sometimes blocked throughout the whole summer by floes massed along this coast by the polar winds. Nevertheless the Norwegian Johannsen succeeded in ascending the Lena to Yakutsk in the steamer *Lena* for the first time in 1878. But it is doubtful whether any regular navigation can ever be established with this river, whose basin belongs entirely to the Arctic Ocean. Yet this region abounds in copper, iron, coal, lead, gold, silver, sulphur, salt, and other minerals. The Vitim sends down auriferous sands in large quantities, and the Lena is skirted for over 900 miles by coal-measures, often cropping out above the surface. In some places the coal-beds, kindled by forest fires, have been burning for years, giving rise to the reports heard from time to time of still active volcanoes in the Lena basin.

Between the Yenisei and Lena basins two other large rivers—the Khatanga, 600 miles long, and the Olenek, double that length—enter the Frozen Ocean in separate channels. The Olenek is 6 miles wide and over 20 feet deep at its mouth; but both alike are practically useless as highways of trade.

This is also to a large extent true of the Yana, Indigirka, and Kolima, flowing east of the Lena from the Verkhoyansk range northwards to the Arctic, as well as of the Anadyr, flowing from the southern watershed to the Bering Sea.

The Amur Basin.

The Amur, whose basin is now politically divided between Russia and China, promises at no distant date to become more important economically than all the Siberian rivers taken collectively. It is formed by the junction of the Shilka and Argun, the former flowing from Mount Kentai in the Khan-ula range mostly through Russian territory, the latter from the south side of the same range, under various names, such as Kerulen, Lukin, etc., mostly through Chinese territory. The main stream is joined on its left bank by the Zeya at Blagevyeshchensk, above which point it is navigable by light craft for several hundred miles from its mouth in the Gulf of Tartary over against the island of Sakhalin.

Even the Shilka, hitherto supposed to be accessible to steamers only as far as Srychensk, has been shown by the surveys in connection with the Siberian railway to be easily navigable to Mitrofanovskaya, 105 miles farther up. The steamer *Kiakhta*, drawing 5 feet, reached this place in 1893 at the average level of the river.

Below the Zeya the Amur is swollen by the Bureya,

also on its left, and by the Sungari and Usuri on its right bank. Between the Zeya and Usuri, a distance of 570

VALLEY OF THE AMUR.

miles, it is accessible to ships drawing 8 feet, and thenceforth to deep-sea vessels, although a serious obstacle to navigation is offered by the bar at its mouth, with a depth of scarcely 13 feet.

COSSACK VILLAGE ON THE AMUR.

On its seaward course the Amur has to force its way through the Khingan range, separating the Mongolian plateau from Manchuria, and farther down through the Sikhota-alin, by which its lower course is deflected almost due north in a line with the Usuri. Between these two points it breaks through the barrier of the Little Khingan, below the Bureya confluence, and again through the rugged Chanyatin uplands. Besides the bar, the shifting sands and intricate channels, both in the river itself and in the shallow waters between its mouth and Sakhalin, offer great obstacles to its navigation, which is further closed by ice for six months in the year. Owing to these difficulties, the attention of the Russians was long directed towards the southern coast region, which they finally secured in 1860, and where they now possess the more or less convenient harbours of Castries Bay, Port Imperial, Olga Bay, America Bay, Victoria Bay (now by them renamed the Gulf of Peter the Great), and Possiet Harbour on the Korean frontier. From source to mouth the Amur has a total estimated length of 2920 miles.

On the south-west frontier of this maritime province is the extensive but shallow Lake Kenka (properly Hanhai), which is 65 miles long and about 25 wide, with an area of 1200 square miles, but is nowhere more than 30 feet deep. This basin, which, notwithstanding its great extent, is a mere reservoir for the rainfall of the surrounding hills, drains through the Sungacha northwards to the Usuri and Amur.

4. *Natural and Political Divisions: West Siberia (Governments of Tobolsk and Tomsk)—The Tundra—East Siberia (Governments of Yeniseisk, Irkutsk, Transbaikalia)—Amur—Primorsk Province—Islands—Sakhalin.*

The conflict between permanent physical conditions and arbitrary political groupings, so common throughout the Russian Empire, is mostly restricted in North Asia to West Siberia. Some portions of this region, which is mainly comprised within the limits of the Ob basin, are attached to the Government of Turkestan, while other portions are included in the European Governments of Perm and Orenburg. To European Russia by a curious fiction are also attributed the Governments of Akmolinsk and Semipalatinsk, besides a part of Turgai, so that Tobolsk and Tomsk are the only West Siberian Governments which are not politically encroached upon. Tobolsk is limited by the eastern slopes of the Urals, while Tomsk comprises the Altai highlands and the upper waters of the Ob and Irtish.

West Siberia.

Seen from the eastern slopes of the metalliferous Ural range, these lowlands seem to stretch away eastwards, like a limitless ocean plain. For some distance beyond Tiumen the land presents the aspect of a heath varied with a few plantations of sickly firs. This is succeeded by the steppe, which as far as the Irtish is diversified with birch and brushwood thickets, interspersed with extensive swamps and shallow basins. But east of the Irtish the country assumes the character of a true steppe, a boundless grassy plain, here and there relieved by a few bushes on the distant horizon. Beyond Omsk it presents the appearance of a prairie with rolling hills, covered with short grass, which affords pasturage for the herds and flocks of the Kirghiz nomads. But the steppe is distinguished from the prairie especially by the numerous lakes, some of considerable size, frequented by wild swans, ducks, and other waterfowl. Towards Semipalatinsk, a feature of the scenery are the square graves of the Kirghiz-Kazaks, made of the trunks of trees, and looking at a distance like little houses or log-huts.

South of this place run the Arkat hills, bare granite masses of grotesque form and rugged aspect, rising to a height of 1200 feet in picturesque outline above the surrounding treeless plains. At Sergiopol the snowy crests of the Tarbagatai range come into view. North of this point the slopes of the Ala-tau afford good pasturage to magnificent herds of oxen, camels, horses, and fat-tailed sheep. Then follow the salt steppes, where animals and wayfarers sink at every step through the saline efflorescence covering the surface. But in the Ala-kul lacustrine district the ground is overgrown with an extensive dense jungle of reeds, affording in winter a slight shelter to the Kirghiz from the fierce snow-storms. A striking contrast to these monotonous wastes is presented by the magnificent scenery of the upland valleys and alpine lakes on the slopes of the Ala-tau.

Proceeding along the Ob valley north from Tobolsk, the impenetrable primeval woodlands are succeeded by dreary bottomless swamps, the true tundra, stretching without a break north and north-eastwards to the Ob and Yenisei estuaries. For eight long winter months the frozen ground is here covered with snow, the glass often falls to 45° below freezing-point, and the cold converts the breath of animals into icy hoar frost. Birds on the wing often fall dead from the skies, the panes of glass start in their sashes, the hardened soil splits into wide and deep fissures, and the very ice on the lagoons bursts asunder. Here the fierce storm often rages for twenty-four hours at a time, during which man and beast remain patiently buried in the snow, as the only means of sheltering themselves from its fury. The heavens are perpetually overcast with dull leaden clouds, the atmosphere is raw and humid, the long gloomy nights are relieved only at intervals by the magnificent phenomenon of the Northern Lights.

Yet in summer the tundra can present even an inviting aspect. Nordenskjöld, who explored the lower course of the Yenisei in the August of 1875, in preparation for his famous expedition by the north-east passage, denies that the tundra presents the aspect of a dreary ice-bound waste, relieved here and there only by a stunted growth of sickly vegetation. Such, according to him, is its aspect at one point only, on the Yenisei, the vegetation being elsewhere, and especially in the islands of the river, of a surprisingly luxuriant character. The fertility of the soil, the boundless extent of the meadow lands, and the abundance of pasturage render the tundra a splendid grazing ground.

Farther south, between Turukansk and Yeniseisk, where the country is overgrown with extensive woodlands of great age, and succeeded near Krasnoyarsk by boundless plains covered with a thick layer of black mould, the tundra merges in a region fully as productive as the most favoured tracts in Scandinavia. Here the natural richness of the soil, combined with the abundance of fish in the rivers, the sparse population, and the absence of markets, renders provisions of all kinds fabulously cheap. The Rev. H. Lansdell, who visited this region in 1879, was offered "live ducks for five farthings each, large fish, called *yass*, for 1½d. a pair, and pike for a farthing each. Milk cost 2½d. a bottle, but young calves in remote villages could be purchased for 6d. each. The belt of rich black earth in the region immediately north of the Altai lets for 3½d. per acre, and from it wheat may be purchased for about one-twentieth its cost in England. Still farther north, in the forest region, rich in excellent timber and fur-bearing animals, meat was bought up wholesale in 1877 at less than a halfpenny per pound; whilst in the tundras the rivers are so full of fish that one of the ordinary difficulties of the natives is to avoid breaking their nets with the weight of the draught. The fish are frozen and sent more than 2000 miles to St. Petersburg, where a very moderate price realises for the fisherman a profit of nearly 100 per cent." [1]

East Siberia.

The course of the Yenisei marks the boundary line between West and East Siberia. Beyond this line the plains are far more diversified by hills, ridges, and even hilly plateaux, which often deflect the course of the streams east and west, whereas in the Ob basin all run north,

[1] Paper in *Proceedings of Royal Geographical Society*, October 1880.

north-west, and north-east. In fact, in East Siberia true lowland plains of great extent are comparatively rare. Even in the Taimur peninsula, between the Yenisei and Khatanga estuaries, the coast ranges are said to attain elevations of from 3000 to 4000 feet, while the Kharaulakh hills between the Lower Lena and the Yana rise in some places to a height of 1300 feet. Farther east and south occur the vast elevated plateaux of Yakutsk, Transbaikalia, the Daurian and Nierchinsk steppes, the Stanovoi uplands, the Amur basin and Kamchatka, filling most of north-east Asia, and reducing the lowland formation to a relatively small area. Here also the Verkhoyansk water-parting runs at an elevation of from 5000 to 6000 feet from the Lower Lena right across to the north-east coast ranges, thus completely separating the head-streams of the Yana, Indigirka, and Kolima from the Lena basin. Speaking generally, while the mean altitude of the southern Altai mountain system falls gradually north-eastwards to the Stanovoi plateaux, the mean altitude of the northern region rises gradually from the Yenisei eastwards to the Bering Sea, here culminating in the Kamchatka peninsula. Even between the Yenisei and Lena basins there is a plateau of palæozoic formation.

In East Siberia, although Kamchatka disappears from the administrative nomenclature, the political divisions otherwise mostly follow the natural lie of the land. Thus the vast Governments of Yeniseisk, Irkutsk, Yakutsk, and Anadyr, this last constituted in 1894, are mainly comprised in the Yenisei, Lena, and Anadyr river basins, while the provinces of Transbaikalia and the Amur correspond with two physically distinct regions, the first stretching from Lake Baikal to the Argun-Shilka confluence, the second comprising the region between the Stanovoi water-parting and the left bank of the Amur. Even the south-eastern divisions of the Usuri territory and the Primorsk

Province answer to two natural divisions, the first comprised between the right bank of the Usuri and the Sikhota-Alin highlands, the second including the strip of coast land between the eastern slopes of these highlands and the Sea of Japan. The new province of Anadyr comprises the north-eastern extremity of Siberia between latitudes 60°-70° and longitudes 164° E. to 170° W. Its population, about 200,000, consists mainly of Lamuts, Koryaks, Kamchadales, Chukchis, and a few surviving Yukaghirs.

To these governments are attached the islands and groups of islands lying off their respective coasts. Of these the largest are the desolate and uninhabited New Siberian or Liakhov Archipelago, north of the Lena delta ;[1] the Bear Islands, north of the Kolima estuary ; Wrangel Land, discovered in 1849 by Kellet on the spot previously indicated by Wrangel, and for the first time circumnavigated in the summer of 1881 by the crew of the American steamer *Rodgers*, not in the vessel itself, as has been stated, but in the small boats belonging to it ; lastly Sakhalin, whose northern extremity is almost connected with the mainland near the mouth of the Amur.

Baron Toll, who made an excursion in May 1893 to the New Siberia group, discovered on the Malyi Liakhov island complete trees of *Alnus fruticosa*, 15 feet long, with leaves and cones, under the perpetual ice in a sweet-water deposit, which contained pieces of willow and bones of post-tertiary mammals—the so-called mammoth layer. It thus appears that during the mammoth period arborescent vegetation ranged as far north as the 74th degree of latitude, that is, at least three degrees farther north than at present. The thick layers of ice seen on

[1] A grant of 14,000 roubles was made by the Government for two Polar observing stations at the Lena delta and on one of the islands of New Siberia during the year 1882.

Kotelnyi island and other members of the Archipelago were also ascertained from their granular structure to date from the glacial period. Amongst the animals met on the islands were numerous white bears and swarms of mice (*Lemmus obensis*), which were in a feverish state of activity, some passing from island to island, some migrating to the mainland, and others coming from the mainland to the Archipelago. The farthest point reached by the explorer was 75° 37′ N. on Kotelnyi island.

Formerly held jointly by Russia and Japan, Sakhalin[1] (Saghalien) or Karasto was by the treaty of 1875 ceded by the latter power to Russia in exchange for the Kurile Archipelago. It has an area of no less than 25,000 square miles, and stretches for 550 miles north and south, with a breadth varying from 15 to 80 miles. It is traversed in its entire length by parallel ridges, of which the loftiest and most continuous is the west coast range, with crests from 3000 to nearly 5000 feet, culminating with La Martinière (Ktönspal) Peak about the centre, 4860 feet. From this chain, which nowhere reaches the snow-line, a few streams flow for short distances, mainly southwards to the coast. Amongst the products the most important is coal, which, although inferior to the English, still commands higher prices than that of Japan or Australia. There are also extensive forests of valuable timber, frequented by numerous fur-bearing animals. The climate and soil are unfavourable to agriculture, but vegetables may be grown and stock-breeding carried on in some sheltered districts. Latterly the Russians have used this island chiefly as a convict settlement for political prisoners, who are sent thither by the sea route

[1] This word is a corruption of the Manchu "Sakhalan anda Khanda"—that is, "Rock of the Amur Estuary"—applied originally to an islet in the mouth of the Amur, and afterwards, by mistake, extended to the island now known as "Sakhalin."

through the Suez Canal. The chief stations are Dui on the west coast, and Mauka Cove towards the south-west extremity. The latter has been chosen as the head-quarters of a company which has obtained from the

SAKHALIN PENAL COLONY.

Government the monopoly of the trade in fish, bêche de mer (trepang), edible seaweed, and other local produce, for a term of ten years.

In 1892 Sakhalin was visited by Professor Krasnov of the Kharkov University, who collected some valuable information regarding its natural history and inhabitants. Owing to the cold currents to which it is exposed, its climate is much colder than most places in the same latitude, the flora resembling that of Spitzbergen, and the conditions of plant life such as prevailed during the glacial epoch in Europe. Noteworthy is the existence side by side of distinct vegetable types, due to variations, not of climate, but of soil and relief. Thus the Polar flora of the summits is succeeded on the slopes down to sea-level by the *taiga,* a forest of conifers (*Abies* and *Picea*), with birch, maple, and mountain ash, while the swampy tracts and lowlands are covered with larch forests in association with *sphagnum* and other peat-bog plants. The inhabitants comprise three distinct groups, the nomad Tunguses of the tundra, the Orochons, who hunt in the taiga, and the Ainu fishermen of the coast-lands, probably the true aborigines. The Russian exiles reap precarious crops of cereals during their enforced sojourn in this inhospitable region (*Annales de Géographie,* July 1893).

5. *Climate: Region of Intensest Cold.*

Amid much diversity, natural in such a wide area, the Siberian may on the whole be taken as the most essentially continental climate on the globe. Here the maximum of cold is reached, not in the Yakutsk district, as is commonly supposed, but in Verkhoyansk, on the Upper Yana, just within the Arctic Circle, where the glass usually falls to 49° C. below freezing-point in January, the mean in Yakutsk being 42° or 43°. "Within the isothermal of – 40° C., a temperature at which the quicksilver freezes, Verkhoyansk alone is included for the whole period from November to February, Yakutsk for December and January only, and Ustyansk (at the mouth of the Yana) for January only, while Tolstoy Noss (at the mouth of the Yenisei) lies beyond this isothermal."[1] It also appears that in these very places the glass rises from 28° to over 38° C., or occasionally as high as 102° F. in July and August, which is about the normal summer heat of most lands lying about the equator. No other region can show such amazing extremes as these, consequently the claim of North-East Siberia to the possession of the most typical continental climate is established. It would seem to be at once colder than the North Pole,[2] and hotter than many uplands under the equator, a condition due to the combination of more cold and heat producing causes than occur elsewhere in the northern hemisphere.

But while the intense heat lasts only for a few weeks, the intense cold prevails for many months, the two extremes being separated by short intervals of broken spring and autumn weather. The result is that in the course of ages the ground has gradually become permanently frozen in many parts of the tundra from about 2 feet below the surface to depths of from 100 to 300 feet, and perhaps even more. It might be supposed that the great elevation of the Altai regions would have the effect of neutralising the difference of latitude, thus rendering the southern highlands as cold as the northern lowlands. But this is far from being the case, and as we proceed southwards the normal temperature rises steadily. The Russians compare the climate of some places in these latitudes with that of Italy, and there can be no doubt that many of the Altai and Amur districts are favoured by a genial healthy climate suitable for the development of agriculture.

[1] *Verhandlungen der Ges. für Erdkunde zu Berlin*, July 1881, p. 275.

[2] The two poles of greatest cold in the old and new worlds oscillate about Verkhoyansk and Cockburn Bay, Adelaide Peninsula (H. W. Klutschek, of Schwatka's Expedition).

The prevailing winter winds are from the south and south-west between the Urals and the Yenisei, but in the Lena basin from the north-west. These icy north-west gales blow steadily for months together, and are felt far beyond the limits of Siberia in the Japanese waters, in the Amur basin, and on the Mongolian plateaux. In summer the rarefaction of the atmosphere causes Arctic breezes to prevail along the western seaboard. But farther east these are succeeded by moist south-east winds from the Pacific, and to this cause the regions east of Lake Baikal are indebted for their abundant rainfall. On the coast-lands the mean exceeds 40 inches, falling westwards to 10 at Yakutsk and to 8 at Kiakhta. In winter the snows are much lighter in the east than in the west, and the Lower Lena and Yana basins, where the cold is intensest, are remarkable for their clear blue winter skies.

6. *Flora and Fauna: The Argali, Marmot, and Lemming—Extinct Mammalia.*

In North Asia the northern limits of timber, while following the coast-line, scarcely anywhere reach the Arctic seaboard. Long before reaching their actual limits, the few stunted larches straggling northwards assume strange distorted forms, trailing rather along the surface than shooting upwards, and often presenting the aspect of withered branches or dead trunks of trees. Nevertheless, in his expedition down the Ob in 1880, Khandachefsky discovered a magnificent forest of large cedars and larch in the valley of the River Nadym, at a point where Petermann places the extreme northern limit of the forest zone.

This forest zone, or "taiga," consisting mainly of species common to Europe, stretches almost uninterruptedly right across the continent, merging everywhere northwards in the tundra. Here the vegetation consists almost exclusively of mosses, lichens, and grasses. South of the forest zone the Ob basin is occupied by the steppe, which in some respects resembles the tundra, both presenting the same cheerless, monotonous aspect, and absence of timber. Even the same species are often found in both, such differences as exist being caused by deficient moisture in the steppe, and deficient heat in the tundra. Sometimes particular species of reindeer and other mosses predominate in the tundra, imparting a pale white or a dull yellow aspect to the scenery for miles and miles along the lower reaches of the great rivers.

In the taiga [1] the prevailing trees are the larch, birch, alder, cedar, and a noble species of pine peculiar to Siberia, which shoots up to a height of nearly 100 feet, with a slender stem seldom exceeding a foot in diameter. But perhaps the most characteristic plants are those producing berries in great variety and abundance. These uncultivated fruits supply food to man and beast, and quantities are preserved for use during the winter.

An English traveller who visited Siberia in the summer of 1893 gives a vivid description of the highly characteristic taiga woodlands. "One's first impression, especially if wandering alone or at night, is the awful loneliness and gloom of the place, where at all times silence seems audible and darkness painfully visible. The air is clogged and damp, the noonday light dimmed to dusk, vegetation matted and dripping, and the pungent smell of must and mould, pervading and permeating everything, pierces to the inmost soul. Thickets of moss-

[1] The term is used somewhat differently in different parts of Siberia. In the Altai it means the wooded uplands abounding in fur-bearing animals; in the north it is applied to the zone of uninhabited woodland tracts bordering on the mossy tundra, which stretches thence to the Arctic seaboard.

SIBERIAN LARCH.

covered, mildewed spruces, interlocked in deadly embrace, vainly wrestling with each other and with pines and cedars for the life-giving light of the genial sun, are woven into one wide wall of appalling thickness. Nor is it sheer denseness alone that renders the taiga impassable; a no less effectual hindrance consists in the wickerwork of storm-felled cedars, birches, and firs, twisted into the general tangle amid mounds of dead leaves, gnarled roots, and musty masses of decaying vegetation. Winds that would sough and wail through the trees of a European forest glide noiselessly by in the taiga. They may bend and break the tops of towering larches and tapering cedars 80 or 100 feet nearer to the clouds, but they stir no gossamer thread in the dismal depths below. The murmurs of the air and the motions of forest and glade are not slumbering only, but frozen and dead; the birds and bees have flown to far-off nests and hives; the winds seem to be pent up in caves, sunlight to have found a gloomy prison, and life an everlasting grave."

M. Cherski's expedition of 1891 to the Indigirka and Kolima basins shows that, despite their high latitude, those regions possess a rich, if not a very diversified, flora in the more sheltered part of the river valleys. The numerous arms of the rivers, often stretching right across the valleys, are adorned with magnificent balsamic poplars and tall willows of striking beauty, while the northern slopes of the valleys, in contrast to the rocky southern sides, are covered in places with white reindeer moss. The meadows of the valleys foster such a rich vegetation, that the traveller visiting this region in June or July is apt to forget the high latitudes he is crossing, especially when the temperature in the sun reaches 113° F. But this tropical heat is soon over, and during the expedition the temperature fell below freezing-point on fourteen days in August, once as low as 19° F., and on six days snow fell (Petermann's *Mitteilungen*, 1892).

In the remote volcanic peninsula of Kamchatka the banks of the inland waters are decked with a clothing of grass, growing with an almost tropical luxuriance, interspersed with bright flowers, alpine rose-bushes, the cinquefoil, and the rare Kamchadale lily. The poplar and the birch grow in clusters on the lowlands, while the slopes of the hills are covered with the sombre foliage of extensive pine-forests.

The prevailing humidity of the Pacific seaboard has also favoured the development of a magnificent vegetation in the Amur basin, where the flora, especially in the islands and along the river-banks, abounds in endless varieties of leafy shrubs and undergrowths. Here also the conifers, oak, elm, ash, walnut, cork-tree, maple, and linden often attain majestic proportions. On the Usuri the ginseng is largely cultivated for the market of China, where this plant fetches its weight in gold, and where it is supposed to be a sovereign remedy against all disorders. The wild vine in some places yields a good grape, and the grasses flourish with astonishing luxuriance; for in this more favoured region the flora includes plants peculiar to the cold, temperate, and warm zones.

This is also largely true of the animal kingdom. In Manchuria the tiger reaches his northernmost limit, and is here associated with the panther, lynx, glutton, and wolf. There are two species of bear, and a transition is effected to the fauna of Siberia proper by the sable, black and red fox, marten, ermine, and other fur-bearing animals. The cedar-groves are here enlivened especially by a species of dark-gray squirrel, whose skin is much prized, fetching large prices on the spot. The ruminants are represented chiefly by the deer, elk, roe, and musk-deer, while there are over 200 species of birds. The Amur, and especially the Usuri River, with Lake Kenka, are incredibly rich in fishes, including the sturgeon, salmon, carp, sterlet, and many other varieties.

This teeming animal life has elsewhere its counter-

SABLE.

part in the prodigious multitudes of marmots and other species of small rodents inhabiting vast tracts from the Tarbagatai right away to Kamchatka and the extreme north-eastern Chukchi lands. The ground in many places is honeycombed with the galleries and subterranean townships of these pretty little troglodytes, who may at times be seen mounting guard in interminable lines on the hillocks at the entrance of their dwellings, suddenly disappearing at the least sound, and as suddenly re-appearing to ascertain the cause of the alarm.

Equally abundant in the north-east are the lemmings. The line of march of these migratory rodents often stretches for miles across the plains between their winter quarters and summer camping-grounds. In Kamchatka

ARGALI.

a lasting alliance has been struck between them and the natives. Whenever the latter are driven by distress to draw from the supplies of their provident little friends during their absence on some distant expedition, they are always careful to replace the stores in more prosperous times. It is also said that, to guard against similar plunder by other less scrupulous marauders, the lemmings conceal their underground granaries with poisonous herbs. So, at least, Krasheninnikov was informed by the natives.

Most of the fur-bearing animals have disappeared from these north-eastern regions, causing many of the old hunting stations to be abandoned. But in Kamchatka the trappers still obtain from 6000 to 8000 sable skins for the Russian market. In other parts of Siberia many species of these animals are also becoming extinct. But here their disappearance is often due as much to the destruction of the taiga as to the skill of the trapper.

LEMMING.

In West Siberia birds are very numerous, and here amongst the more characteristic species are the golden eagle, the white-throated alpine lark, and the gray-headed wagtail. In this region the Arkat hills still afford a refuge to the Argali (*Ovis ammon*), a magnificent mountain sheep, with enormous thick and twisted horns over three feet long. This species, whose original home seems to be the Central Asiatic plateau, resembles in its habits the steinbok and chamois, and, owing to its extraordinary speed and velocity, is very difficult to bring down.

Of domestic animals, the most useful are the fat-tailed sheep and camel in the steppe, the reindeer in the tundra, and the yak in the Upper Yenisei basin, where it reaches its northernmost limits.

Siberia was in former epochs the home of a large species of rhinoceros and of the mammoth, some specimens of which have in recent times been found preserved by the ice in an almost perfect state.[1] Vast quantities of fossil ivory from these animals, amounting at one time to 40,000 lbs. yearly, have been obtained in the archipelago of New Siberia, and parts of the Arctic seaboard.

7. *Inhabitants: Table of the Siberian Races.*

Excluding the Bashkirs, who dwell chiefly west of the Ural River, and the Kara-Kirghiz and Kirghiz-Kazaks, whose camping-grounds lie chiefly south of the Aralo-Caspian and Ob water-parting, the native inhabitants of Siberia scarcely number 750,000 altogether. And even these, few as they are, seem to be mostly in a process of more or less rapid extinction or absorption in the advancing Slav element. Certainly Siberia belongs henceforth to the Russians, in the same sense that Australia has become a new home of the English race. They have already occupied a continuous broad zone stretching from Europe across West Siberia and along the southern highlands to Lake Baikal, and thence through Transbaikalia and down the Amur and up the Usuri to the Pacific seaboard. Here they are firmly established at Nikolayevsk and Vladivostok, the extreme northern and southern points of the Primorsk Province. They have also occupied both banks of the Yenisei throughout its

[1] A rhinoceros by Pallas in 1771 on the Vitin, a mammoth by Adams in 1799 on the Lena, and others during the present century.

entire course, most of the Ob and Irtish, the Lena down to Yakutsk uninterruptedly, besides numerous detached stations on the Lower Lena, in Kamchatka, Sakhalin, and elsewhere. Large portions of the really arable lands are thus already held by Russian agricultural colonies, and great efforts are now being made by the Government to direct the migrations of the peasantry from Europe to the Amur basin. Certain tracts, such as the distinctly steppe region of West Siberia, will doubtless continue to remain in the hands of the natives, for they are uninhabitable except by nomad tribes. But all the broad lands available for cultivation will be occupied by the Slav race.

None of the natives have any vitality except the Yakuts of the Lena basin and the Kirghiz of the West Siberian steppes. The Ostiaks of the Ob basin and the Yukaghirs of the Indigirka and Kolima Rivers are actually dying out, and a similar fate threatens to overtake the Giliaks of Sakhalin and opposite mainland, as well as the Samoyedes of the Lower Ob and Yenisei.

With the exception of the still unclassified "Hyperborean" group, all the aborigines belong to various branches of the Mongolo-Tatar ethnical and linguistic family. From the subjoined table of these races the Bashkirs and Kirghiz are omitted, the main sections of these races being included in Europe and the Aralo-Caspian basin respectively:—

MONGOLIAN STOCK.

Races.		Religions.	Population.
Kalmuks	Zungars Targuts Khoshods Turbets Chorasses Teletzes	Buddhists and Shamanists	20,000

Races.		Religions.	Population.
Buriats (East Branch)	Kudara, Selenga, Khorinsk, Barguzin	Shamanists, Buddhists, and Christians	250,000
Buriats (West Branch)	Tunka, Verkolensk, Olkhon, Kuda, Ida, Balagansk, Alarsk		

MANCHU STOCK.

Tunguses	Lamuts, Oroches, Orokhos, Chapogirs, Golds, Menegrs, Manguns, Samagirs, Ngatkons, Nigidals, Negdas, Tazi, Olenes	Shamanists and nominal Christians	80,000

FINNIC STOCK.

Samoyedes	Yuraks, Tagurs, Abators, Koibals, Soyots, Motors, Karagasses, Kamastes, Tagvis	Shamanists and nominal Christians	35,000
Ugrians	Ostiaks	Shamanists	25,000
	Voguls	Nominal Christians	4,500
Mixed Finno-Tatars	Darkhats, Soyons	Buddhists	15,000
	Assan, Arinzi, Kotti	Shamanists	5,000

TURKI STOCK.

Yakuts	Christians and Shamanists	200,000
Red Tatars, Black Tatars	Mostly Christians	80,000
Teleuts	Sunnis	
Kumandes	Christians	

UNCLASSIFIED SUB-ARCTIC RACES.

Koriaks	Pagans	5,000
Chukchis	Pagans and nominal Christians	12,000
Yukaghirs	Shamanists	1,600
Kamchadales	Nominal Christians	3,000
Ankali, extinct?	Pagans.	...
Giliaks		5,000
Ainus		3,000
Eskimos		500

SLAV STOCK.

Races.	Religions.	Population.
Great Russians .	Orthodox	4,500,000
Little Russians .		
Poles . . .	Roman Catholics mostly .	

SUNDRIES.

Races.	Religions.	Population.
Chinese . . .	Buddhists.	10,000
Manchus . . .		
Koreans . . .		7,000
Japanese . . .		400
		5,262,000

GILIAK.

The Buriats.

The Mongolian race is in Siberia best represented not by the Kalmuks, but by the less known Buriats, who have been long settled on both sides of Lake Baikal. Previous to the Russian conquest all were still addicted to the old Shamanist religion of Siberia. But towards the close of the seventeenth century those dwelling east of Lake Baikal adopted Buddhism, while most of the others conformed to the Orthodox Church. Like most Mongolian peoples, the Buriats are of a decidedly phlegmatic temperament, betraying such an inborn disinclination to work that they often need the stimulus of actual hunger to exert themselves in any way. They are stolid, reserved, sullen, and uncourteous to strangers. Through the Russians they have acquired a passionate love of drink and tobacco, and children eight or nine years old are now often met with Chinese pipes in their mouths. The Buriats are in other respects a harmless, peace-loving people. Amongst them murder is rare, and highway robbery unknown, although they are still prone to acts of petty theft. Formerly nomads and stock-breeders, they have recently become successful

BURIAT GIRL.

agriculturists, and also show a marked capacity for industrial pursuits, often proving more skilful than their Russian teachers.

Beneath an outward show of Buddhism and Christianity, the Buriats, like so many other Siberian peoples, are still at heart genuine Shamanists. The Shamanistic cult, which is based entirely on oral tradition, and which is little removed from nature-worship, was formerly universally diffused throughout Siberia. But it could scarcely hope long to resist the attacks of the Buddhist propaganda, supported as this was by a zealous priesthood and a rich religious literature.

Of the Buddhist Sacerdotal order there are three degrees in Siberia, the two first alone bearing the title of Lamas.

Notwithstanding their ignorance, the Lamas have betrayed a fanatical zeal in the cause of Buddhism, everywhere suppressing Shamanistic practices, and even successfully resisting the spread of Russian Christianity amongst the aborigines of East Siberia.

The Tunguses.

Conterminous on the north with the Buriats are the Tunguses, who occupy an enormous domain in East Siberia, stretching from the Yenisei to the Pacific seaboard, and at two points reaching northwards to the Frozen Ocean. In the Lena basin this domain is largely encroached upon by the Yakuts; but the coast-lands from the Amur nearly to the Arctic Circle are still almost exclusively held by the various divisions of the widespread Tungus family. The Tunguses contrast most favourably not only with the sluggish Buriats, but with all the other races of Siberia. Travellers are never wearied of extolling their many admirable qualities; and there can be no doubt that they are one of the very noblest types of mankind. They are cheerful under the most depressing circumstances, persevering, open-hearted, trustworthy, modest yet self-reliant, a fearless race of hunters, born amidst the gloom of their dense pine-

TUNGUSES.

forests, exposed from the cradle to every danger from wild beasts, cold, and hunger. Want and hardships of every kind they endure with surprising fortitude, and nothing can induce them to take service under the Russians, or quit their solitary woodlands, where they cheerfully face the long and harsh winters, when the snowstorm often rages for days together.

The dress of the Tunguses is picturesque, and even elegant, especially when contrasted with the coarse and slovenly garb of the Buriats. "Surprising resemblances in the designs of the materials seem to show that the Tunguses must at one time have maintained constant intercourse with Japan."[1]

Most of the Tunguses in the Baikal district have been baptized; but Russian orthodoxy has scarcely penetrated below the surface. They look on the rites of the Church as mere formalities, practising them only under compulsion, or in the presence of the Russians. When engaged in the chase, or remote from the European settlements, they are still true nature-worshippers. Hemmed in and continually encroached upon, especially by the Russians and Yakuts, the Tungus race seems destined to ultimate extinction as a distinct nationality.

The Yakuts.

The domain of the Yakuts, who are the most energetic and versatile of all the Siberian peoples, lies mainly on both sides of the Middle and Lower Lena, with isolated settlements on the left bank of the Lower Indigirka and Upper Kolima. This is the north-easternmost point reached by the Turki race, of which the Yakuts are a distant branch. During their migrations eastwards the Yakuts have become largely intermingled, especially with the Tunguses. Their Turki type has thus become so profoundly modified, that their original kinship with the Western Turki peoples, from whom they are separated by a vast interval, is now attested chiefly by their speech. At the same time, there is perhaps some exaggeration in the oft-repeated statement that the Lena Yakuts and the Osmanli of Stambul can easily converse together.

[1] Reclus, vi. p. 359.

While all the other aborigines of Siberia seem to be dying out, the Yakuts are actually increasing in numbers.[1] They have been not inaptly described by Wrangel as "men of iron," and more inured to cold than perhaps any other people in the world. Their territory includes both Yakutsk and Verkhoyansk comprised within the zone of intensest cold in the old world. Yet they seem to be almost indifferent to the rigours of a climate where the glass falls in winter to nearly 50° below freezing-point. In a temperature of −32° R., Kennan met them airily arrayed in nothing but a short shirt and a sheepskin, lounging about, joking or gossiping, as if they were enjoying the balmy summer zephyrs of some favoured temperate zone. They are at the same time extremely industrious, skilful artisans and agriculturists, and probably the most intelligent traders in North Asia. From their preternatural cleverness in driving a bargain the less quick-witted Russians have named them the "Jews of Siberia," and, unless it be the Chinese, they certainly yield in this respect to no other Asiatic people.

In their greater frugality the Chinese have also the advantage over the Yakut, who, with all his inherent energy and powers of endurance, seldom works except under pressure of actual want.

Although mostly baptized, the Yakuts are no better Christians than the Tunguses and the other "converts" to Russian orthodoxy in Siberia. Beneath the outward parade of Christianity they are not merely Shamanists but true nature-worshippers at heart. With many curious rites they conjure the powers of nature, filling mountain, stream, and valley with many good and evil spirits, whose numbers have been increased by additions from the Calendar and pandemonium of the Russians.

[1] From about 50,000 in the beginning of the century to 200,000 in 1880.

Above all there is doubtless a supreme being; but he is too far off to hear their prayers, or too good to need their supplications. It is the evil ones who require to be propitiated, as no harm can come from the good spirits. The two principles of good and evil took part in the creation, the former making the earth small and level, the latter coming and tearing it up enraged, whence the hills and the valleys. And the valleys became river-beds, and the great lakes and seas gathered round about the high mountains.

The Chukchi, Koriaks, and Kamchadales.

Beyond and partly overlapping the Yakuts and Tunguses is the interesting group of "Hyperboreans," filling the Chukchi and Kamchatka peninsulas, and occupying a portion of Sakhalin and of the opposite mainland about the Lower Amur. Since Nordenskjöld's expedition round the north-east coast, the Chukchis, who give their name to the north-eastern peninsula, have again been the subject of much controversy. But there can be little doubt that W. H. Dall is right in affiliating them to the Koriaks, who probably form a connecting link between them and the Kamchadales. All these tribes, together with the neighbouring Yukaghirs, would be readily grouped with the Mongolo-Tatar peoples, but for their speech, which differs in its structure fundamentally from the Ural-Altaic linguistic family. But it will be seen in chapter vi. (Japan), that the Ainus, if not also the Giliaks, stand on a different footing, and must be separated altogether from both connections.

The Koriaks, probably the parent stock of all the sub-Arctic races except the Ainus, possess many commendable qualities, and are especially noteworthy for their gentle and kindly disposition. A harsh word is never uttered against their women, and the children are treated with extreme tenderness. Hence all the more surprising seems the long prevalent practice of despatching their nearest kindred when enfeebled by age or infirmities. But this inconsistency is more apparent than real. The weak and aged are both alike incapable of performing the offices or undergoing the hardships inseparable from the nomad state. Hence to these untutored children of nature it seems a merciful act to spare them a lingering death by this means.

Essentially distinct from the Koriaks are the Lamuts, who dwell on the shores of the Sea of Okhotsk, and who are evidently a branch of the Tungus or Manchu family.

The Kamchadales, or aborigines, of the Kamchatka peninsula differ both in speech and appearance from the neighbouring Koriaks. They are favourably spoken of by Kennan and other travellers who have visited them, but since the year 1780 they have been reduced to one-half their former number by disease, famine, and other troubles. In summer they spear the salmon as they ascend the stream, and cultivate a little rye, potatoes, and turnips, besides which they keep some cattle and barter their furs with the Russians for tea and sugar. The interior of their houses is scrupulously clean, the walls, roof, and floor being planked over with rough but spotless birch boards, while the windows are adorned with chintz curtains and the walls hung with American engravings; but the doors are so low that ingress has to be effected on all fours. All have long been Christians, and the little Byzantine church is never missing in the centre of the villages, which invariably stand amidst clumps of trees on the banks of streams abounding in fish. Some of the northern islands of the Kurile Archipelago are inhabited by the Kamchadales.

The Ostiaks, Samoyedes, and Voguls.

In West Siberia nearly all the still surviving aborigines are members of the widespread Finnish race. The

OSTIAKS.

Finns are supposed to have come originally from the Altai and Sayan highlands, where they are still represented by the Soyots. These Soyots, by many regarded as the parent stock of the race, occupy a considerable area about the head-waters of the Yenisei, on both sides of the Russo-Chinese frontier, reaching from the Tanuola range northwards to the Krasnoyarsk district, and from the Upper Yenisei eastwards to the Buriat domain about Lake Baikal. They are extremely skilful artisans, and seem to have inherited the arts of the so-called "Chudes," an extinct prehistoric race, traces of whose culture are still met in various parts of Siberia.

The other branches of the Finnish stock in Siberia are the Ostiaks, Samoyedes, and Voguls. The Ostiaks are scattered in isolated groups along the Ob basin northwards to the estuary, and eastwards to the Yenisei between Yeniseisk and Turukhansk. In this wide domain of some 400,000 square miles they scarcely number 25,000 souls altogether, and seem to be everywhere either rapidly dying out or becoming absorbed amongst the surrounding Russian settlers. Their old national organisation is completely broken up, and they have almost ceased to dwell in settled abodes since the destruction of their villages and strongholds by the early Russian invaders of West Siberia. They formerly paid tribute in peltries, but with the gradual disappearance of the fur-bearing animals the taxes have been raised in specie. Such is the depressed condition of the race, that it has been proposed to distribute the children amongst the Russian peasantry, and leave the adults to die out in the course of nature.

Even after death the prospects of the Ostiaks are not of the brightest; for although there is a "third world," where there are no more bodily ailments, they never reach this stage, but are confined to the "second world," a far less happy abode lying somewhere beyond the Ocean, away north of the Ob estuary. Their belief in Shamanism is still unshaken, and nowhere else does the wizard, or medicine-man, enjoy more influence than among the Ostiak tribes. The brave man may possess bodily strength, they say, but the Shaman has the words of wisdom. The strong man may hurl the dart, but its course is directed by the Shaman, through whom alone it hits or misses the mark.

A still more primitive people are the Samoyedes, whose territory lies mostly within the Arctic Circle from

SAMOYEDES OF ARCHANGEL.

the head-waters of the Khatanga westwards to the Kanin peninsula. They are usually represented as dwelling on the Arctic seaboard; but the eastern or Tavgi branch do not appear to have anywhere quite reached the coast, which they hold to be the rightful domain of the "white bear people."

Some of the Samoyedes are baptized, but all alike are true Pagans, or idol-worshippers. Their gods are carnivorous, and are fond of raw flesh, which is accordingly thrust between their teeth at stated times. The Khatanga tribes keep entirely aloof; but some of the others have already become assimilated to the Russians, while others on the Ket and other eastern tributaries of the Ob are spoken of as "Ostiaks."

The eastern slopes of the Urals are occupied by the Voguls, who reach southwards to Yekaterinburg, and eastwards along the Konda valley nearly to the Lower Irtish. Reduced merely to a handful, the Voguls were formerly a powerful people, representing the primitive stock whence came the Magyars of Hungary.

"Like many other Finnish peoples, the Voguls have their family *totems* tattooed on their heads, arms, and legs.

"The Voguls are probably the least sociable of the Siberian aborigines. In summer they live in isolated family groups; in winter they pitch their tents or build their huts far apart from each other. Even the family spirit seems but slightly developed. The hunter may have one or more wives according to his means, but the least disturbance dissolves the union, and the husband will then often live quite alone, accompanied only by his reindeer and dog."[1]

8. *Topography: Omsk—Tobolsk—Yekaterinburg—Tomsk—Beresov — Obdorsk — Smeinogorsk — Barnaul — Semipalatinsk—Krasnoyarsk—Irkutsk—Kiakhta—Vladivostok.*

The Siberian towns claim consideration rather for their prospective than their actual size and importance. Many hamlets, consisting of fifty or sixty log-huts, such as Turukhansk or Okhotsk, figure in large letters on the maps, either because they are the official centres of

[1] Reclus, vi. p. 340.

administration, or because they are the only stations or settlements occurring for hundreds of miles in the almost uninhabited regions of East Siberia. Excluding Yekaterinburg, which is comprised in the European Government of Perm, there are only three towns altogether with upwards of 40,000 inhabitants—Irkutsk, Omsk, and

OKHOTSK.

Tomsk; and beside Yakutsk, there is one only in the whole of East Siberia, Krasnoyarsk, whose population exceeds 10,000. Yet some of these towns cover vast spaces, being laid out with broad, straggling streets, and low wooden houses in the midst of extensive plantations and waste grounds. Most of them are concentrated in the Ob basin, which alone contains about four-fifths of the entire population of North Asia. All, except Tiumen, are comparatively new, dating since the Russian conquest,[1] and none of them can boast of any "monuments" in the European sense. The dreary monotony of log-cabins and wooden huts is seldom relieved by anything beyond a few whitewashed houses and public buildings, such as the official quarters, barracks, convict prisons, and the like, all designed on a uniform plan, and imposing only because of their mean surroundings.

TOBOLSK.

In the Ob basin the largest place is Omsk, capital of West Siberia, which takes its name from the Om, standing at the confluence of that river with the Irtish. Here the Government buildings are of brick, and there are several

[1] Even Sibir, capital of the Tatar kingdom, overthrown by Yermak in 1581, has been washed away, with the bank of the Tobol River on which it stood. But Tumen (Tiumen), which appears on Herbertstein's old map (1549), is still a flourishing place.

churches, including one for the Kossaks, one for the Roman Catholics, and a third for the Protestants, besides a large mosque for the surrounding Bashkirs and Kirghiz. Lower down the Irtish, Tobolsk presents from a distance a really picturesque appearance, perched as it is on a bluff facing the junction of the Tobol from the north-west. From Tobolsk the steamers ascend the Tobol to the Tura, on which stands Tiumen, which claims to be the oldest place in Siberia. Above the Tura the Tobol is joined by the Isset from the Urals, about the head-waters of which stands Yekaterinburg, centre of the important mining industries of this region. Standing almost on the verge of the two continents, Yekaterinburg presents more the aspect of a European city than any other in Siberia. Here are a meteorological observatory and the "Society of the Ural Naturalists," besides extensive porphyry, malachite, jasper, rock-crystal, and other ateliers, whose products adorn the palaces and museums of every city in Europe.

In this district are several large mining stations, the general *entrepôt* for which is Irbit, at the junction of the Rivers Irbit and Nitza. A mere village during most of the year, Irbit becomes the Siberian Nijni Novgorod during the annual fair, when it is visited by 15,000 to 20,000 dealers, whose wares exchange hands to the value of from £1,750,000 to £2,000,000.

In the eastern parts of the Ob basin the chief place is Tomsk, on the Tom above its junction with the Ob. Tomsk is the centre of a very large local trade, and some of its streets, with their bright and well-stocked shops, present quite a cheerful aspect. Here was laid, in 1880, the first stone of the Siberian University, which with its future botanic garden and other branches is destined to render Tomsk the intellectual centre of North Asia. Here begin the extensive gold-fields discovered in 1830, which, before the opening of the Californian El Doradoes, yielded more of the precious metal than the whole of America.

Some 640 miles below Tobolsk, on the verge of the forest and tundra zones, lies the little town of Beresov, on the banks of the Sosva, near its junction with the Ob.

In the Altai uplands the chief mining stations are Serianovsk, Smeinogorsk, and Verkhniy Pristen. At the latter place the ores are shipped on a peculiar kind of craft called "karabass," and floated down the Irtish to Ust-Bukhtarminsk. The road beyond Ust-Kamenogorsk to Smeinogorsk crosses a cultivated hilly district, dotted over with several large villages. Here is probably the oldest mine in the Altai region. But the operations have been lately suspended, and replaced by the smelting works for the ores brought down from Serianovsk, Rybinsk, and elsewhere. Not far from Smeinogorsk is the famous imperial stone-polishing establishment of Kolivan, where the finest porphyries, jaspers, and marbles are dressed, but only for such large objects as chimney-pieces, monumental vases, tables, slabs, and the like.

From Kolivan the way leads through the Baraba steppe north to Barnaul, the chief town in the Altai highlands. This place, which lies on the left bank of the Upper Ob, in an extremely fertile and flourishing district, possesses several scientific institutions and technical establishments.

South-west of Barnaul lies the busy agricultural town of Semipalatinsk, most of whose inhabitants are Siberian Tatars, distinct from the Kirghiz, partly Sunnis, partly Christians, with seven mosques and two churches. Here the broad, sandy streets, lined with low wooden houses, give the place the appearance of a city of dunes. The road leading thence southwards to Lake Balkhash passes by Sergiopol, whence a view is commanded of the snowy Tarbagatai range.

IRKUTSK.

In the Yenisei basin the first place of any consequence crossed by the main route from Tomsk is Krasnoyarsk on the Upper Yenisei, in a fertile district, where the streams wash down auriferous sands. Below the junction of the Yenisei and Angara lies the little town of Yeniseisk, centre of a vast administration, and near the head of the Angara, where it emerges from Lake Baikal, stands the city of Irkutsk, capital of East Siberia. In some respects Irkutsk is the most important, as it is the largest, city in Siberia. Frequently wasted by fires, it has always risen rapidly from its ashes, and, thanks to its vital position on the trade and military route through Kiakhta to China, it must always remain an important place. Irkutsk is also a great centre of the fur trade, and amongst its public institutions are a gymnasium, library, theatre, and a flourishing geographical society.

In Transbaikalia the main highway leads up the Selenga valley to the Chinese border, where stand the well-known trading towns of Kiakhta on the Russian and Maimachin [1] on the Chinese frontier. Through these places pass the great tea, silk, and rhubarb caravans from China to Russia. But Kiakhta has lost much of this trade since the opening of the Suez Canal, which has developed a sea-borne traffic between the Chinese free ports and Odessa on the Black Sea.

In the north-eastern regions Verkhoyansk on the Upper Yana and within the Arctic Circle is noteworthy as perhaps the coldest place on the globe. But here the most considerable town is Yakutsk on the left bank of the Middle Lena above its junction with the Aldan. This is the proper capital of the Yakut nation, the most enterprising and prosperous of all the indigenous races.

[1] That is, *Mai-Mei-Chin* = "The Chinese Mart."

VLADIVOSTOK.

There are no large towns in the Amur province, Blagoveshchensk, the capital, having only 20,000 inhabitants. But on the coast are the two important naval and trading stations of Nikolayevsk and Vladivostok, the former at the mouth of the Amur, the latter close to the Korean frontier on the Sea of Japan. But Nikolayevsk has lost much of its importance as the port of entry of the Amur, owing to its severe climate and the intricate navigation of the river, which is usually blocked with ice for six months in the year. Vladivostok—that is, "Ruler of the East"—was founded lower down the coast in 1860 to obviate these inconveniences. Considerable sums have already been spent on its docks, piers, arsenals, and fortifications, with the intention of making it the chief naval station on the Pacific seaboard.

In his official report (Foreign Office, *Annual Series*, 1892) Mr. W. C. Hillier anticipates a great future for Vladivostok, which is rapidly becoming a commercial terminus of great importance. There can be no doubt that, as the construction of the Trans-Siberian railway progresses, the trade of this port must largely increase. Much of the brick-tea now forwarded to Siberia through Tientsin and Kalgan will eventually be shipped from Hankow to Vladivostok, while the extension of the railway cannot fail to open up new districts, at present very sparsely inhabited, of which Vladivostok must become the trading centre. The imports from Korea consist at present almost exclusively of cattle and food-stuffs; but a native passenger traffic is already being developed, and everything tends to show that on the completion of the trans-continental trunk line this hitherto little-known Russian port will become one of the chief centres of trade on the Pacific seaboard.

9. *Highways of Communication: The Trakt—Railway Projects.*

Few regions present fewer obstacles than Siberia to the general movement of the population. Doubtless many parts of the taiga are almost impenetrable, and the great rivers run mainly in the direction of the meridian. But beyond the limits of the dense forest zone the open steppe and boundless rolling plains stretch with little interruption from the Urals to the Pacific seaboard. Even of the great rivers, the Middle Amur flows east and west, while many of the tributaries of the others follow the same direction. A great navigable highway, broken only by two short portages between the Ob, Yenisei, and Lena basins, thus affords a natural line of communication across North Asia to within a short distance of the east coast by the Lena and Aldan, and quite to the coast by the Amur. This circumstance, combined with the sparseness of the population, explains the surprising rapidity with which the whole land was overrun by the Kossaks within twenty years of Yermak's second expedition, resulting in the capture of Sibir in 1581. Since then the country has been traversed with comparative ease in all directions by naturalists and scientific explorers, such as Gmelin (1733-42), Pallas (1771-72), Lesseps (1787-88), Wrangel (1821-23), Erman (1828-30), Castren (1842-43), Middendorff (1843), Radde (1855-59), Venyukov (1856), Krapotkin (1865 - 66), Finsch (1876), Baron Toll (1893-94), and many others. Yet the survey of the seaboard can scarcely be said to have been completed till Nordenskjöld successfully made the north-east passage in 1878-79.

From Yekaterinburg, the chief station on the Perm railway, the great caravan route runs across the

Tobol and Ishim valleys to Omsk. Here one branch follows the Irtish valley southwards through Semipalatinsk and Sergiopol to Lake Balkhash and the Ili valley. But the great Siberian trunk line is continued from Omsk for 400 miles across the Baraba steppe to Kolivan on the Ob. Here it is deflected southwards to Tomsk, whence it runs due east as far as Krasnoyarsk on the Yenisei, and thence north-east to Irkutsk on the Angara. From Irkutsk the communication is maintained both by steamer across Lake Baikal and by land round its southern extremity to the Selenga valley, where the trade route runs southwards to Kiakhta on the Chinese frontier. Another line is continued eastwards across Transbaikalia and over the Yablonovoi range down to Khita on the Shilka River. Here the road mainly follows the course of the Shilka across the Nierchinsk steppe to the Amur and down the main stream through Blagoveshchensk to the Usuri confluence, where it ramifies northwards along the Lower Amur to Nikolayevsk, southwards up the Usuri valley and over the Sikhota-alin coast range down to the Pacific at Vladivostok.

The *trakt*, as the great trunk line from Perm to Kiakhta is called, has been one of the chief instruments in developing trade and diffusing civilising influences throughout Siberia. It is traversed by long lines of waggons and sleighs, which will often make from 40 to 50 miles a day. Along the route the various halting stations have gradually grown into considerable centres of population, generally consisting of a single line of two-storied houses from one to two miles long.

From Tomsk the old route to Yeniseisk is continued north-eastwards to Yakutsk, whence one road branches northwards over the Verkhoyansk range down to the Yana valley at Verkhoyansk, while a second runs due east across the Aldan valley and over the Stanovoi plateau

BLAGOVESHCHENSK.

down to Okhotsk on the Sea of Okhotsk. From Verkhoyansk a track followed by Müller leads to Nijnekolimsk on the Arctic, and thence across the Chukchi country to the Gulf of Anadyr, in the Bering Sea. Okhotsk also communicates round the head of the Sea of Okhotsk and down the Kamchatka peninsula with Petropavlovsk. This line was traversed by Lesseps in 1777-78, and again in 1865-66 by Kennan, who also explored the region between the Gulf of Anadyr and the Sea of Okhotsk.

The railway from Perm over the Urals to Yekaterinburg, now connected with the European system, forms the first section of the great North Asiatic trunk line, which is intended ultimately to run from the Urals right across Siberia to Vladivostok on the Pacific, with a branch to Pekin or Port Arthur, at a total estimated cost of about £26,000,000. The central section from Siberia to the Amur basin is now (1896) creeping round Lake Baikal, a narrow road having been cut through the cliffs, which here rise from the margin of the lake to a great height. But the cost of the section from Irkutsk round the south side of the lake to Transbaikalia has been estimated at no less than £2,500,000, while its construction will necessarily take a great deal of time. Hence it has been decided to meanwhile run a short line of 54 miles from Irkutsk to Listvinichnaya on the lake, and to establish between the port and the opposite shore a regular line of steamers, the communication being kept open during the winter months by a powerful steam icebreaker capable of cutting through five or six feet of ice. At the same time the success of Captain Wiggins's project (see below) has proved that the heavy plant for the railway may easily be brought from Europe up the Yenisei and its Chulym and Angara affluents, which might be greatly improved for navigation at a slight outlay. It is therefore proposed to build at once the line 112 miles long between Achinsk and Krasnoyarsk in order to connect the two great navigable basins of the Ob and Yenisei, and thus facilitate the construction of the link between Krasnoyarsk and Irkutsk. On the West Siberian section, between Chelyabinsk and Omsk, most of the earthworks are finished, and the rails have been laid over nearly half the total distance of 500 miles. Rapid progress is also being made on the section of 326 miles between Omsk and the Ob, as well as on the Ob-Krasnoyarsk line, and on the Usuri section, of which the first 70 miles, between Vladivostok and Nikolskoye, were opened for passenger and goods traffic in 1894. The whole line, which is expected to be completed by the year 1905, will have a total length of 5000 miles, exclusive of the connections with China.

Amongst the Siberian highways must now be included the north-east ocean route round North Cape and through the Kara Sea to the Yenisei estuary, and thence by river craft to Yeniseisk. Despite the difficulties and uncertainties of the Kara Sea navigation, this route, open, of course, only during the short summer season, may be regarded as fairly established by the persevering efforts of Captain Joseph Wiggins, who made his first successful voyage in 1874, and who since then has crossed the Kara Sea nine times. In 1893 he conveyed the first Russian Government river steamers that have ever reached the Yenisei waters, and at the same time succeeded in landing at Yeniseisk and Krasnoyarsk, some 300 miles farther up the river, the first section of rails for the Trans-Siberian railway, returning with a full cargo of Siberian produce. "All this," the enterprising pioneer concludes, "should now surely prove the Kara Sea to be a commercial route."

10. *Administration: Education—Industries.*

Excluding the portions attached to Europe and Turkestan, the whole of North Asia comprises the two great administrative divisions of West and East Siberia, whose capitals are Omsk and Irkutsk respectively. Each of these is subdivided into a number of governments and provinces, which in their turn are distributed into circles and districts. The military system still largely prevails in the Amur Government, which is divided into Kossak "regiments" and "battalions." But elsewhere the administration is mainly modelled on that of European Russia. The municipal, judicial, and ecclesiastical departments are all theoretically based on the same uniform plan. But owing to the vast distances and the difficulty of communicating in winter with the central authorities, the local officials and commanding officers enjoy almost absolute control in their several jurisdictions. In some places the natives hardly understand the existence of the "White Tsar," and know of no higher power than the district magistrate. But in the more settled parts, and especially in the Tobolsk, Tomsk, and Irkutsk Governments, these functionaries are fully as responsible to the higher authorities as those of European Russia.

In some respects the people enjoy even greater personal liberty than in the west. In Siberia there are no nobles or specially privileged classes; serfdom never gained a footing in the land, and through the increasing traffic with California, ideas of freedom and independence, unknown to the Western Mujiks, have already penetrated into Siberian society. "As we advance eastwards," remarks a Russian writer, "the freer and more independent do we find life and opinions among us."

Education is still in a very rudimentary condition. In some places the Russian settlers have even forgotten their mother-tongue and become assimilated to the surrounding aborigines. In East Siberia, with a population of over 1,700,000, there were only 283 schools, attended by 8610 scholars, in 1870, and in the whole of Siberia there were only two periodicals, a weekly and a monthly, in 1878. Nevertheless, a certain intellectual life has been fostered in the larger centres of population. There are geographical and other learned societies in Irkutsk, Yekaterinburg, and elsewhere, and in 1880 imperial consent was at last given to the establishment of a "Siberian University" in Tomsk, from which much is expected.

Meanwhile agriculture is in such a backward state that the crops scarcely yield sufficient for the local consumption. Stock-breeding, however, is conducted on a very extensive scale, and notwithstanding the ravages of the "Siberian plague," said to have first broken out in the Baraba steppe, vast herds of cattle and of horses, almost in a wild state, are bred on the rich pasturages of the southern plains and upland valleys. On the other hand, both the chase and the deep-sea fisheries have fallen off with the gradual disappearance of the fur-bearing animals and of the large cetacea from the northern waters. The mining industry also, although still of primary importance, has suffered by the competition of the Californian, Australian, and South African fields. Of other industries, perhaps the most important is distilling. Vast quantities of coarse spirits are produced from grain and potatoes, and retailed in the taverns which abound in all the towns, and especially in the mining districts. Returns of these various resources will be found in the subjoined tables.

11. *Statistics.*

Areas and Populations.

	Provinces.	Area in sq. miles.	Population (1889-94).
Western Siberia	Tobolsk	539,659	1,313,400
	Tomsk	331,159	1,299,729
Eastern Siberia	Irkutsk	257,061	421,187
	Transbaikalia	236,868	545,338
	Yakutsk	1,533,397	280,200
	Yeniseisk	987,186	458,572
Amur Regions	Amur	172,848	87,705
	Primorskaya (Maritime Province)	715,982	102,786
	Sakhalin	29,336	19,644
		4,833,496	4,538,561
Parts of W. Siberia attached to European Russia		150,000	1,820,000
Parts of W. Siberia attached to Turkestan (Kirghiz Steppe)		756,000	2,000,000
	Total Siberia	2,739,496	8,358,561

Approximate Population of Siberia according to Races.

Mongol Stock.		*Finnish Stock.*	
Tunguses	80,000	Voguls	4,500
Buriats	250,000	Samoyedes	25,000
Kalmuks	20,000	Ostiaks	25,000
Chinese, Manchus	50,000	Soyots	8,000
Koreans	5,000		
Sub-Arctic Races.		*Slav Stock.*	
Chukchis	12,000	Great Russians, Little Russians, Poles	5,500,000
Koriaks	5,000		
Kamchadales	3,000		
Giliaks	5,000		
Ainus	3,000		
Yukaghirs	1,600		
Turki Stock.		*Sundries.*	
Yakuts	200,000	Gipsies	5,000
Tatars	80,000	Dolgans	500

Chief Towns in Siberia.

Irkutsk	50,000	Blagoveshchensk	20,000
Omsk	55,000	Semipalatinsk	12,000
Tomsk	42,000	Smeinogorsk	12,000
Yekaterinburg	37,000	Vladivostok	15,000
Tobolsk	22,000	Kiakhta	11,000
Barnaul	17,000	Yeniseisk	8,000
Krasnoyarsk	13,000	Vyernyi	22,000
Tiumen	35,000		

Growth of Population in Siberia.

1796	1,193,145	1870-73	3,340,362
1816	1,540,424	1884	4,093,535
1869	3,327,627	1894	8,358,000

Products of the Siberian Mines (1891).

Gold	63,432 lbs.	Platinum	3,000 lbs.
Silver	30,000 ,,	Copper	2,866 tons.
Lead	290,000 tons.	Iron	820,000 ,,
Steel	40,000 ,,	Coal	46,000 ,,
Salt	35,000 ,,	Graphite	?

Yield of the Siberian Gold Mines (1726-1894)	£148,000,000
Yield of the Silver Mines since eighteenth century	35,000,000
Present average annual yield of gold	3,400,000
Present average annual yield of silver	220,000

Exiles to Siberia from 1823 to 1858.

Men, 238,482; Women and Children, 42,844; Women and Children voluntarily accompanying their friends, 23,285; total, 304,618.
Total during the last 250 years, about 1,000,000.
Present yearly average, 14,000 to 15,000.
Condemned to exile (1891): 14,315 men; 1003 women; total, 15,318.

Russian Free Emigration to Siberia.

1885	9,680	1889	30,140
1887	13,900	1892	100,000
1888	26,130	1894	180,000

Public Instruction in Siberia exclusive of the Urals.

Elementary Schools (1894), 1480; Attendance, 51,000.
Higher Schools, 75; Attendance, 8700.

THE GREAT RIVERS OF SIBERIA.

	Length. Miles.	Area of Drainage. Sq. miles.	Nav. Waters. Miles.
Ob-Irtish . . .	3400	1,420,000	9000
Yenisei-Angara . .	3300	1,180,000	5000
Lena-Vitim . . .	3280	1,000,000	6000
Amur-Argun . . .	3066	800,000	6000

Steamers on the West Siberian Rivers (1844), 1; (1860), 10; (1870), 22; (1880), 37; (1894), 107.

EASTERN ASIA

CHINESE EMPIRE, JAPAN

CHAPTER V

CHINESE EMPIRE

1. *Boundaries—Extent—Area.*

By the expression Chinese Empire will here be understood all the lands, either absolutely administered from Peking, or indirectly forming part of the Chinese political system. In this chapter will therefore be treated not only China proper, Mongolia, Manchuria, Kulja, and Kashgaria, but also Tibet, which, notwithstanding a certain more or less real autonomy, is practically controlled in its foreign relations by Chinese diplomacy, and Korea, which, till the disastrous war of 1894-95, was in the same position.

The region thus defined, besides the political unity derived from this circumstance, is further united by the bonds of race and religion. For the vast majority of its inhabitants belong to various modified forms of the Mongolic type, and constitute various branches of the Buddhist religious world.

Compared with the other great states of the world, China takes the foremost rank in respect of population, while in extent yielding only to England and Russia. It occupies the whole of Central and East Asia, the Indo-Chinese peninsula alone excepted. For by the expression

Central Asia should properly be understood the great continental tablelands confined north and south by the Altai and Himalayan mountain systems, and stretching from the Pamir—that is, from the converging point of these systems—eastwards to China proper. On these tablelands of the Pamir, Tibet, and Mongolia rise the great continental rivers—Oxus, Sir, Indus, Brahmaputra, Yang-tse, Hoang-ho, Amur, Ob, and Yenisei—which flow west, south, east, and north, to the Aralo-Caspian basin, the Indian, Pacific, and Arctic Oceans. Excluding the Pamir, which is at present a sort of neutral land between the three empires of British India, Russia, and China, all converging at this point, these central plateaux, constituting the true heart of the continent, and determining its great water systems, form politically an integral portion of the Chinese Empire. Consequently to China alone belongs Central Asia, although the expression has found a place in the official language of Russian bureaucracy.

Thus comprising Central and East Asia, the Chinese Empire is almost everywhere clearly delimited, on the north and north-west by Asiatic Russia, on the south and south-west by British India, on the south-east by Indo-China, and on the east by the Pacific Ocean. From the Kizil-art, the water-parting of the Oxus and Tarim basins, about 75° E. long., it stretches across 53 degrees of longitude for a total distance of 3000 miles to the east coast of Korea in 128° E. long.; and from the great northern bend of the Amur on the Siberian frontier across 34 degrees of latitude for 2400 miles southwards to the island of Hainan. Within these limits it has a total area roughly estimated at 4,500,000 square miles, with a population of perhaps 350,000,000.

An estimate based on official returns for 1842 gave 405,000,000 for China proper, to which probably 20,000,000 should be added for the rest of the empire.

But since then enormous losses were caused not only by the wars of the Taipings and Muhammadan Dungans in the south and north-west, but also by the inundations and shiftings of the Hoang-ho, and the terrible famine by which the northern provinces have been wasted in recent years.

2. *Relief of the Land: The Kuen-lun Mountain System—The Nan-shan, Khingan, and Nan-ling Ranges—The Cross Ridges—Plateaux and Depressions.*

The great frontier mountain systems of the Himalayas and Altai, enclosing the central plateaux on the south and north, and converging westwards round the Pamir, as well as the Tian-shan, lying partly in Russian and partly in Chinese territory, have been described in previous chapters. Our information with regard to the internal systems, especially in the western regions, is still extremely defective. In a general way it may be stated that in the west—that is, in Tibet and Mongolia—the great ranges run mainly west and east, and assume somewhat the character of bold escarpments to the great central tablelands, which stretch at different elevations from the Himalayas northwards to the Altai. But in the east—that is, in China proper—the direction is rather north-east and south-west, and even north and south. Here also the tendency is, especially on the Tibetan and Indo-Chinese frontiers, to broaden out into extensive and irregular highland regions, in which the general direction of the main ridges is indicated by the course of the great rivers flowing from the Tibetan plateau to the Chinese and Indo-Chinese seaboards.

Thus the great Kuen-lun (Kuen-luen) system, rooted westwards in the Pamir plateau, runs under diverse names, such as the Tuguz-daban, the Altyn, and Kilien-

KOK-SU, TIAN-SHAN.

shan (Nan-shan), for hundreds of miles along the northern edge of the Tibetan plateau to the Tsaidam and Kuku-nor districts towards the Chinese frontier. In the western section, between Chinese Turkestan and north-west Tibet, the Kuen-lun rises to altitudes of over 20,000 feet about the Yurung-Kash and Keria head-streams of the Khotan-darya, and it is here crossed by some extremely difficult passes, such as Russian Pass (19,000 feet), and that of Lubashi (17,500), both for the first time utilised in 1889-90 by Captain Grumb-chevsky in his daring attempts to penetrate from this direction into the interior of Tibet. On reaching the little lake Gugurtlik, beyond the Lubashi Pass, this intrepid explorer discovered that at that point the plateau itself stood at a height of over 17,000 feet.

After surveying the Mustagh-ata group, Bogdanovich passed in the same year, 1889, south from Yarkand to Mumuk in the valley of the Tiznaf, a large affluent of the Raskam-darya, that is, the main eastern head-stream of the Yarkand. Here he crossed the main range of the Kuen-lun, also near its western extremity, by the hitherto unvisited Takhta-Korum Pass, at an altitude of about 17,000 feet. The ascent took eight hours of hard climbing, towards the summit huge mountains converg-ing on every side to form a vast amphitheatre entered by a giant staircase of natural steps of great width, slightly inclined and studded with blocks of fallen rock. The character of the scenery agrees well with its Persian name, *Takhta-korum*, "Rocky Throne." In summer the pass is quite free from snow; but the descent on the south side is very abrupt down to the Kulan-aghil tributary of the Yarkand. Beyond this valley rises a snowy range, which was also crossed by the Kokelan Pass (17,000 feet), leading down to the Yarkand near its source by the Malgum-bash gorge and the valley of the Bazar-dereh, a raging torrent, 30 to 40 yards wide. After nearly losing his life in an attempt to cross the Yarkand at this point, the daring explorer had to retrace his steps by the Takhta-Korum Pass.

After surveying the Central Tian-shan regions, the Pievtsoff expedition of 1889-90 passed from Yarkand southwards to the northern slopes of the Kuen-lun. From Nia, which gives its name to one of the numerous intermittent streams descending to the Tarim basin between Khotan and Cherchen, a determined attempt was made in the spring of 1890 to penetrate over the border range into the Tibetan tableland. A preliminary excursion made late in the previous autumn by Roborofsky as far as the Astik and Uzu mountains, had led to the belief that the plateau might easily be reached with a caravan from this direction by skirting the Astyn-tagh, or "Russian Range," as it was renamed by Prjevalsky. But this belief proved to be delusive, and the efforts made to advance beyond the Kuen-lun southwards nearly ended in disaster.

From Kara-sai, at the foot of Russian Range, two parties were sent to explore the surrounding regions, one under Kozlof striking south-east and east, the other under Roborofsky proceeding along the northern slope of Russian Range towards the Kerian River. Leaving the Uzu-tagh on the left, Roborofsky travelled over 40 miles by the Saryk-tuz valley to the Kanbulak gold mines, 14,000 feet above the sea. Near the Sin-bulak spring, source of the Saryk-tuz, which is a head-stream of the Tolun-Khoja flowing north to the Tarim basin, the "Tsar-Liberator" and other peaks of Russian Range rise to a height of 20,000 feet. Southwards the range breaks off abruptly towards the Lake Shor-kul depression, which was reached by a pass nearly 17,000 feet high, but falling with a gentle incline down to the Tibetan plateau. The lake is

5 miles by 3, its brackish waters being encircled by flat, marshy shores, thickly covered with saline efflorescences. From this point the party, following the Shor-kul valley, reached the Kerian River, which, however, was found to be dry, its sources being still ice-bound. Roborofsky was consequently obliged to return in all haste from this arid waterless region to Kara-sai, a total distance of 134 miles, having satisfied himself that the route was impracticable for a caravan, owing to the great elevation and general lack of fodder. Nor was Kozlof more fortunate, the easterly route followed by him also proving too rugged and barren for a camel caravan.

Other excursions made during the summer of 1890 towards the south and east were attended by the same results. The water-parting between the Tarim and Tibetan drainage areas was again passed at an altitude of nearly 17,000 feet; but the whole region was found to be extremely barren, rugged, and uninhabited, and consequently quite impracticable even for small expeditions. A river was reached on the plateau flowing first east and then south-east, at an elevation of 16,000 feet. "Beyond it rose another of the slaty ridges, which we ascended, and from its summit had a splendid view of the mountains folding the Kerian River in their stony embrace, and stretching away to the south-east, and then suddenly falling away to the east. We could see for 20 miles to the south of us, as far as a distant ridge, the intervening expanse being filled with bare, weather-worn heights and serrated ridges, all of one character. It was a monotonous, dreary landscape, devoid of human beings and animal life of any kind, with those everlasting dark, slaty ridges, worn by time and weather, all trending in one direction. The region is evidently rainless, for there are no water-channels here, and the heights of the mountains are scored by the winds, not by aqueous agencies.

The atmosphere is excessively dry, judging from the total absence of moss on the soil and rocks. Snow, however, falls daily, but is swept away by the winds and evaporated by the dryness of the air. Wherever drifts collect and thaw in the sun the ground is damp; but we came upon no springs, lakes, or other natural reservoirs. The winds from the north-west and north-east blow with terrific force, mostly from 11 to 12 o'clock in the day, and at 8 in the evening, sometimes at night, when the frost exceeds – 10° Centigrade [14° F.]. . . . I had advanced altogether 120 miles (46 south of Uzu-tagh), and had ascertained that the country was impassable for beasts of burden. The enormous altitude and constant storms, the absolute want of fodder, and the rocky surface, injuring even shod horses, render a passage in this direction impracticable. This country, it may therefore be assumed, will long remain unexplored by Europeans" (V. Roborofsky, *Proceedings of Royal Geographical Society*, 1891, p. 102).

The Altyn-tagh section of the Kuen-lun system was one of the most surprising results of Prjevalsky's expedition in 1877 to the Lob-nor. For this snowy range rises abruptly to an elevation of 13,000 to 14,000 feet within 120 miles of the lake, where it was formerly supposed that the sands extended for several degrees of latitude southwards to the scarp of the Tibetan plateau. The discovery is of great importance as helping to explain many hitherto unintelligible passages in Chinese records in connection with the wars and migrations of the Huns and Mongols.

But the direct route of the old silk caravans along the north foot of the Altyn Range, between Lob-nor and Saitu (Sachu), had not been traversed since the time of Marco Polo till the year 1893, when it was, so to say, rediscovered by Mr. and Mrs. Littledale. The track followed ran from Abdal, near the mouth of the Tarim, nearly due east to about 93° 40′ E., where it turned north-east to Saitu. It lay for the most part along the northern slopes of the range, crossing numerous projecting spurs by passes from 7000 to over 11,000 feet high, above which the crests still rose 2000 or 3000 feet higher. The path was almost everywhere extremely rugged, water and fodder for the pack animals being generally very scarce, while deep ravines had constantly to be surmounted in the upland valleys. These ravines are formed by the erosive action of the rains furrowing deep channels in the loess which covers the slopes in many places. Snowy peaks of great altitude rose above the track at intervals, and the range was found to increase in altitude eastwards. About the meridian of Saitu (95° E.) it trends round from east to south-east, running in this direction under the name of the Humboldt Range (the Hai-shi-goa of the Chinese) towards the Kuku-nor depression. This section, crossed by passes from 11,000 to over 16,000 feet, is flanked on both sides by other snowy ranges, such as the Shah-hull-gar, all of which seem to converge at the Katin-la Pass (15,800 feet), above which the Tungo or Zagit Peak towers to a height of 18,000 feet. Beyond this point the system is continued still to the south-east round the north side of Kuku-nor all the way to Hsi-ning, within the Chinese frontier, here merging in the Nan-shan and Ala-shan systems. In the western section of the Altyn Mountains Mr. Littledale procured the specimens of the wild camel which are now preserved in the Natural History Museum, London.

The Tsaidam plains are skirted on the south by the parallel Shuga (15,600) and Burhan-bota, or Burhan-Buddha (15,800), ranges, which seem to branch off from the Altyn south-eastwards. Still farther south rises the great Dang-la range, which under the 33rd parallel runs at a mean elevation of probably 20,000 feet across eight degrees of longitude (90°-97° E.). The Dang-la (Tang-la) Mountains, which were skirted in 1892 by Mr. Rockhill, may be regarded as the true climatic parting-line for this part of Central Asia; they completely intercept all the moisture-laden clouds which drive before the south-west monsoon, and which the Himalayas fail entirely to arrest. Hence the northern slopes are a comparatively dry, arid waste, whereas the southern plateau, the Naktsang of the Tibetans, stretching 600 miles west to the Pamirs, is for nearly half the year deluged with rain, hail, or snow. The high rugged range to the east of the Dang-la and to the south of the Soloma (Upper Hoang-ho), plays the same part in determining the climatic relations of east Tibet (Rockhill, *The Land of the Lamas*, p. 174).

The Tsaidam ranges are continued far into China proper, between the Hoang-ho and Yang-tse valleys, under the names of the Hsi-king-shan, Tsing-ling, Funiu-shan, and Mu-ling. German orographists group these ranges collectively as the Eastern Kuen-lun, of which they regard the Burhan-bota[1] as the central, and the Tuguz-daban and Altyn as the western section. The Kilien-shan, or Nan-shan, thus sinks to the position of a subordinate northern offshoot of the Kuen-lun system, which stretches from the Kizil-art (Eastern Pamir range), with many interruptions, across North Tibet and Central China, for 2700 miles eastwards to the Lower Yang-tse-kiang.

[1] The Burhan-bota, which, notwithstanding its great elevation, nowhere reaches the snow-line, runs for 130 miles along the northern edge of the Tibetan plateau, and forms the southern limit of the marshy Tsaidam steppe. Prjevalsky writes *Burkhan-Buddha*, Rockhill more correctly *Burhan-bota*, the *kh* of *Burkhan* being the usual Russian transliteration of the simple aspirate *h* beginning a word or syllable; cf. *Khamil* for *Hami*.

The Nan-shan is again continued north-eastwards, partly along the great wall of China, through the snowy Kuliang and Liang-chu, the Ala-shan[1] (11,600), the Khara-Narin-ula, In-shan,[2] Munni-ula (9000), Sirung Bulik, Suma-hada, Shara-hada, and other parallel ridges, to the head-waters of the Lohan, at the converging point of Mongolia, Manchuria, and China, north of Peking. Here the system is gradually contracted till it forms a junction with the volcanic Great Khingan range, which runs between Mongolia and Manchuria, due north to the Amur, near the confluence of the Argun and Shilka.

Little was known of the Khingan system till 1887-88, when it was traversed four times and surveyed throughout almost its whole length by the Russian traveller, M. Garnak, and also visited by M. Ressine. Throughout his long wanderings Garnak crossed the crest of the range several times, and followed the eastern slopes facing Manchuria as far as the territory of the Solons and Dahurs. Numerous natural history specimens were collected by his brother, who accompanied him, and these explorations, supplementing the rough preliminary surveys of Krapotkine and Fritsche, give a good general knowledge of the Khingan uplands.

East of the Khingan, Manchuria is occupied in the north by the Ilykhuri-alin and Duss-alin, skirting the

[1] The Ala-shan mountains rise abruptly above the left bank of the Hoang-ho, and run for 150 miles north and south between Kan-su and the Ala-shan country, Mongolia. They culminate with Mount Bugutin (11,600 feet), but nowhere reach the snow-line.

[2] "The natives do not know this name, and have their own names for different parts of the range. In a wider sense the term In-shan applies to all the mountains from the northern bend of the Hoang-ho, through the Chakar country to the confines of Manchuria" (Prjevalsky's *Mongolia*, i. 153). Mr. Rockhill, however, thinks that "In-shan" may be "a defective transcription of Ta-ching-shan," the only name he heard used. He adds that west of Pao-tu "the range is called Wula-shan, then Lang-shan as far as Ala-shan" (*Geo. Jour.*, May 1894, p. 359).

right bank of the Amur, and by the Shanyan-alin (10,000 to 11,000 feet), forming the frontier towards Korea. This peninsula is mountainous throughout, and especially on the east side, where a coast-range, here and there flanked by parallel inner ridges, forms a southern extension of the Sikhota-alin coast-range of the Russian Maritime Province. Geologically the Kuen-lun, or at least its western section, is of far older date than the Himalayas. The prevailing rocks are syenitic gneiss and more recent triassic formations, whereas in the southern range is comprised the whole series between the palæozoic and eocene deposits. Hence the Kuen-lun, rather than the Himalayas, must be regarded as the eastern extension of the Hindu-Kush, and the true backbone of the continent in this direction.

In China proper the provinces of Shan-si and Pe-chi-li are traversed by the Siwe-shan, Man-tu-shan, Tao-tsu-shan, running south-west and north-east from the Hoang-ho. Parallel with this system are the Utai-shan, Luyen-shan, Mian-shan, Niao-ling, and other ridges, filling the whole of Shan-si, and continued beyond the Hoang-ho by the Ming-shan and Sung-shan through North Ho-nan Here a junction is effected with the Funiu-shan section of the Eastern Kuen-lun, which at this point forms the boundary between the provinces of Ho-nan and Hu-pe.

Our knowledge of the geological structure of the northern provinces, and of the Hoang-ho basin with its vast loess deposits, has been considerably enlarged by the observations of Lóczy, the geologist attached to the Szechenyi expedition (1877-80). Extremely thick and widely distributed lacustrine deposits have been discovered in the north-west, which are proved to be of Pliocene age by the mammalian remains contained in them. Evidence is thus afforded that Tibet itself must at that epoch have been covered with numerous flooded depressions of great extent. Baron von Richthofen, who has studied Lóczy's reports published in 1893, accepts the view that most of the deposits which he called "Lake Loess" in Northern China belong to this formation. Mr. Rockhill also remarks that the Nan-shan Mountains, which he crossed in 1892, offer "a most admirable illustration of the mode of formation of the loess, and of the continual redistribution going on, and the formation of new deposits under the action of the torrential rains which for months every year deluge this section of the country."

During his journey of 1887 along the Central Asian trade route, Col. M. S. Bell had many opportunities of studying the loess formation, to which the basins of the Hoang-ho and its Wei affluent are mainly indebted for their great fertility. From Khavailu (Huo-lu-hien) in Shansi for 150 miles to Sze-tien, the track crosses continuous loess hills stretching from the Pekin-Kalgan road south-westwards to the Hoang-ho. It extends altogether for about 1400 miles along the great Central Asian route, and is regarded by Richthofen as a subaerial formation developed in a region of inland drainage, representing the accumulated residuum of countless generations of herbaceous growths, combined with much material which was distributed over the surface by wind and water and fixed by vegetation. "The loess is a solid friable earth of brownish-yellow colour, and when triturated with water not unlike loam, but differing from it by its highly porous and tubular structure; these tubes are often lined with a film of lime, and ramify like the roots of plants. Amongst the constituents very fine sand and carbonate of lime predominate next to the argillaceous basis. It spreads alike both over high and low ground, smoothing off the irregularities of the surface, and its thickness often considerably exceeds 1000 feet. It is not stratified, and has a tendency to vertical cleavage. The loess is full of fossil land-shells, and contains bones of land quadrupeds, but no remains of either marine or of fresh-water shells. It is very fertile and requires little manure" (Richthofen, quoted by Bell, *Geo. Proc.*, 1890, p. 59).

All the loess districts are subject to two serious drawbacks, dust storms and bad roads, both due to the light, friable nature of the soil. On his journey through Central Asia in 1887, Captain Younghusband noticed these evils in the region about Kalgan towards the Mongolian frontier north-west of Peking. Here "the inhabitants were attempting in a half-hearted way to cultivate the fields, which were constantly being covered with layers of dust by the horrible sandstorms, which occurred almost daily at this time of the year (April). The country is of the formation called loess, a light, friable soil which crumbles to dust when the slightest pressure is put on it. In consequence of this the roads are often sunk 30 or 40 feet below the level of the surrounding country, for when a cart passes along a road the soil crumbles into dust, the wind blows the dust away, and a rut is formed. More traffic follows, more dust is blown away, and gradually the roadway sinks lower and lower below the surrounding level; for the Chinese here, as elsewhere, never think of repairing a road" (*Geo. Proc.*, 1888, p. 490).

South of the Lower Yang-tse the whole of south-east China is occupied by extensive and nearly parallel chains, such as the Shi-shan (16,000 feet), the Ja-ling, Ta-yu-ling, Yung-nien-ling, Tung-lo-ling,[1] and Pu-ling, whose

[1] The Tung-lo-ling, an easterly section of the Nan-ling, separates Hunan and Kiang-si from Kwang-tung, but does not form a true water-parting between the Yang-tse and Canton basins, for it is pierced by a stream rising north of it and flowing south to the Kiang or North River of Canton. It is crossed by the important Che-ling (1200 feet) and Mei-ling passes.

normal direction is also from south-west to north-east. This system merges through the Nan-ling range with the southern and south-western highlands of Kwang-si and Yun-nan on the Indo-Chinese frontier.[1] Here we enter one of the least-known regions on the globe, on which some little light has been thrown by Gill, Riley, Desgodins, and Baber. But its thorough exploration is needed to solve the many obscure questions connected with the sources and water-partings of the Yang-tse, Min, Mekhong, Salwin, Irawadi, all of which flow for long distances in close proximity through the narrow longitudinal valleys formed by the Cross Ranges stretching from the Yun-nan highlands along the Tibeto-Chinese frontier between the Brahmaputra and Yang-tse basins northwards towards the Kuen-lun system.

These "Cross Ridges," as Blakiston calls them, are obviously in a geological sense an eastern extension of the Tibetan plateau itself, which has here been cut up into parallel chains running mainly north and south. The beds of the running waters, to which the Ranges would seem to owe their existence, lie still at elevations of from 8000 to 10,000 feet above the sea, and the great trade route from Lassa through Batang to West China maintains a normal elevation of no less than 12,000, with occasional passes nearly 17,000 feet high. The ranges between the River Kinsha (Yang-tse) and its tributaries, the Yalung and Min, rise above the snow-line which Gill here fixes at from 14,000 to 15,000 feet. The Nenda, or "Holy Mountain," east of the Upper Kinsha valley under the parallel of Batang, is some 20,000 feet high, and sends down glaciers to all the surrounding valleys. Farther east the snowy peaks of

[1] To the whole of this south-eastern system, which in the Ping-ya-shan rises above the snow-line, Richthofen gives the collective name of Nan-shan.

Surong, running north-west and south-east, are nearly if not quite as high, while above a parallel chain east of the Yalung River, the Jara, or "King of Mountains," commands all the surrounding heights by 5000 feet. This chain is continued northwards to the Bayan-Khara system, where some of the crests may possibly rival those of the Himalayas themselves. Amongst them are the Ngomi-shan, ascended in 1879 by Riley; the Siwelung-shan, or "Snow Dragon"; the "Seven Nails," supposed by Gill to have an altitude of 19,000 to 20,000 feet.

Between the Tarbagatai and Zungarian Ala-tau lies the depression of the so-called "Zungarian Strait," through which access is gained from Turkestan along the Sassik-kul, Ala-kul, Ebi-nor, Sir-nor, and other eastern extensions of the Balkhash lacustrine system, into the Mongolian plateau. In the same way between the Barluk-Orkochuk and Little Altai lies the valley of the Black Irtish, which again gives access to North Mongolia through Lake Ulungur and River Urungu. For the valley of this river sweeps round the south-eastern extremity of the Little Altai to the Kobdo plateau, where it has its source. The Kobdo plateau itself stretches from the Little Altai beyond Lake Ubsa to the Tanuola range, by which it is separated from the valleys of the Upper Yenisei and Selenga. From this point the North Mongolian plateau is broken by no other prominent range until we reach the Great Khingan, by which, as already stated, it is separated from Manchuria.

From this rapid survey it appears that China proper is by no means a vast lowland plain formed by the alluvia of the twin rivers Hoang-ho and Yang-tse. It is a distinctly highland region almost everywhere occupied by vast mountain systems, except along the lower courses of the great streams and on the east coast. And even here the lowland formation is broken in the upland peninsula of Shan-tung, projecting seawards between the old and new channels of the Hoang-ho, and culminating at its extremity in the Kuan-in-shan (2900 feet).

It also appears that the great Central Asiatic plateau consists in reality of several distinct sections differing enormously in elevation and extent from each other. These sections are grouped round the central basin of the Tarim, which is in fact rather a depression than a plateau, falling to little over 1600 feet above sea-level. South of it the land rises in successive stages from 3000 to 6000, 10,000, and 15,000 feet, the probable mean altitude of the Tibetan plateau, at once the most elevated and extensive on the globe. Above this vast tableland the intersecting ranges attain altitudes of from 20,000 to 25,000 feet, culminating in the southern scarp of the Himalayas with peaks ranging from 26,000 to 29,000, the highest summits on the surface of the earth.

North of the Tarim basin the land also rises in terraces of 3000, 6000, and 15,000 feet, here culminating with the Tengri-khan (25,000), central and highest point of the Tian-shan. Beyond the Tian-shan the ground again falls gradually to about 1500 feet in the Zungarian depression (Tian-shan Pe-lu), north of which it attains a height of 7000 or 8000 feet in the Kobdo plateau. This elevation is maintained in North Mongolia eastwards to the head-waters of the Amur. But in the central parts the Gobi desert stretches from Lob-nor at a mean height of probably not more than 3000 feet to the Khingan range. Lastly, the closed basin of the Kuku-nor between the Nan-shan and Burhan-Bota ranges stands at an altitude of not less than 10,500 feet above sea-level.

3. *Hydrography: Inland Drainage, Lob-nor and Ili Basins—Seaward Drainage, The Hoang-ho, Yang-tse-kiang, Pei-ho, Liao-ho, and Si-kiang Basins—Kuku-nor and the Tibetan Lacustrine System.*

Those water systems of Central and East Asia which are altogether comprised within the limits of the Chinese Empire are few in number, and seldom of great extent. Excluding the already-described Amur basin, now shared between China and Russia, not more than five large rivers find their way in independent channels to the Pacific coast, and of these two only, the Hoang-ho and Yang-tse, attain the proportions of great continental streams. The inland drainage also, apart from the numerous small lacustrine closed basins of Tibet, is represented chiefly by the Ili flowing beyond the frontier to Lake Balkhash, by the Ike-aral and Ubsa-nor basins of the Kobdo plateau and the Lob-nor of Kashgaria.

Of these inland systems that of the Lob-nor is by far the most extensive. This lake or flooded morass, the true position of which was first determined by Prjevalsky in 1877, receives through the Tarim River nearly the whole drainage of Chinese Turkestan. Here the surrounding Tian-shan, Kizil-art, Karakorum, and Kuen-lun send down numerous streams, including the Ak-su, Ugen-darya, Shah-yar, Kashgar, Yarkand, and Yurang-kash (Khotan), all of which are collected by the Tarim and carried through a still imperfectly-explored course to the Kara-buran, or west end of Lake Lob. The Tarim also receives from the north the discharge of Lake Bagrach (Bostang-nor) through the Koncheh-darya (Kaidu-gol), which forces a passage through the intervening Kuruk-tagh ridge. But the Cherchen-darya, rising in the Tuguz-daban (Western Kuen-lun), flows through the sands intermittently from the south directly to the lake at the Tarim confluence.

Of all these affluents of the Tarim, the largest and in every way the most important is the Yarkand-darya, the course of whose upper tributaries has not yet been quite settled. But despite the discrepancies between the statements of the latest explorers, Grombchevsky and Dauvergne (1889-90), it is certain that the farthest sources of the Zarafshan, its main head-stream, lie on the northern slope of the Karakorum range, in long. 78° E., lat. 37° N., from which point it flows first north-west, then north and north by east to Kusharab, where it enters the plains after a course of about 200 miles through the western Kuen-lun Mountains. At Chang-jangal it is joined by the romantic river Oprang, whose valley, surveyed in 1889 by Younghusband, winds amid a series of huge glaciers between the Mustagh and Aghil ranges. Below Chang-jangal the Zarafshan (Raskam) appears to flow between the Kichik-tung and Arpatallek ranges, west and east, northwards to Langar (above Kusharab), where it was crossed and found to be a copious stream by Dauvergne. Near Kusharab it is joined on its left bank by a considerable affluent, which has its rise at the Wakhijrui Pass close to the source of the Oxus, and flows thence through the Tagh-dum-bash Pamir northwards to Tash-kurgan, where it turns abruptly east to its junction with the Zarafshan.

Below Kusharab the united waters flow as the Yarkand-darya mainly east by north through Yarkand to their junction with the Kashgar-darya, which descends as the Markhan-su from the slopes of the Kizil-Agyn in the Trans-Alai range. The Markhan-su flows mainly east to its junction with the Kizil-su ("Red River"), which descends from the southern slopes of the Tian-shan, and which is by some regarded as the true upper branch of the Kashgar-darya. In any case, the Kizil-su retains its name below the Markhan-su confluence as far as the plains, where it becomes the Kashgar-darya. It is the only stream which permanently reaches the Yarkand-darya from the Tian-shan highlands, and the Yarkand itself, although a copious river throughout its middle and upper course, rapidly loses volume by evaporation and the extensive irrigation works developed along its banks on the arid plains of Chinese Turkestan.

East of the Yarkand the Takla-Makan desert is traversed by the Khotan, which is formed by the junction of the Yurangkash and Karakash, at the Koshlash camping ground, about 68 miles below the town of Khotan, on the Yurangkash. The Khotan, whose course was followed by Mr. Carey in 1885 all the way to the Tarim confluence, flows almost entirely through a desert region, where it runs out, leaving in the dry season nothing in its bed except some pools here and there. Two miles below Khotan cultivation ceases, and, with the exception of one or two patches at Yangi Arik and Tawakal, 10 and 40 miles respectively below Khotan, no land under tillage is met all the way to the Shah-yar rice-fields below Kuchar. "There is plenty of land to all appearance suitable for the plough, but the water-supply is considered insufficient to irrigate a larger area than is already tilled" (A. D. Carey, *Geo. Proc.*, 1887 p. 734).

Below the Khotan confluence, the hitherto confused hydrographic system of the Tarim has been carefully explored and cleared up by Mr. Carey. This accurate observer shows that in this section of its course the main stream ramifies into two branches, one of which is the Ugen-darya or North Channel, which is consequently not an independent river as described by previous explorers. It is skirted on its left (north) bank by the Shah-yar, which descends from the Tian-shan uplands, and which farther east takes the appropriate name of Inchiki, or "Thread-like River." All three watercourses—Inchiki, Ugen, and Tarim, as the South Channel is called—flow in a parallel easterly course to the Kultokmit-kul lake or swampy lagoon, where they converge in a single channel. "The general character of the country from Shah-yar to Kultokmit-kul, the point at which the two

SCENE ON THE UPPER HOANG-HO.

branches of the Tarim reunite, is a dense reedy swamp, with occasional sheets of water in the area reached by the flood waters, bordered by a desolate saline desert" (Carey).

At Kultokmit-kul, which is the frontier station towards the Lob district, the Tarim trends round sharply to south by east. Here the united waters, which the natives still call the Yarkand-darya, from the chief head-stream, flow in a broad channel from 380 to 400 feet wide, with a mean depth of about 2·0 feet. Lower down it throws off the Kiok-ala-darya, which, after a course of 75 miles in an independent channel 150 feet wide, again joins the main stream 60 miles above its mouth. Much of its water is drawn off to irrigate the surrounding "tara," or fields, whence the name of *Tarim*, now applied to the river itself.

The lake consists of two sections, Kara-buran, about 18 miles long, and Kara-kurchin, or Chon-kul, 50 to 60 miles long, and nowhere more than 12 wide. Both are connected by the channel of the Tarim, and seem to be little more than 3 or 4 feet deep except at the junction, where the Tarim is 14 feet deep and 125 wide, with a velocity of 170 feet per minute. The whole basin is little more than a flooded morass, choked with reeds, and gradually disappearing eastwards in saline marshes. But the lake itself was found by Prjevalsky to be fresh, and well stocked with carp and marena (*Coregonus marœna*). "The whole of Lob-nor is equally shallow, only here and there occur occasional pools, 10 to 13 feet deep. . . . But the fact that almost all the lakes of Central Asia show signs of desiccation is well known" (Prjevalsky's *Lob-nor*, p. 100).

The term Lob-nor is applied by the natives to the whole course of the Lower Tarim, the lake itself generally taking the name of Chon-kul, or "Great Lake."

The Ili is formed at the head of the Kulja valley by the junction of the Tekes and Kunges. From this point it flows through Kulja westwards beyond the Russo-Chinese frontier to Iliysk, where it trends north-westwards to its delta at the south-eastern extremity of Lake Balkhash. The Ili is thus partly a Russian and partly a Chinese river, and its valley forms one of the weak strategical points of the Chinese Empire, for the upper course of the Kunges leads beyond the Narat Pass (9800 feet) between the Odon-kura and

Katun-daba, spurs of the Tian-shan, eastwards to the Mongolian plateau. From the sources of the Tekes in the Muzart (11,600 feet) to its mouth, the Ili has a total length of 750 miles, of which about 450 are navigable to a point 50 miles above the town of Kulja.

Apart from the already-described Brahmaputra, the seaward drainage of the Chinese Empire is mainly represented by the Hoang-ho and Yang-tse-kiang (" Yellow " and " Blue " rivers), which jointly drain an area of probably not less than 1,360,000 square miles. Both differ from the other great continental streams, inasmuch as they have their farthest sources not merely within the scarp or near the edge, but in the very heart of the great central plateau. The Hoang-ho was always supposed to rise in the springs known to the Chinese as the Sing-su-hai or " Starry Sea," on the marshy Odon-tala plain (" Thousand Springs ") in the Tangut country, south of the Burhan-Bota, or Central Kuen-lun range. But the exact spot was approximately determined in 1884 by Prjevalsky, who, after a march of about 70 miles across a barren plateau 14,000 or 15,000 feet high due south of the Burhan-Bota Mountains, came upon the sources of the Hoang-ho on the Odon-tala plain. Here the main stream is formed by two mountain torrents converging from the south and west, and flowing eastwards in two or three channels 70 to 90 feet wide and 2 feet deep at the fords. After a course of 12 miles it passes through the two lakes Jarin and Orin 14,000 feet above the sea in 35° N. lat., 97° E. long. The Hoang-ho, which takes its name from the yellow loess formation occupying most of its basin, is remarkable not only for its extremely circuitous course, but also for its tendency to break through formidable barriers and to shift its channel from epoch to epoch. Thus after leaving Lake Orin it makes almost a complete circuit round the Amne

Machin Mountains, and then turning north pierces the Dzun-mo-lun range (10,800 feet), as if bent on forcing a passage to the Kuku-nor basin. But failing to burst the rampart of the Kuku Mountains, it is here deflected eastward to Lan-chau in Kan-su close under the Great Wall of China. At this point it makes a tremendous bend along the east slope of the Ala-shan and the south foot of the In-shan round the Ordos peninsula, north, east, and south to the Funiu-shan, or eastern extension of the Kuen-lun system.

This bend was known only at a few isolated points before the year 1893, when Mr. and Mrs. Littledale on their eastward journey through Asia struck the Hoang-ho at Lan-chau. From this place, which stands at the head of the navigation, they descended the river, partly by raft, partly by scow, all the way to Bautu near the north-east corner of the bend. The trip took 25 days, and the stream, although shallow, was found free of obstructions, except above Ning-hsia, where the raft was nearly wrecked by some dangerous rapids. Here the stream rushes between nearly vertical rocky walls at one point scarcely 20 yards apart; "the river every few minutes made sharp turns, boiling and surging, and at the bends the water would be heaped up against the rocks higher than my head as I stood on the raft, and then flow away in a succession of whirlpools. Fortunately after a couple of hours the gorge came to an end and the river widened out and went at a more sober pace" (*loc. cit.* p. 468). The river was generally about a quarter of a mile wide, flowing between banks overgrown with bush or willows, beyond which nothing was to be seen but the drifting yellow sands of the desert. Yet between Ning-hsia and Bautu there was a good deal of traffic, scows laden with wool or grain drifting with the current or returning up stream towed by their crew usually of five men.

At Tung-kwan, converging point of the three provinces of Shensi, Shansi, and Ho-nan, the Hoang-ho is joined on its right bank by its great tributary the Wei, which has its source far to the west in the Kansu uplands, and which flows thence in a nearly due westerly course through the rich and thickly peopled loess region of Shensi. At the confluence the main stream turns abruptly east to Lung-men-kow in Ho-nan, where begin the extraordinary shiftings of its lower course, which for their vast extent and destructive character are altogether elsewhere unparalleled. "In all our ordinary maps the Hoang-ho enters the sea in lat. 34° south of the great peninsula of Shan-tung. This was its true course down to some thirty years ago (1853), and for six centuries before that. But in the earliest times of which the Chinese have record, the Hoang-ho discharged into the Gulf of Pe-chi-li —that is, north of Shan-tung and its mountains—and it continued to do so, though with sundry variations of precise course, till the thirteenth century. Before the latter period the river had occasionally thrown off minor branches to the south of Shan-tung, but it then changed its course boldly to the latter direction, and so continued till our time. The tendency to break towards the old northern discharge had long existed, and was resisted by a vast and elaborate series of embankments. These gave way partially in 1851; following floods enlarged the breach, and in 1853 the river resumed its ancient course, the plains of Pe-chi-li, and now enters the gulf of that name in lat. 38°" (Col. Yule).

The whole of this region from the Yang-tse along the Grand Canal across the old, and down to the new course of the Hoang-ho, has been explored by the English engineer G. J. Morrison. "For some little distance down the new course seems to have resisted all attempts to confine it within reasonable bounds. It has overflowed the low-lying country, and presents the appearance of a lake with numerous shoals and channels between. None of these had more than 2 feet of water throughout their entire length, and this part of the river can hardly be said to be navigable. After about 25 miles, however, there is a change for the better. Embankments have been built along both sides, which protect the country from floods. The authorities have also attempted to confine the river by planting trees. But the river has eaten away the banks, in some places leaving the trees growing, in other cases carrying them off and depositing them in heaps elsewhere, thus forming serious and dangerous obstructions in the river. At one point we passed through what appeared to be more like a flooded plantation than a river, and although the current of at least 4 miles per hour was with us, it took about two hours to go 2 miles." [1]

According to Mr. E. L. Oxenham, the best plan for regulating the discharge would be to replant the Shansi and Shensi hills, which are at present bare of timber. A more equable rainfall would thus be ensured, and less earth would be swept down by torrents. It is the vast quantities of silt deposit that make the Yellow River so difficult to control and the level of its bed so difficult to preserve. The cultivation of the barren desert (by irrigation) which the river at its northern bend flows through would also be highly beneficial (*Consular Report*, 1887).

In proportion to its length, estimated at from 2500 to 2600 miles, the Hoang-ho receives fewer large tributaries than perhaps any other river in the world. Its upper course is joined by the Ta-tung-ho flowing from the southern slopes of the Nan-shan across the Kuku-nor eastwards to its left bank. This river, which flows through a very mountainous region, has not yet been

[1] Paper in *Proceedings of Royal Geographical Society*, March 1880, p. 148.

thoroughly explored, although crossed at several points by Prjevalsky during his excursion to Kuku-nor.

This great lake, although approached from the south by the Hoang and on the north by the Ta-tung, is none the less a closed basin, standing 10,500 feet above sea-level in the Tangut country, which is a sort of debatable land between China, Mongolia, and Tibet. The lake, which has the form of an ellipse over 200 miles in circumference and 2300 square miles in extent, is very salt, and of an exquisite dark-blue colour, compared by the Mongolians to blue silk. The shores are flat and shelving, but towards the west is a rocky island with a temple inhabited by ten lamas, who have no means of communicating with the mainland during the summer. But in winter pilgrims cross over the ice with presents of provisions for the hermits.

Although now a closed basin, the Kuku-nor seems to have formerly communicated westwards with a long-vanished lake, the largest in all Tibet, which filled the whole of the Tsaidam plain between the Nan-shan and Burhan-bota north and south. This swampy region is now traversed by several streams, the chief of which is the Bayan-gol ("Rich River") or Tsaidam, which receives the overflow of Lakes Tosu-nor and Alang-nor from the south, and which flows for over 300 miles in a north-westerly direction, at last losing itself in the Dabsun-nor marshes. Here is a depression between the Altyn and Nan-shan, through which the former Lake Tsaidam must have sent its superfluous waters north-west to the Lob-nor basin. The gradual isolation of all these basins affords one of the most striking illustrations of the process of desiccation that has been going on throughout Central Asia from the remotest times. First the Kuku-nor fails to reach the Tsaidam; then the Tsaidam ceases to communicate with Lob-nor and ultimately dries up, while Lob-nor sinks to a mere reedy morass some 3 or 4 feet deep, and the thickly-peopled plains of the Tarim basin are converted into the Takla Makan sandy waste.

Although not an absolute desert, Tsaidam, that is, the "Salt-swamp," is one of the most desolate regions in Central Asia. "Occasionally," writes Mr. Rockhill, who skirted its southern margin in 1892, "we saw Mongol tents on some dry spot in the swamp, and each time with renewed wonder that human beings could live in such a place as the Tsaidam, where there is but swamp and sand, willow, brush or briars; where mosquitoes and spiders thrive; where the wind always blows, the heat of day is intense, and the cold of night piercing. Fear alone holds them there—fear of the Golok and Panaka [marauders] prevents them from occupying the rich pasturages along the clear streams in the mountains to the south of their God-forsaken plain" (*Geo. Jour.*, May 1894, p. 368).

In the Shang district south-east of Tsaidam this explorer discovered the long narrow Lake Tosu-nor standing at an altitude of a little over 13,000 feet. This basin, which runs east and west for a distance of at least 35 miles, with a breadth of from 2 to 2½ miles, discharges through the Yogor-gol (Bayan-gol) north-westwards to the Tsaidam swamp. A visit was also paid to the smaller lake Alang-nor, which lies much farther west, and which sends its overflow through the Alang-gol to the Yogor-gol. The Alang-nor, which is intersected by the 97th meridian, stands about 1000 feet higher than the Tosu-nor.

Some 70 miles south of the Shuga range the Hoang-ho basin is separated from that of the Muriu-ussu, or Upper Yang-tse-kiang, by the Baian-kara-ula ("Rich Black Mountains"), which under various names runs about 450 miles east and west without anywhere reaching the snow-line. But towards the north-west this range is connected with the Kuen-lun by the snowy Gurbu-naiji, whence flows the Napchitai-ulan-Muren, a chief head-stream of the Yang-tse. The confluence of these rivers in 94° E. long. and 34° 50′ N. lat. marks the limit of Prjevalsky's expedition to the Kuku-nor in 1871-73, when he was obliged to retrace his steps, unable either to reach Lassa or the actual source of the Yang-tse. At the confluence he found this river already 750 feet broad, and flowing in a bed over a mile wide, which, "as our guide assured us, is entirely covered with water during the rainy season in summer, when it sometimes even overflows the banks" (ii. 221). The first ford, even after the subsidence in autumn, lies 20 miles above the confluence, and the actual source of the main stream is in the Dang-la mountains, a ten days' march farther up about 34° N. lat., 91° E. long. From this point the Muriu-ussu, or "Winding Water," flows east and south to its junction with its great tributary, the Min, at Siu-chau-fu, and the Chinese, who regard the Min as the true head-stream, call the other branch Kinsha-kiang ("Gold-sand River"), from its source to the confluence.

The windings of the Yang-tse are no less remarkable and somewhat analogous to those of the Hoang-ho. The upper courses of both are separated only by the narrow Bayan-khara water-parting, and their waters have at various epochs been intermingled in a common delta, or at least connected by numerous natural and artificial channels in the coast province of Kiang-su. But while the normal direction of each is from west to east, in their middle course they are deflected for hundreds of miles north and south respectively, the Hoang sweeping round the Ordos peninsula to the foot of the In-shan mountains, the Yang-tse penetrating southwards far into the Yun-nan highlands. Here their channels diverge as much as 15 degrees of latitude (26°-41° N. lat.), enclosing an intervening space fully 1000 miles long, and stretching from South Mongolia nearly to the frontiers of Burma. From the source to the mouth of the Yang-tse the distance in a straight line is scarcely more than 1800 miles, but owing to these astonishing meanderings its total length is estimated by Von Tillo at no less than 3158 miles, with a drainage estimated by Blakiston at 750,000 square miles.

From the junction of the Min on the Yun-nan frontier, where the Kin-sha becomes the Yang-tse, the main stream flows with many windings, mainly east-north-east, through the great provinces of Se-chuen, Hu-pe, Ngan-whei, and Kiang-su, to its delta in the Tung-hai, or Eastern Sea, over against the southern extremity of Japan.

Throughout the greater part of its course the Yang-tse is fed by a vast number of tributaries, some of great size, and jointly affording a navigable water highway of not less than 12,000 miles. Its upper course is joined by the Yalung (Yarlung) or Niachu, which flows from the Bayan-khara slopes parallel with the Muriu-ussu. Lower down comes the Min-shan (Min) or Wen from the north, which at low water is navigable for 200 miles to Sintsin-hien, nearly 2000 miles from the sea. But below the junction the main stream is obstructed by several rapids in its passage through the "Cross Ranges" down to the central plains. From Pingshan to the coast, a distance of 1760 miles, the total fall is about 1500 feet (Blakiston), but very unevenly distributed. The descent is very rapid in the romantic hilly region on the Se-chuen and Hu-pe frontier, where several magnificent gorges follow in quick succession. Here the stream is contracted to a few hundred yards, and rushes in some places at the rate of 10 or 11 miles per hour through its deep rocky bed. Between Kwei-chow and Ichang the chief "tan" or

THE MI-TAN GORGE. UPPER YANG-TSE.

rapids extend over 100 miles; but below the stupendous Lon-kan and Mi-tan defiles the hills suddenly recede, the great river broadens out to a uniform width of 2600 feet, and a depth of 20 to 30 feet at low water.

A conspicuous feature in the Min basin is the sacred Mount Omei, first described by Colborne Baber, and ascended in 1890 by Mr. A. E. Pratt during his second botanical excursion from Shanghai to Ta-tsien-lu on the eastern border of Tibet. Omei, which lies within two days' march of Kia-ting-fu on the Min, rises abruptly above the surrounding plain to an absolute height of 11,100 feet, and is accessible only from one side. The others are extremely steep, one being "a precipice nearly a mile and a third high, the highest sheer declivity, perhaps, in the world." On the mountain, which is constantly visited by many thousands of pilgrims, there are from sixty to eighty Buddhist temples attended by about 2000 priests. Omei is crowned by a temple standing close to the edge of the great precipice. "The present structure is of wood, but the original was of bronze, and tons of bronze slabs with the image of Buddha are still lying on the ground. They appear to be Indian work, and were originally covered with gilt. At one of the lower temples we saw a life-size brass elephant. It looked like Indian work, and had probably been cast in sections" (Pratt, *Geo. Proc.*, 1891, p. 338).

Throughout its lower course the Yang-tse is lined on both sides, but especially on its right bank, by numerous shallow lakes or reservoirs, which during the floods are filled up by the overflow of the main stream and its tributaries. These lakes, of which the largest are the Tung-ting-hu in Hunan and the Po-yang-hu in Kiang-si, thus serve to regulate the inundations and prevent the widespread ruin often produced by this cause in the Hoang-ho basin. The Tung-ting has an area of at least

2000 square miles, and receives the Yuan-kiang, Lo-kiang (Tse-kiang), and Heng-kiang, and other rivers from the south, which collectively drain an area of 80,000 square miles in the province of Hunan. From this basin the two adjacent riverain provinces are respectively named Hu-pe and Hunan—that is, "North of the Lake" and "South of the Lake."

Below the Tung-ting the Yang-tse receives its great affluent, the Han-kiang, flowing from the Tsing-ling (Eastern Kuen-lun) through Shensi and Hu-pe south-eastwards to its left bank at Hankow. Although obstructed by rapids at some points, the Han is throughout navigable, and in summer accessible to steamers for a distance of 600 miles. It flows through a magnificent region, well watered, extremely fertile and healthy, and abounding in vegetable and mineral resources of all kinds. But in its lower course it flows in a bed higher than the surrounding lands, so that during the floods the whole country from the confluence to Lake Tung-ting is sometimes transformed into a vast inland sea.

Beyond Hankow the great Lake Po-yang plays the same part for the province of Kiang-si that the Tung-ting does for Hunan. Its chief influent is the Kia-kiang, which flows from the Tung-lo-ling range through Kiang-si to a large delta on the south-west side of the lake. This basin, which has an area of 1800 square miles, is studded with islands, and although in many places covered with forests of reeds, its northern section is very deep and skirted by lofty hills. The wooded headlands, inlets, and islets are everywhere interspersed with towns, villas, towers, pagodas, crowning every eminence, and rendering this one of the most charming regions in China.

Below the Po-yang the Yang-tse flows in a majestic stream through the hilly and flourishing province of

Ngan-whei, and by the great cities of Ngan-king and Nan-king to its delta, which fills a large part of the province of Kiang-su. Here it has a mean discharge of perhaps 735,000 cubic feet per second, ranking in this respect, according to Guppy,[1] next after the Amazons, Congo, and La Plata. Its basin occupies nearly one half of China proper, comprising some of the richest lands in the world, with boundless material resources of every kind, and especially including vast coal-measures. The riverain provinces, where were sown the first seeds of Chinese culture, have for ages supported a teeming population, variously estimated at from 100 to 200 millions. The main stream itself and its navigable tributaries are everywhere crowded with many hundred thousand junks and boats of all sizes, on which a floating population of millions pass their whole lives. A fire which broke out at the port of Wuchang in 1850 destroyed 700 junks, several thousand smaller craft, and proved fatal to no less than 50,000 of this riverain population.

The tides penetrate for over 200 miles up the estuary, which in many places is over 300 feet deep and 60 miles wide at its mouth. Here, however, the navigation is much obstructed by numerous islands and shifting sands, with scarcely more than 14 feet at low water in the deepest channels. But at the flow, vessels drawing 18 to 20 feet easily pass up, and the danger most to be dreaded is perhaps the dense fogs often enveloping the whole estuary and neighbouring seaboard. At the same time, the sedimentary matter, brought down at the rate of perhaps 6300 million cubic feet yearly, is constantly accumulating at the mouth, raising the sandbanks and enlarging the islands. Tsungming, the largest of these, formerly washed by the tides, now supports a population

[1] *Nature*, 20th Sept. 1880.

of 2,000,000 industrious peasantry and fishermen on an area of about 400 square miles.

Great changes have taken place in the lower course of the Yang-tse, which formerly threw off branches from various points above Nan-king eastwards to the large Lake Tai-hu, and thence southwards to Hang-chow bay. The whole of the Shang-hai peninsula thus formed a portion of the delta, which also extended through countless channels northwards to the Hoang-ho, when that river discharged into the Yellow Sea. Here the connection is still maintained even with the present channel of the Hoang-ho by means of the famous Imperial or Grand Canal, which runs through a district where land and water seem to be inextricably intermingled, mainly northwards from the Yang-tse across both the old and new beds of the Hoang to the Pei-ho.

The basin of the Pei-ho, although vastly inferior in extent to those of the Hoang and Yang-tse, derives a special importance from the fact that within its limits is situated Peking, capital of the empire. It is formed by a large number of streams, such as the Pei-ho, Hwen-ho, Tsu-ho, Hu-to-ho, Chang-ho, and Wei-ho, which flow mostly in independent channels to within a comparatively short distance of the coast. Here they all converge at Tien-tsin, whence the united stream flows in a broad navigable channel to the Gulf of Pe-chi-li. It is probable that the various branches of this river system formerly found their way in separate channels to the coast; for all the lowland Pe-chi-li plains form a marine basin, which has been slowly filled in with the alluvia brought down by these rivers from the Shansi and Mongolian frontier highlands. They are still subject to extensive inundations, covering an area of about 6000 square miles, and representing, as it were, the former extension of the Gulf of Pe-chi-li towards Peking and the surrounding hills.

These inundations are caused by the swollen waters of the head-streams, which, being unable to find room in the common channel below Tien-tsin, overspread its banks far and wide, and present a continuous sheet of water stretching from the gulf inland nearly to Pao-ting. Amid this waste of waters nothing is visible except the towns and villages perched on eminences dotted over the plains; the river-banks are washed away, the streams shift their course, the crops are destroyed, and the people left a prey to famine and disease. Yet in an area of scarcely 60,000 square miles the province of Pe-chi-li, mainly comprised in the Pei-ho basin, was estimated to have a population of close on 37,000,000 in the year 1842. At present it probably amounts to less than half that number. The inhabitants have been forced to emigrate in hundreds of thousands to Manchuria and Mongolia, owing to the disastrous floodings of the Pei-ho.

According to Guppy, the Pei-ho drains an area of some 56,000 square miles, has a mean discharge of 7500 cubic feet per second, and brings down to the Gulf of Pe-chi-li about 80,000,000 cubic feet of alluvial matter every year.

In the neighbouring province of Manchuria the drainage is mainly through the Sungari and Usuri to the Amur. But the southern portion between the Khingan range and Korea drains through the Liao-ho southwards to the Liao-tung Bay, which is a northern extension of the Gulf of Pe-chi-li. In its upper course the Liao takes the name of Sira-muran-pira (Shara-muren), which with its chief affluent, the Lohan-pira, rises in the highland district where the Mongolian escarpments merge in the Great Khingan range. The Liao pursues a tortuous semicircular course from these uplands round to its mouth at the head of Liao-tung Bay. It is navigable during the floods only in its lower course within the Chinese province of Shing-king; but since the Russian occupation of all the seaports north of Korea this river has acquired great importance as the only seaward outlet of Manchuria.

In the extreme south-east of China proper the Nan-ling with its eastern extensions forms the water-parting, broken at one point, between the Yang-tse and the Si-kiang or "West River." The Si-kiang basin thus occupies nearly the whole of Kwang-si and Kwang-tung, besides parts of Yun-nan and Kwei-chow, and also encroaches both on Tong-kin in the south and beyond the Tung-lo-ling range on Hunan in the Yang-tse basin. Here the Pe-kiang ("North River of Canton") flows from the north side of the range due south to its junction with the main stream west of Canton. This river has its farthest head-streams in the Yun-nan and Kwei-chow highlands, whence it flows due east to its delta and estuary in the China Sea.

The North River between Lo-chang and Shao-chow "has some shallows, which would be impediments to steam navigation; but from Shao-chow the only troubles would be the freshets, which cause the water to rise quickly, and the river to run swiftly; but as far as I could learn there is nothing to prevent regular steamboat traffic being carried on as far as Shao-chow" (G. J. Morrison). Thus affording direct communication between the Yang-tse basin and the southern provinces, the Pe-kiang has at all times occupied a position of primary importance as a military and trade route. In this respect it ranks far before the Si-kiang itself, which has yet a navigable course of nearly 900 miles. Below its junction with the Koli-kiang it penetrates through a series of magnificent gorges into the province of Kwang-tung, where it has a depth in some places of from 50 to 150 feet. The tides are felt for a distance of 180 miles inland, and at high water most of the countless channels and branches of the delta are navigable. In the delta, which has an area of over 3200 square miles, the chief channel is the east branch, known as the Canton River, which is joined from the east by the Tung-kiang. Below the confluence the stream is contracted to a narrow bed, commanded by Forts Humen—that is, "Tigers' Throats"—whence this part of the river takes the name of Bocca Tigris, or simply the Bogue. Immediately below the forts it broadens out to the estuary of the Shu-kiang, or "Pearl River," said to be so called from another stronghold, the Hai-chu, or "Pearl Fort," familiarly known as the Dutch Folly.[1] The network of channels in the delta is inhabited by an enormous floating population, estimated at about 350,000, most of whom pass their whole lives on the water.

On his overland journey from Tong-kin to Canton in 1891, Mr. A. R. Agassiz struck the upper Si-kiang at Lang-chow near the Franco-Chinese frontier in Kwang-si, and followed the course of the river from that point all the way to the estuary. The Tso-kiang, or "Left River," that is, the southern fork of the Si-kiang, is formed by the junction at Lang-chow of the Kao-ping-ho from Yun-nan, and the Sung-chi-kiang, which sweeps round from Kwang-tung by Langson to the confluence, thus twice crossing the frontier. It took seven days to descend in a small boat from Lang-chow to Nan-ning, a short distance below the Yu-kiang, or "Right River," that is, the northern fork of the Si-kiang, which also has its rise in Yun-nan. It is navigable from the town of Pe-se, where it enters the province of Kwang-si, and it had long been one of the chief highways to Yun-nan till 1890, when much of its traffic was lost by the opening of the Red River in Tong-kin. At the confluence of the Right and Left Rivers, the united waters take the

[1] That is, the "Dutch Fort," the English word *fort* being pronounced *foli* in "Pigeon English."

name of Si-kiang for the rest of its course to its mouth at Macao. Nan-ning, six hours below the confluence, ranks as the third largest city of Kwang-si, and does a large trade with Canton despite the heavy transport charges and the high duty levied on goods forwarded by the river. Below Nan-ning the Si-kiang is a broad deep stream, well suited for steam navigation for many miles at a stretch, but here and there obstructed by rapids. Beyond Hsun-chow, where it is joined on the left bank by the Hun-shui-kiang from the north-west, the main stream "is a splendid expanse of water, which might be rendered navigable to steamers by a slight expenditure of engineering skill." Boats take two days to descend from Hsun-chow to Wu-chow at the junction of the Tan-kiang from the north, about 200 miles above Canton. The whole journey from Lung-chow, a distance of some 600 miles by water, took twenty-seven days, although the traveller was informed that under favourable conditions it is occasionally made in fifteen days. As far as Wu-chow there is no obstacle to the ascent of steamers drawing 8 feet, and the absence of steam this observer attributes partly to the opposition of the large number of people engaged in the local carrying trade, partly to the patriotic sentiment of the Chinese, who object to the sight of a foreign flag on another river of China, as well as on the Yang-tse-kiang. They abstain from running their own vessels lest foreigners should demand the same privilege (*Geo. Proc.*, 1891, p. 259).

Besides the lakes already mentioned in connection with the river systems, there are few bodies of still water in the empire anywhere except on the Tibetan plateau. Here, although mostly of small size, lakes are extremely numerous, and the lacustrine character of this region becomes more evident with every fresh exploration. More than half of the Kachi tableland is dotted over

LAKE NAMTSO (TENGRI-NOR).

with closed basins, which are probably the remains of inland seas formerly draining through openings in the frontier ranges. In the west the largest would seem to be the Ike-namur and Bakha-namur, which form the central basins of a lacustrine system, stretching for over 120 miles south-west and north-east. Many of these lakes were visited in 1874 by Nain Sing, who found that they were the remains of far more extensive basins, some of which were already reduced to mere swamps or quagmires covered with a saline efflorescence. Most of the lakes are salt or brackish, while some are still perfectly fresh. The Dangra-yum, or "Mother Dangra," 86° N., 31° E., is 180 miles in circumference, and is dominated on the south by Mount Targot-yap, or "Father Targot."

East of the Dangra-yum several other lakes are said to discharge their waters northwards to the Chargut-tso (Bower's Chargat-cho ?), which is itself supposed to communicate with one of the great rivers flowing to the Indian Ocean. But better known is the Tengri-nor lying in the south-east corner of the Kachi plateau, within 60 miles of Lassa. It is 50 miles by 15 to 24, and of unknown depth. A favourite place of pilgrimage is the convent of Dorkia on its west side, commanding a superb view of its blue waters and of the surrounding snowy peaks. The Tengri is not a closed basin, as had been supposed, for it discharges through the Nak-chu, flowing from its north-west corner to the outlet of the Chargut-tso. North of it lies the Bul-tso, or "Borax lake," covering an area of 24 square miles, whence formerly came much of the so-called "Venetian" borax.

But before the recent explorations no clear idea could be formed of the hydrographic relations. From Nain Sing's journey of 1874-75 the central plateau was known to be a lacustrine region dotted over with numerous lakes; but much ignorance continued to prevail regarding the drainage of these flooded depressions. We now know that Tibet, taken as a whole, comprises three distinct drainage areas: East Tibet, from about 91° E. long., draining to the Indian and Pacific Oceans through the great rivers of China and Indo-China; West Tibet, from about 80° E. long., draining in the opposite direction to the Indus, and to a small extent through the Ganges and the San-po to the Bay of Bengal; Central Tibet, mainly between 80° and 90° E. long., that is, the lacustrine plateau, whose true character as a region of inland drainage without any seaward outflow has been determined by Captain Bower's journey. This explorer's route lay for the most part to the north of Nain Sing's, and he was thus able to follow for hundreds of miles the great chain of lakes heard of by the Indian surveyor. On the central plateau there appear to be no well-defined watersheds, so that rivers may be met flowing in almost any direction, but all discharging into the closed lacustrine basins. Some of the lakes are very large, and one especially, doubtfully named the Garing Cho, is described as "a noble sheet of water, stretching out east and west to an enormous distance, more worthy to be called an inland sea than a lake." Yet, like all the other lakes of this region, it showed unmistakable signs of having been formerly much more extensive than at present. It lies in the Naksung district under 32° N., 89° E., at an altitude of 15,420 feet, somewhat to the north-west of the Tengri-nor basin surveyed by Nain Sing. Near the western extremity of the chain, due east of the Lanak Pass leading from Ladakh into Tibet, stands a still more elevated sheet of water, the Horpa Cho (17,930 feet), probably the highest lake on the surface of the globe. Beyond the Horpa Cho follow the Charol Cho and the Aru Cho, the latter a magnificent basin

17,150 feet above sea-level, lying in a grassy district frequented by "wild yak and antelope in incredible numbers. This peaceful-looking lake, never before seen by European eye, seems given over as a happy grazing-ground to wild animals—a veritable sportsmen's paradise" (Bower). Having no outflow, all these basins were found to be saline or brackish, except one extremely picturesque lake studded with islands, ramifying in all directions amid the upland valleys, and overshadowed on the south side by a cone-shaped snowy peak. It lay to the south of the Garing Cho, from which it was separated only by a narrow neck of land. But the explorer failed to ascertain its name, and he everywhere had the greatest difficulty in determining the correct nomenclature of the region traversed by him. "It is almost impossible to get the correct names of places or lakes in Tibet, as every Tibetan lies on every occasion on which he does not see a valid reason for telling the truth. Sometimes I have asked half a dozen men separately the name of a lake, and received half a dozen different answers." This vast lacustrine zone appears to lie mainly between 30° and 35° N. lat., and between 80° and 92° E. long., and the innumerable basins dotted over its surface are probably the scattered remains of a continuous sheet of water, a vast inland sea which sent its overflow eastwards to the Pacific Ocean.

On the frontier of Tibet and Kashmir the three lakes, Pangkong, Mognalari, and Noh, form an almost continuous basin, stretching nearly west and east, but at different levels, and apparently in the line of an old watercourse. They wind along with a present depth of about 150 feet, and a total area of 210 square miles north-westwards, to a point where they formerly communicated through the little River Tankseh with the Shayok, a tributary of the Upper Indus. But with the gradual subsidence of their waters the emissary ceased to flow, and they now form a closed basin, saline in the Pangkong or lower section, but still fresh higher up.

Towards the south-east of Tibet lies Lake Palti (Yamdok-tso), usually described as ring-shaped, with a large island filling most of its basin, and rising 2300 feet above its surface. But the native explorer who visited it in 1875 ascertained that the supposed island is really a peninsula connected by a narrow tongue of land with the south side of the lake. The water, described by Manning (1811) as slightly brackish, this explorer found to be perfectly fresh.

4. *Natural and Political Divisions: Tibet—The Tarim Basin (Chinese Turkestan) — Mongolia — Zungaria and Kulja—Manchuria—The Great Wall—The Gobi and West Mongolia—South-East Mongolia—Korea—China Proper—Islands: Hainan, Macao, Hong-Kong.*

In the Chinese political system the great administrative divisions correspond to a large extent with the main physical regions. Tibet and Korea, off-lying members of the system in the extreme south-west and north-east, are marked off from the rest of the empire by their geographical position no less than by their political status. In the same way the remote western regions of Kashgaria and Zungaria—the former occupying the Tarim basin, the latter embracing the upper courses of the Ili and Irtish — are held by military tenure, while the nomads of the vast Mongolian plateau are ruled through their respective hereditary Khans. Since the conquest (1644) of China proper by the Manchus, the north-eastern region of Manchuria has been brought into more intimate relationship with the Hoang-ho and Yang-tse basins. Here also the great provinces are generally disposed in harmony with the respective areas of drainage. Thus Pe-chi-li is mainly comprised in the Pei-ho water system, while Hu-pe, Hunan, and Kiang-si are each occupied by the basins of large rivers flowing to either bank of the Yang-tse.

Tibet.

Tibet, taken in its widest sense, comprises the whole region between the Kuen-lun and Himalayas, which is at once the most elevated section of the Central Asiatic plateau and the loftiest tableland on the face of the globe. It forms a mass of irregular uplands sloping gradually eastwards, but scarcely anywhere falling below 12,000 or 14,000 feet above sea-level. Northwards, the Kuen-lun escarpment falls rapidly towards the low-lying plains of the Tarim basin, while on the west and south the massive sweep of the Himalayan system forms a natural barrier towards British India. The eastern boundary has been crossed only at a few points, but here also the "Cross Ridges," roughly answering to the Yung-ling of Chinese geographers, serve as the frontier line towards China proper.

In recent years the work of exploration has been prosecuted with great vigour by several travellers, who, although unsuccessful in their main object of reaching the capital, have considerably advanced our knowledge of the Tibetan plateau, and especially of the northern and central parts. The researches of Dalgleish and Carey, Bonvalot and Prince Henri of Orleans, Pevtsof, Bower, Rockhill, Younghusband, Littledale, and others have opened up new fields of exploration in hitherto inaccessible regions. They have determined the continuity of the Kuen-lun orographic system across 20° of longitude, and revealed the trend and structure of its main branches. They have shown the lacustrine character of the central tableland, and traced nearly to their sources some of the great rivers which here take their rise. They have at the same time thrown much light on the climatic conditions of these arid upland regions, which, nevertheless, are found to contain an extraordinary abundance of animal life. Their researches have also revealed the existence in former times of a long chain of flourishing oases extending along the northern foot of the Kuen-lun, indicating the route followed in mediæval times by the Chinese silk caravans, and the productive gold-fields of Northern Tibet have likewise been brought to light.

A very fair idea may now be formed of the relief and general configuration of the great Tibetan plateau, which in many parts attains an altitude of from 16,000 to 18,000 feet, and is traversed by vast mountain ranges mainly with an east and west trend, rising 6000 and 8000 feet still higher. Captain Bower, who entered from the west (Ladakh), and passed right through eastwards to China by a route not previously followed by any European since the time of the Jesuit Père Desideri (1715), travelled in this direction for five months (June to September 1891) without ever once camping at a lower elevation than the summit of Mont Blanc, or say 16,000 feet. This explorer distinguishes two essentially different regions—the whole of Central and Northern and nearly the whole of Western Tibet, to which he gives the name of Chang, and South-east Tibet. The Chang consists of a lofty tableland with hills mostly of a rounded form, and snowy ranges disposed normally in the direction from east to west. Very little of the Chang is inhabitable throughout the year, and most of the treeless, grassy tracts are too remote from the winter pastures of the nomads to serve as summer grazing-grounds.

Hence the camping places are met chiefly round the margins of the plateau, which, at least south of the Dang-la range, enjoys a sufficient rainfall and snowfall to support a short, crisp, and exceedingly nourishing grass, such as so often grows on soil covered with snow during the winter months. Hence the surprising abundance of wild animals, such as yak, Tibetan antelope, gazelle, goa, and kiang, although it is difficult to understand how they survive the long winter months, when the thermometer falls to —15° F., and when the cold is intensified by the prevailing high winds. Bird life is less plentiful, but it includes some extremely rare species, such as two kinds of eared pheasant, the snow partridge intermediate between the true partridge and the grouse, the blood pheasant, the Tibetan sand grouse and bar-headed goose. Amongst the insects, which are not numerous, are the burrowing bee, and butterflies ranging up to 17,600 feet. Of the 115 species of flowering plants collected by the Bower expedition, one was found at an altitude of 19,000 feet, probably the highest at which any flowering plant has yet been met.

The Chang is necessarily a nomad domain, sparsely occupied by wandering tribes, who are partly of Mongol, partly of Tibetan stock, but who are all alike described in far from flattering language by recent explorers. Some, such as the Chukpas of the central region, may be called professional marauders, periodically raiding the cattle or levying blackmail on the more peaceful populations, and plundering the caravans along the trade routes, which radiate in various directions from Lassa. But the "peaceful" nomad populations themselves appear to be little better, and according to Captain Bower, "the character of all these nomads is much the same—greedy, faithless, and suspicious. Their suspicions do not attach only to foreigners; every camp seems to view every other camp as not only a possible but a very probable enemy."

In its physical aspects South-east Tibet presents a marked contrast to the Chang wilderness. Thanks to a more copious rainfall, and to a much lower mean elevation, falling in places to 12,000 and even 11,000 feet, it enjoys a more genial climate, while the surface is scored by deep river valleys, and diversified with steep, well-wooded hills. Hence agriculture takes the place of pastoral pursuits, and the nomads dwelling in tents are replaced by a settled population living in houses and cultivating various industries. But in other respects these semi-civilised Tibetans are little better than their nomad neighbours—"faithless, immoral, cowardly, and untruthful; to those they are afraid of they are servile, but to those they are not afraid of insolent. Their physique is distinctly good, and they appear to be able to stand any amount of cold and hunger; less industrious and skilful than the Chinese, they are still an active, lively people" (Bower).

Bower estimates the whole population of Tibet at not more than 8,000,000, half under the rule of the Deva Zhung, that is, the central government of Lassa, and half in Chinese Tibet, mostly under their own chiefs and practically independent, or owning only a nominal allegiance to Lassa. He attributes the scanty population to three main causes—the custom of polyandry, which is largely, though not universally practised; the large number of Buddhist monks (lamas), nearly half a million, who, though probably only nominal celibates, are forbidden to marry; lastly, the vast extent of wilderness unsuitable for settlement, and capable of supporting only a few scattered nomad tribes besides wild yak and antelope. In fact the great bulk of the inhabitants are concentrated in the settled south-eastern districts, and of these all directly subject to the Lassa government are at least nominal Buddhists. But "the Buddhist religion, as seen in Tibetan countries, has nothing in common with the pure morality preached by Gautama Buddha. The doctrines of the founder are too abstract for the average Tibetan mind, and this has led to innovations which have developed until the grossest superstition, little better than African fetishism, and bearing hardly any resemblance to the original precepts, is all one meets in the stronghold of Buddhism" (Bower).

The little-known northern division of Kachi is occupied in the west by the Hor of Turki stock, in the east by the Sok of Mogul stock, and has hence been sometimes called Hor-Sok-pa. Here are the numerous chains of lakes discovered by Nain Sing, which very possibly drain eastwards to one or more of the great rivers of Indo-China. Still farther south the high and apparently continuous Ninching-tangla ridge divides this basin from that of the San-po, or Upper Brahmaputra River. Here is the true Bod-yul, or Land of the Bod—that is, of the Tibetan race; and here is situated Lassa, capital of Tibet, together with the more fertile and thickly-peopled portion of the country.

Bod-yul comprises eight divisions:[1] — 1. Nari-Khorsum (chief town, Gartokh), embracing all Western Tibet as far as the Mariam La Pass (82° 30′ long.); 2. Dok Thol (chief town, Sarka-Jong), extending from the Mariam La Pass to the Kalha Pass (87° E.); 3. Chang (chief town, Shigatze), extending eastward to the Khamba La Pass; 4. U (chief town, Lassa), stretching thence eastward to the twelfth stage on the southern route from Lassa to Peking; 5. Monhuil, embracing the Tawang

[1] Markham (*Tibet*) speaks of four only — Nari (Ari), U, Isang, and Kam, treating Hor as distinct, and omitting reference to Dok Thol and Monhuil.

district between Bhutan and the Daphla (Lhopa) country; 6. Kham (chief town, Tsiamdo), between U and Se-chuen; 7. Hor-Sok, or Kachi, occupying the northern parts as far as the Kuen-lun, and over which the Central Government has scarcely a nominal jurisdiction; 8. Jyade (Rgya-sde), the "Chinese Province," extending east and west over 200 miles between the Dang-chu River and Tsiambo, with a breadth of 60 or 70 miles from the Dang-la range in the north to Larego, Shobando, and other districts in the south governed from Lassa. Of Jyade little was known before it was visited by Mr. Rockhill during his second Tibetan journey in 1892. Yet, next to the province of Lassa, it is certainly the most interesting region in the whole of Tibet. Its inhabitants, who are practically independent both of China and Tibet, have from remote times professed the Binbo religion, a form of Shamanism or devil-worship, which the Buddhists of Lassa have long in vain tried to stamp out. In the seventeenth century, when the Chinese began to extend their rule over Tibet, they stopped the incessant warfare between the rival religions by forming a separate province of all the Binbo districts, and placing it under the jurisdiction of the Emperor's Amban (Ambassador) at Lassa. This is the present province of Jyade, whose thirty-six debes (chiefs) are appointed by the Amban. Most of the country is pasture-land above the timber-line (here about 13,500 feet above sea-level), and though poor, the people are light-hearted and freer in their manners than the priest-ridden natives of Lassa. "The importance of the Binbo religion has not, I believe, been heretofore suspected. All along the eastern borderland of Tibet, from the Kuku-nor to Yun-nan, it flourishes side by side with the lamaist faith. In Jyade, where there are certainly 50,000 people, it rules supreme, and in all the southern portions of Tibet not under the direct rule of Lassa, its lamaseries may be found. So it seems that this faith obtains in over two-thirds of Tibet, and that it is popular with at least a fifth of the Tibetan-speaking tribes" (Rockhill).

The total area of the eight Tibetan divisions has been roughly estimated at some 560,000 square miles; but with the north-eastern Kuku-nor district, Tibet covers a space of about 675,000 square miles, and 800,000 square miles if we include the districts in Kashmir and Se-chuen (West China) occupied by peoples of Bod stock.

The Tibetans are essentially a commercial people, to such an extent that most of the officials and head lamas of the monasteries are said to keep agents and carry on trade on their own account. There are also many Muhammadan and other foreign traders settled in Lassa, the great emporium of the country. From Northern China come silks, gold lace, precious stones, and carpets; from Mongolia and Kachi, leather, saddlery, sheep, horses, salt, and borax; from Se-chuen, tea, cotton goods, porcelain; from Tawang, Bhutan, and Sikkim, rice, indigo, coral, pearls, sugar, spices, and Indian wares; from Ladak and Kashmir, saffron, silk, and Indian produce.

Among the chief exports are gold and silver, productive mines of which are found in various parts. The richest gold-field is that of Thok Jalung, north-east of Gartokh, in 32° 30′ N. But the great staple is wool, of which vast quantities, and of the finest texture, might be produced on the boundless grassy plains and mountain slopes of Bod-yul. Tibetan musk is highly esteemed, but so great is the demand that it reaches the coast in a very adulterated state. Salt abounds everywhere, and is obtained chiefly by solar evaporation in shallow basins.

Of the imports specially important is the brick-tea, consisting of the coarser leaves and twigs, and described by Baber as the merest refuse. It is first pounded into moulds, and then broken into "bricks" 9 or 10 inches long by 7 wide and 3 thick, conveyed by mules and carriers over the lofty passes into the country. The annual import from Ta-chien-lu to Batang, the Tibetan emporium, is estimated at 10,000,000 lbs., valued at £160,000. The tea is apparently paid for mainly by Indian rupees, which have entered the country in vastly-increased quantities of late years, and are said to have now become the currency of Tibet.

Chinese cottons are also largely imported, but at Lassa are supplanted by Indian goods. Silks of bright colours fetch their weight in silver, and English woollen cloths are in great demand notwithstanding the competition of Russian goods. Desgodins saw numerous packets on their way to the salt-works, bearing the name of a Halifax maker, and he adds that scarlet is the favourite colour, although a good golden yellow would sell well. Mr. Edgar reports that the demand for indigo is very great, and that the profit varies from 50 to 100 per cent.

Altogether it is evident that the Tibetan trade offers special advantages to English and Indian traders, were it possible to establish free communication. But ever since the war between China and Nepal in 1792 the passes to India have been closed. Even the native surveyors, sent forward disguised as traders, have had great difficulty in passing through. The Tibetans say that this policy is due to orders from Peking; but there is reason to believe that the real obstacle is at Lassa.

A first step, however, towards breaking down Tibetan exclusiveness was taken in 1894, when Mr. J. Hart, of the Imperial Chinese Customs, induced the authorities at

Lassa to agree to an Indo-Tibetan commercial treaty to come into force in May 1895. A station is to be opened at Yatung, on the Tibetan side of the frontier, where an agent of the Indian Government will be accredited. All articles except war materials, intoxicating and narcotic drugs, are to be admitted free of duty for the first five years, during which Indian tea is excluded. But after that period the treaty may be modified and tea allowed to enter subject to the same duty as Chinese tea imported into Great Britain. At Yatung will also be stationed an officer of the Chinese Maritime Customs to regulate the trade, as in the treaty ports in China. The Tibetans are expected to take in exchange for wool, gold, borax, and other local produce, Manchester goods, hardware, crockery, and especially patent medicines, their faith in foreign drugs being unbounded.

The Tarim Basin (Chinese Turkestan).

The basin of the Tarim River forms the western section of the relatively low Central Asiatic plateau, which the Chinese geographers styled the Han-hai, or "Dried-up Sea." From the Pamir to Lob-nor it stretches west and east about 900 miles, with a mean breadth of 500 between the Tian-shan and Kuen-lun ranges. Its prevailing character is that of a vast sandy plain enclosed north, west, and south by a horseshoe-shaped rampart of the loftiest and grandest mountains, rising in ridges of 18,000 to 20,000 feet, with peaks shooting up to 25,000 and even 28,000 feet. From the snows and glaciers of these highlands rush down the streams which flow through the common bed of the Tarim to Lob-nor.

The open approach to the west afforded in former times easy means of access to migrating nomad tribes and military expeditions from China and Mongolia.

Along the northern and southern edges of the basin there lie at intervals the remains of fertile oases, such as the Lob (Lop), Charchand (Cherchen), Kiria, Khotan on the southern route, and Hami (Khamil), Kuchar, Aksu on the north, some of which were of great extent and importance. But the sands driving before the winds in ceaseless billows from the eastern Gobi have gradually encroached on the cultivated lands, swallowing up populous and flourishing cities, memorials of which are still found in the gold and silver ornaments, and even in the "bricks of tea" constantly exhumed at certain spots. Extensive ruins of cities are known to exist in the Lob district.

The whole basin comprises four natural divisions—highlands, lowlands, desert, and swamp or lacustrine tracts. The first include the plateaux and deep valleys of the encircling ranges, barren hill-slopes, rich pastures on the more level portions, but with a general deficiency of vegetable and animal life. The lowlands comprise the strip of country intervening between the mountains and the desert. This is the only permanently-settled and cultivated portion of the country. Although the soil is naturally poor, it is extensively irrigated and brought under cultivation in the vicinity of the Khotan, Yarkand, Kashgar, and Aksu Rivers, where there are some exceptionally fertile tracts.

But most of the Tarim basin consists of an undulating plain of shifting sands, sloping gradually eastwards from 4000 to 2000 feet, and traversed by the various rivers flowing to the Tarim. The banks of these streams are fringed by strips of fir, poplar, willow, and tamarisk forest, interspersed with dense growths of tall reeds and grass. But all the rest is an inhospitable waste, with a deep coating of loose sandy or saline soil, impracticable alike to man and horse.

SWAMPS OF THE TARIM.

The swamps and lakes Lob, Bagrach, Yeshil-kul, and Karga, are formed by the overflow of the rivers, and are unhealthy tracts overgrown with reeds and swarming with waterfowl. "The poplar woods, with their bare soil, covered only in autumn with fallen leaves, parched and shrivelled with the dry heat, withered branches and prostrate trees encumber the ground.

"But cheerless as these woods are, the neighbouring desert is even more dreary. Nothing can exceed the monotony of the scenery" (Prjevalsky, *Lob-nor*, p. 60).

The whole of the Khotan district, and especially the neighbouring Kuen-lun mountains, abound in gold, silver, iron, lead, copper, antimony, salt, saltpetre, sulphur, soda, and coal. Gold and precious stones are chiefly found in the beds of streams, and the Kappa, Sorghak, and other auriferous districts of Khotan are said to employ 7000 hands, with an annual yield of nearly 80 cwts. Its transport to India is generally a very lucrative venture, bringing in profits of from 20 to 24 per cent.

Silk is cultivated in Khotan, but notwithstanding its good quality, a defective method of reeling renders it of little use for the export trade. It is employed with wool and gold thread in the manufacture of the Khotan carpets, which are made from patterns usually handed down from master to pupil. A renowned product and former article of manufacture is jade, found only in Khotan and the northern Kuen-lun valleys. Here there is a plentiful supply, especially from the quarries in the Karakash valley and south of Khotan. It was carried far and wide in mediæval times, and jade implements have been picked up even in Western Europe. The Yu-moun, or "Jade Gate," in the Great Wall in north-west Kansu, seems to have been named from the jade caravans which passed that way to China. This interesting historical industry was for a time suspended after the expulsion of the Chinese from Kashgaria in 1864. Although much reduced, the city of Khotan had still an estimated population of 30,000 when Mr. Carey passed through in 1885. The ruins of the wall of an ancient and much larger city, which included the sites of the present Muhammadan and Chinese quarters, "were distinctly traceable at many points" (*ib.*).

The Takla-Makan wilderness was skirted on its west side by the Pievtsoff expedition of 1889, which followed the Yarkand-darya route southwards to Nia and the Kuen-lun highlands. Both banks of the Yarkand are here fringed by broad belts of well-watered and wooded tracts from 17 to over 20 miles wide, beyond which in all directions stretches "the limitless kingdom of the desert sands, which hide in their bosoms things curious and unknown. Many cities, once flourishing, happy, and well populated, lie buried there. The dwellers on the desert border sometimes venture themselves amongst the sands in search of valuables, which they dig up in the ruins of the ancient towns buried in sand; but they never go farther than three or four days' journey. The limitless and mysterious nature of the unknown and awful waste that has become the cemetery of a once prosperous land frightens people, and the time is far off when the daring European will traverse the desert in many directions and discover to the world the secrets hidden by the sandy ocean of the desert of Takla-Makan, as the natives call it. Wild camels are apparently its only inhabitants. Nearer the river tigers have trodden paths through the woods and reeds, and mercilessly wage war against the boar and morals (deer). There, too, on the edge of the desert one meets occasionally the light and timid antelope, besides wolves, foxes, hares, and small rodents, which are found almost everywhere" (Lieut. Roborofsky, *Geo. Proc.*, 1890, p. 31). Even the inhabitable river-side zones are occupied only in summer by a few scattered groups of Dolon nomads; for these tracts are infested during the hot season by gnats, gadflies, mosquitoes, tarantulas, scorpions, and water-snakes which prey on the fishes of the streams.

Zungaria—The Great Wall.

The term Zungaria, unknown to the Chinese, derives from the Zungars, a branch of the Kalmuks, or Western Mongolians, who suddenly acquired great power early in the eighteenth century. Their empire stretched east and west from Hami to Lake Balkhash, and they were strong enough to invade Tibet and sack its capital in the year 1717. But after a chequered history of some sixty years they fell as rapidly as they had risen above the political horizon. Their overthrow by the Chinese in 1757 was attended by the most frightful massacres, in which the whole nation perished, leaving behind it nothing but the name which Western writers still continue to apply to the region at one time forming the centre of their power.

Zungaria, which is administratively connected with, but physically separated from, Kulja (Upper Ili valley), occupies the whole region between the Central Tian-shan and the Western Altai. It has no natural frontier towards Mongolia, with which it everywhere merges imperceptibly, and which it resembles in its main physical features. Towards the west it is not bounded so much as intersected by the Ektag-Altai, the Tarbagatai, and the Alatau, which with their eastern extensions run rather east and west than north and south. Thanks to this disposition of the ranges between the Altai and Tian-shan, the Central Asiatic tableland, elsewhere enclosed by continuous and mostly impassable mountain barriers, here opens through no less than three distinct depressions down to the Aralo-Caspian basin. Between the Ektag-Altai and the Tarbagatai lies the Upper or Black Irtish valley, continued right into Mongolia by the Urungu River, and nowhere rising more than 2500 feet above sea-level (Sosnovsky). But far deeper is the southern depression between the Tarbagatai and the Ala-tau, which is itself divided into two sections by the intermediate Barluk-Orkochuk ridge also running east and west. Between this ridge and the Saura, or eastern extension of the Tarbagatai, runs the second approach, which passes by the town of Chuguchak, and which, although less open, is more frequented than the others. Lastly, the third and southernmost passage is clearly marked by the Ayar-nor, Ebi-nor, and the undecided steppe rivers, all formerly presenting a continuous waterway communicating eastwards with the Central Asiatic mediterranean (Gobi), and connected westwards through Lakes Ala, Sassik, and others, with Lake Balkhash—that is, with the Aralo-Caspian basin.

The physical complexity is reflected in the ethnical confusion especially of the Ili valley, which has been the common battle-ground of rival races and conflicting creeds for ages. Kulja, as the Upper Ili valley is now called, is naturally by far the richest land in the empire beyond the limits of China proper, and has at times supported vast populations dwelling in numerous large cities and thriving towns scattered over its fertile and highly-cultivated plains. But the frequent revolts, first of Zungars, then of Dungan and Taranchi Muhammadans, in which momentary success on either side was invariably followed by wholesale extermination, have in recent times converted these magnificent lands into a howling wilderness. The victims of the successive Zungarian and Dungan insurrections, extending over more than a century, must be reckoned literally by millions, and the scene of desolation now presented by the ruined cities and wasted plains of unhappy Kulja baffles all description.

Kulja, which was temporarily occupied by Russia from 1871 to 1880, forms a triangular space some 26,000 square miles in extent, wedged into the very heart of the Central Tian-shan, and opening down the Ili valley towards Semirechinsk and Lake Balkhash. Its population has been reduced from over 1,000,000 to little more than 100,000 in 1880, and in the whole of Zungaria, with an area, including Kulja, of 146,000 square miles, there are less than 500,000 inhabitants.

Recently a great part of extra-mural China has been merged in a vast administrative division, which, under the designation of Sin-Kiang, includes the whole of Kashgaria, Ili, Zungaria, Outer Kansu, and the highlands extending from Urumtsi eastwards to and beyond Hami. Hung-Miotza, that is, Urumtsi, has been chosen as the seat of government, and efforts are being made to introduce an efficient military organisation with a view to future aggression on the part of Russia. At Urumtsi "the Chinese have concentrated their chief military strength, and here they are building a new city, and what they think to be an impregnable walled town; unfortunately they do not possess the military knowledge to see that they are occupying an indefensible site. The merchant, however, sees in the new Sin-Kiang province a territory fairly rich in the precious metals, producing cereals in abundance, besides cotton and silk, of a healthy and enjoyable climate, rich in all fruits—grapes, figs, apricots, melons, etc., and vegetables—well watered by cool streams, and supporting over two millions of a hardy race of excellent physique; affording two secure lines of communication, both cart-roads—the Pe-lu and Nan-lu

—the natural and only lines of conquest and of commerce leading from the west to the east, from Russian Central Asia to the central and south-western provinces of China. True it is that cultivation is almost limited to the narrow strips of rich oases, excepting in the extreme west (Kashgaria, Yarkand, Khotan); but they suffice to meet all military requirements of good communications, can be stocked with supplies of food for man and beast, and pass through large towns capable of supplying all the refitment requirements of an army. Its peoples are easily ruled, are not fanatical, and will, with assured peace, increase rapidly in numbers, and constitute a rich store whenever emigration may be directed into neighbouring and under-populated regions—Kansu, for instance. It is certainly a possession not to be despised, one whose inhabitants have never had an independent past, but whose possible dependent future must surpass its most palmy days of old under Chenghiz Khan, when it attained its greatest degree of prosperity" (Colonel Mark S. Bell, *Geo. Proc.*, 1890, p. 91).

Mongolia.

The swampy Lob-nor district offers little interruption to the sweep of sandy wastes which stretch continuously across 40 degrees of the meridian from below the cities of Kashgar and Yarkand eastwards to the Great Khingan range. The western section of this inhospitable wilderness as far as Lake Lob takes the name of the Takla-Makan desert; the eastern, thence to Manchuria, that of the Great Gobi or Shamo desert. But the whole forms essentially one geographic unit, and is by some writers spoken of simply as the Eastern and Western Gobi, or even as Eastern and Western Mongolia. For while the whole of this region forms on the one hand the true primeval home of the great Mongolian branch of the human family, it is, on the other, often difficult, at times even impossible, to say where Mongolia begins and Gobi ends.

Taken thus in its widest sense, Mongolia comprises the whole northern section of the Central Asiatic plateau between the Kuen-lun and Altai mountain systems. Towards the west the Tian-shan projects midway between these ranges eastwards to the Urumtsi depression, thus dividing the western portion into a northern and southern region roughly indicated by the Chinese expressions Tian-shan Pe-lu, and Tian-shan Nan-lu—that is, the northern and southern Tian-shan routes. By the use of these terms the Chinese people showed from the earliest times a surprising appreciation of the disposition of the land in Western Mongolia. In their eyes the Tian-shan, itself mostly impassable, clearly indicated the routes to be followed in order to penetrate into the Western world. But whereas the Nan-lu led, so to say, to a cul-de-sac at the eastern foot of the Pamir, the Pe-lu gave direct access through more than one depression to the Aralo-Caspian basin. Hence the vast importance to China of the extreme north-western portion of Mongolia, now commonly but most inconveniently spoken of as Zungaria. For within this region are comprised all the natural openings which either through the Balkhash or the Irtish basins lead from Central to Western Asia.

From Zungaria Mongolia proper stretches south of the Nan-shan highlands, eastwards to the Khingan range and almost to the gates of Peking. North of the imperial capital the great commercial route is soon reached, which crosses the Gobi desert and Mongolia to the Siberian frontier town of Kiakhta. A two days' trip from Peking towards the Great Wall brings the traveller to the "fortified city" of Chang-piu-chao, which on a closer inspection proves to be a mere village surrounded by mud walls. A few hours beyond it lie the five mighty gates of the valley of the imperial tombs. In this sandy plain, enclosed by an amphitheatre of lofty mountains, stand the colossal tombs of thirteen Chinese emperors disposed in crescent form at the foot of the wooded hills. Farther on lies the wild and frowning gorge of Nang - kao, through which formerly flowed a rushing torrent, its narrow bed here confined between steep rocky banks. An interminable line of massive walls, flanked at intervals by turrets and battlements, is carried over the crests of the craggy heights, following snake-like all their sinuosities as far as the eye can reach. At the first glance it becomes evident that this is the Great Wall of China, and after penetrating farther into the rugged valley we perceive two parallel lines running close together over the summits of the rocky hills, and sharply defining their outlines against the horizon.

A little farther on rises the barrier of ramparts separating China from Mongolia. The buttresses and apertures of the bastions are somewhat out of repair, but at this point little trace of decay or damage can be detected in the Great Wall, which rises suddenly to the right and left, broken at regular intervals by square towers, and, like a huge snake turned to stone, winding away over the summits of the highest ranges. Repeatedly repaired, rebuilt, and even altered in its general direction, little if any of Shi Hoang-ti's original structure now remains. But such as it is, with all its windings and the double and triple lines erected at certain points, it has a total length of 2000 miles, or one-twelfth of the circumference of the globe. Near Lan-chau in Kansu, where it was crossed by Colonel Bell in 1887, the Great Wall is merely a mud bank "six to eight feet high, and but a few feet thick; it is often wanting." Captain

Younghusband also, who crossed it at two points near the other end in the same year, aptly remarks that the inner branch north of Peking, "under the eyes of

THE GREAT WALL.

the Emperor, is a magnificent structure built of immense blocks of granite. It is some 40 or 50 feet in height, and wide enough at the top to drive two carriages abreast on, winding up and down the steep hill-side, over the summits and across the valley far away into the distance, and the credulous European tourist who comes out here to see it, imagines that it extends thus for hundreds and even thousands of miles. But where I passed through it next, scarcely one hundred miles from Peking, it had dwindled down to a miserable mud wall, not 20 feet in height, of no thickness, and with gaps in it often from a quarter to half a mile in width. The gateway there was very typical of a feature of the Oriental character. There were massive doors, and a lofty gateway, two guns pointing down the road, and a detachment of soldiers to collect customs duties, while twenty yards to the right was a gap in the wall wide enough for a brigade in line to pass through" (Capt. Younghusband, *Geo. Proc.*, 1888, p. 489).

The Gobi and West Mongolia.

West Mongolia proper, comprising the lowest plateau between the Altai and Tian-shan and the eastern section of the region between the Tian-shan and Kuen-lun, has a mean elevation of probably not more than 2000 feet. Farther east the waterless and treeless plains of Gobi stretch from the Tola, a head-stream of the Selenga, south-eastwards to the Darkhanola range, which rises to an elevation of 5000 feet. So far the land does not yet assume the aspect of a true desert, for the hill-sides are still overgrown with scrub from 2 to 3 feet high, and the plains covered with grassy tracts supporting numerous herds of cattle. But here begins the extensive depression which reaches to the Mandal Pass, 3700 feet high. At the Olong Baishing ruins the land falls to a still lower level, and here is seen the so-called "Rocky Girdle," a natural rampart of syenite stretching in a straight line east and west, and forming a clear landmark between North and Central Mongolia. South of this line begins the true desert of Gobi—the Shamo of the Chinese—

the lowest points of which are found at Ergi, Ude, Durma, and Shabadurghuma. The higher grounds are in some places strewn with rubble and blocks of porphyry and jasper, besides chalcedony and carnelian, interspersed with saline plants. The depression itself consists not so much of drift sand as of a sandy soil charged with alkalies, evidently the bed of a former marine basin, where still flourish the Arundinaceæ and nearly all the species common to the Caspian Sea. South of Durma the land again rises to the level of the shores of this dried-up mediterranean, attaining at Tsagan-Balgasu an elevation of 4550 feet, a height corresponding exactly with that of the northern edge of the basin at Urga. The plateau attains its greatest elevation towards the east, where it is cut off from Manchuria and the plains of Pe-chi-li by the intervening Khingan range and the highlands, stretching thence south-westwards to the In-shan mountains.

The region stretching north of Ulia-sutai to the Kobdo plateau was explored by Mattussovski in 1870.

This traveller also visited Lake Ike Aral, one of the largest in West Mongolia. Here he ascertained that Lake Kirghiz in the north-east of Kobdo, although of small size, forms nevertheless the centre of the water system in this region, receiving the overflow of all the surrounding lakes and rivers.

Our knowledge of Mongolia has also been greatly enlarged by the remarkable journey undertaken in 1872 by the English traveller, Ney Elias, from Peking westwards to the Russian Altai. Beyond Kalgan he reached the Belgian missionary station of Si-yun-tse, where wheat, oats, millet, and especially the poppy, are cultivated. The poppy seemed to be the chief inducement for the Chinese to settle here, but no reliable data could be procured respecting the opium trade, which, notwithstanding

PLAINS OF MONGOLIA

the high duty, is said to form the most lucrative business in Mongolia. The route to Kwei-hwa-chang lies for over 140 miles across a somewhat hilly pasture-land, and about 40 miles farther on a pass 5900 feet high leads down to a valley whose soil consists of a brown-yellow loess, intersected by numerous clefts and fissures, often 30 feet deep. Beyond the hills these crevasses even serve as regular dwellings for the people. Kwei-hwa-chang consists of two towns, and enjoys an extensive trade in tea, flour, millet, and the wares in demand amongst the Mongolians. From this point the traveller visited Hokow on the Hoang-ho, a small but busy place near extensive beds of a hard, slaty coal.

From Kwei-hwa-chang two routes—a government road and a caravan track—lead to Ulia-sutai. But the Mongolian steppe presents little variety for the traveller either way. The general aspect of the desert consists of low hills with intervening valleys and plains, rather stony than sandy, here and there intersected by low, rocky ridges, and mostly destitute of vegetation. The best water is found near the hills, where it is always sweet, while that of the plains is often brackish.

On 8th October Elias reached the River Onghin, which, after a south-easterly course of about 100 miles, loses itself in the desert. Proceeding westwards along the slopes of the rugged red and gray granite Kangai hills, he reached the Tui and the Baitarik, the largest of the Kangai rivers. Here the country is wild and barren, although frequented by wild asses and ponies in herds of from twenty to thirty each. On 25th October he camped on the left bank of the Chagan-tokoi, which flows south-west and west parallel with the Sirke range. This range forms an important geographical feature of the land, some of its crests rising from 3000 to 4300 feet above the general level of the surrounding plains. To the north-west lie the hills whence flow the Ulia-sutai and the Buyanta, and which are crossed by a pass 7450 feet high. From Ulia-sutai the traveller made his way to Kobdo by the River Yabkan and Lake Ike-Aral. A pass over 9000 feet high leads from Kobdo to the Chinese frontier town of Suok, whence a second high but easy col in the Altai brings to the River Chu and the town of Biisk.

In 1886 the Gobi was crossed along a new route by the Russian explorer, G. N. Potanin, who started from Goltai and followed the course of the River Eszin, which flows across the desert, and in summer sends down a considerable volume of water to the lacustrine depressions. Along its upper and middle course the Eszin is flanked on both sides by long ranges of moderately elevated chalk and sandstone hills, which in many places are covered by the blown sands of the desert.

Farther down the Eszin ramifies into two branches, one of which trends eastwards and expands in the now nearly dried-up Lake Sugu-nor. The other arm takes a westerly course to the great salt lake Gashun-nor, which lies in one of the most desolate parts of the wilderness, where no grass or fresh water is anywhere to be found for a distance of fifty miles. After exploring the western shore of the Gashun-nor, M. Potanin struck northwards, crossing the eastern spurs of the Altai system, which here develops four parallel ranges running in the direction from west to east, and enclosing broad valleys traversed by watercourses which are flushed after heavy rains. The northernmost of these ranges appears to culminate in the snow-capped Ichi-Bogdo, beyond which the traveller made his way by Lake Orok-nor up the Tui valley to the Kuljussai Pass and thence down to Lake Ugli-nor near the main highway between Ulia-sutai and Urga. Four other important caravan routes were crossed in the region between the Eszin River and Lake Ugli-nor.

During their three years' wandering in the Gobi (1884-86) M. Potanin and his associates surveyed several thousand square miles of hitherto unknown land, determined numerous positions by astronomic observations, and made vast natural history collections, including over 1500 botanical specimens and some 15,000 insects.

South-East Mongolia.

For much valuable information regarding South-East Mongolia we are indebted to Col. Prjevalsky, who visited the Ordos and Ala-shan regions on his journey to Kukunor in 1871. Proceeding in a south-easterly direction from Kalgan, this intrepid explorer came upon the Inshan range, skirting the northern bend of the Hoang-ho. Even before leaving the Kiakhta caravan route, a change is perceptible in the aspect of the country. The hills become higher and more craggy, while grass becomes more scanty, the pasturages being succeeded still farther west by extensive waterless valleys, where the nomads are entirely dependent on the wells dug at intervals along the route. The highest ranges are the Shara-Hada and Suma-Hada, wild and rugged uplands, where the traveller discovered the wild Ovis Argali in flocks of as many as fifteen together. Farther on the Muni-ula range, over 7000 feet high, forms with the Hoang-ho a well-defined landmark in the distribution of birds and mammalia. From these mountains the city of Bautu was reached, a large and busy but dirty place on the Hoang-ho, near the hill where the wife of Jenghiz-Khan is supposed to lie buried.

From Bautu the route lay across the Bagakhatun, southernmost and largest branch of the Hoang-ho, to the Ordos country, where the population is entirely confined to the Hoang-ho valley for about 70 miles west of

Bautu. On the left bank of the river lies the Ala-shan region, mostly a dreary lifeless waste of shifting sands, destitute alike of vegetation, birds, and mammals. The small tracts, where the sand is mingled with the loam and alkalies, produce a scanty but peculiar vegetable growth. The Ala-shan range rises some twelve miles to the west of Din-yuang-ing, capital of the province and residence of a native "van" or prince. The range, about 140 or 150 miles long, rises everywhere abruptly above the Hoang-ho valley, and presents a decidedly alpine character, culminating southwards with Mount Bayan-Tsumbur, 10,600 feet high.

Farther north lies the domain of the Urutes, occupying all the country between the Ordos and the territory of the Chakhar and Khalkha Mongolians in Ala-shan. Here the land is undulating and even hilly, rising steadily to a height of 5900 feet, or 2300 above the Ala-shan plains and 2500 above the Hoang-ho valley.

It was during this journey that Prjevalsky witnessed the somewhat rare spectacle of a steppe fire near Lake Dalai, north of Kalgan. "Towards evening a small light was visible on the horizon, which in the course of two or three hours became a long line of fire advancing rapidly across the open plain. A solitary hill in the centre was soon enveloped in flames, and appeared like a great building burning above the rest. The lake resounded with the loud cries of startled birds, while all was still and quiet on the plain."[1]

On the same occasion this traveller paid a visit to the famous temple of Bathar Sheilun, in the Sirung Bulik mountains, a little north of Bautu. It is "picturesquely situated in the midst of wild rocky scenery, and regarded as one of the most important in South-East Mongolia. The gorgeous shrine is four stories high, and surrounded

[1] *Mongolia*, i. p. 108.

by a cluster of houses inhabited by 2000 lamas, whose numbers are increased to 7000 in summer by the pilgrims who visit the temple, many coming from great distances. We ourselves saw near Lake Dalai a Mogul prince on his way to pray here. He had a large quantity of goods and chattels, and was followed by a train of several hundred sheep to supply him with provisions on the road."[1]

Manchuria.

The Great Wall is continued westwards across the northern bend of the Hoang-ho to the neighbourhood of Su-chau about 39° 30′ N., 99° E., and eastwards round to the Gulf of Liao-tung, thus completely enclosing China proper and part of Tibet from Mongolia and Manchuria. The hilly region of Manchuria stretches from Northern China northwards to the Amur, and from the Great Khingan range eastwards to Korea and the Usuri River.

Manchuria, that is, the Land of the Manchus, usually bears the Chinese name of Tung-san-sheng, the "Three Eastern Provinces," these being Liao-tung or Feng-tien in the south, washed by the Yellow Sea; Kirin in the centre, and Helung-kiang, "Black Dragon River" (the Amur), in the north. The whole region has an area of about 380,000 square miles, with a population approximately estimated at twenty-two millions, distributed over the three provinces as under:—

	Population.	Capital.
Liao-tung	12,000,000	Mukden
Kirin	8,000,000	Kirin
Helung-kiang	2,000,000	Tsitsihar
Total	22,000,000	

[1] *Mongolia*, i. p. 155.

Although the administration is essentially of a military character, all the chief appointments being held by Manchu military officers, the country has long since lost

MANCHU SOLDIER.

its distinctive nationality, and may already be regarded as practically an integral part of China proper. The southern province always has been essentially Chinese; Kirin is being rapidly assimilated, and even in the north the Manchu race has been absorbed in all the settled districts. The surface is mainly mountainous, or at least hilly, with volcanic chains ranging from 3000 to 6000 feet of mean elevation running in all directions without any clearly-defined orographic system. The only extensive level tracts are the fertile alluvial basin of the Liao

CHINESE FARM ON THE AMUR.

River flowing south to the Yellow Sea, and that of the Nonni affluent of the Sungari, which drains a great part of the gently-rolling northern steppe lands.

Notwithstanding its proximity to the Pacific Ocean, the climate of Manchuria is distinctly continental, with a great range of temperature from $-48°$ F. in winter in the northern districts to 87° or even 90° in the shade in summer. But the cold season is bracing and healthy, and the summer delightful, except in the marshy tracts subject to the horrible plague of gadflies, mosquitoes, and other winged pests. The soil is highly productive, yielding most of the crops grown in China, besides a great variety of oleaginous beans, the oil extracted from which forms the staple export of Manchuria. The forests abound in useful timber, such as the pine, walnut, oak, and elm, besides the highly-prized ginseng, fetching from £10 to £20 an ounce in the Chinese market. The country abounds in minerals, although mining operations have hitherto been restricted to mere washing of auriferous sands. "In one spot we found iron and gold within a few miles of one another, and we were told that there was also a silver mine close by. There is also abundance of very good coal and peat. A good deal of gold is exported, but mining is strictly contrary to the law, and the day before we arrived at Sansing [on the Sungari] a man was executed for it" (H. E. M. James, *Geo. Proc.*, 1887, p. 537).

The Sungari, which rivals the Amur itself in the volume of its waters, is the main artery of Chinese Manchuria. In 1886 Captain Younghusband and Mr. James ascended this river to its source in the Chang-pei-Shan Mountains, which they found to be only 8000 feet high, instead of 12,000, or even 15,000, as was formerly supposed. Manchuria enjoys a healthy climate, with a fertile soil and great mineral wealth, so that it is quite capable of receiving the superabundant populations of North China. These industrious agricultural colonists have gradually migrated in such numbers to the Sungari valley, that the aboriginal Manchu tribes now form the minority of the population. Since 1864 the Rev. A. Williamson has made several important expeditions into this region, which has also been traversed in recent times by the Russian Archimandrite Palladius, by Captain Younghusband, Mr. James, the Rev. J. A. Wylie, and others.

At Sang-Sing the Sungari is joined from the right by the Hurka (Khurkha), whose banks are thickly peopled by Chinese settlers. From Ninguta on this river a road leads over the ridge separating the Hurka from the Siufun, a small coast stream flowing to Victoria (Peter the Great) Bay near Vladivostok.

Russian steamers now occasionally ply on the Sungari, and have even penetrated to Tsitsihar on the Nonni. They have also entered the Hurka, though this river is so shallow and rocky that Ninguta can be reached only in small boats. The frontier towards Korea is continued from the Shan-yan-alin range by a narrow strip of neutral and uninhabited territory southwards to the Yellow Sea. But the so-called wooden palisade traced on the maps from the eastern extremity of the Great Wall to the Upper Sungari has long ceased to exist, at least as a distinct boundary line. It never could have possessed any strategical importance, being quite incapable of defence, nor is it possible any longer to make out its general direction from the few straggling stumps of trees, which are now all that survives of the original stockade.

Captain Younghusband speaks in glowing language of Manchuria, "a noble country well worthy of being the birthplace of the successive dynasties which, issuing from it, have conquered all the countries round, and of that dynasty which to-day holds sway over the most populous empire in the world. The fertility of the soil is extraordinary; the plain country is richly cultivated, and dotted over with flourishing villages and thriving market towns, and the hills are covered with magnificent forests of oak and elm. The mineral resources are at present undeveloped, but coal and iron, gold and silver, are known to be procurable. The climate is healthy and invigorating, but very cold in winter, when the temperature varies from 10° below zero Fahr. in the south to 40° or more below zero in the north. . . . Every year thousands of colonists from the northern provinces of China are flocking into these districts, opening up new tracts and building large thriving villages and towns. Unfortunately brigandage is very rife in Northern Manchuria; nearly every one carries arms of some sort, and one never sees small hamlets or detached farmhouses, because the people have, for their own protection, to collect together in large villages and towns. And the Chinese must be careful lest a neighbouring power, actuated of course by the purest of motives in the interest of the advance of civilisation, does not take upon itself to stop this brigandage. With regard to the population of Manchuria, perhaps the most noticeable point is the paucity of Manchus inhabiting the country. The original inhabitants seem to have gradually drained off to China proper, and their places are now being taken by emigrants from the provinces of Shantung and Chihli" (*Geo. Proc.*, 1888, p. 487).

Western Manchuria has been far less explored than the other portions of the province. Here the Khingan range is crossed by passes 3800 feet high, leading to the Mongolian plateau. Nor do these mountains everywhere form an effective barrier between the two countries; for they have long been invaded by Mongolian tribes, which have encroached far beyond this natural barrier into South-West Manchuria. At the same time, both Mongolians and Manchus are being gradually absorbed or displaced by the Chinese immigrants, so that the whole of Manchuria threatens soon to become ethnically an integral part of China proper. The Khingan range must then resume its position as the natural frontier towards Mongolia.

For much information on this borderland between the two countries we are indebted to the Russian astronomer Fritsche, who travelled in 1873 from Peking to the Russian frontier station of Staro-Zurukhaituyevsk on the Argun.

In eighteen days he reached Bei-lei-gu, beyond which point the Chinese begins to merge in the Mongolian population.

Soon after leaving He-shui, he reached the Shara-muren, here a turbid stream flowing between sandy banks about 1500 feet above sea-level.

About 47° 7′ N. and 118° E., crossing the Cholotu-davan at an altitude of 4000 feet, he reached the west side of the low border range, which here runs south-west and north-east in a rolling steppe, 3000 feet above the sea, but gradually sloping down to the River Argun at a level of 2000 feet.

Apart from the thinly-scattered "Yurtas," and the Chinese city of Khailar, along the whole route through the lands of the Barin, Ude-Michin, Khalka, and Solon Mongolians, between 43° and 50° N., the explorer passed only seven Lama monasteries, which in some parts of Mongolia seem to take the place of towns. The Mongolians of this region are governed by their own princes, the head-governor alone of the above-mentioned vassal-lands being a Manchu appointed from Peking and residing in Khailar. Here the Chinese traders are numerous, from this centre distributing their tea, tobacco, bread, saddles, yurts, and other wares, at little isolated stations scattered over the steppe.

Korea.

The peninsula of Korea, projecting southwards between North China and Japan, must in some respects be regarded as an independent section of the Asiatic mainland. It stretches in a south-westerly direction from 42° 31′ to 34° 40′ N., and from 125° to 129° E., between the Yellow Sea and Sea of Japan west and east; while the northern frontier towards China and Russian Manchuria is marked by the course of the Yalu and Tuman rivers. With the numerous islets on the south and south-west coast, its superficial area is estimated at about 82,000 square miles, and most of this area is of a distinctly highland character. The surface rises continually eastwards, attaining in the east coast ranges an altitude of from 7000 to 8000 feet. Since 1882 several expeditions have been made to the interior, which consists mainly of wooded hills, with intervening fertile valleys, and in the north vast lava fields, in some places 40 miles long and 100 to 140 feet deep.

The ordinary native name of the country is Tsyo Syeun, with an alternative Keirin or Korai, whence the current form Korea. Politically it constitutes an autonomous hereditary monarchy, divided into eight "tao," or provinces. But of all modern States it had till recently maintained the most exclusive isolation. The barriers of seclusion, however, were at last removed in the year 1882, when a treaty of amity and commerce was made with England, and since then with the United States, Germany, and other Western powers. The three ports of Fu-san, Chemulpho (Jinsen), and Won-san (Jen-san) are now open to the trade of the world, and much information regarding the resources of the country has already been collected by Carles, Douthwaite, Cavendish, Goold-Adams, and other recent travellers.

Above the bleak northern highlands towers the mighty Peh-tan-shen (Paik-tu-san), or "White-crested Mountain," forming a conspicuous landmark towards Manchuria, whose height Chinese geographers have estimated at 20 li, or about 7 miles! But this tremendous altitude was reduced to less than 9000 feet by Mr. James and his party, who first ascended the White Mountain in 1886, and more accurately fixed at 8900 feet by Captain Goold-Adams, who repeated the exploit in 1891. The White Mountain is a long-extinct volcano, whose crater is now flooded by a charming little lake six or seven miles in circumference, the "Dragon Prince's Pool" of the Manchurians, who hold the spot in the highest veneration as the reputed cradle of their race, or rather of the royal tribe which at present rules China.

The mountain, which lies about the water-parting of three distinct river basins, rises above a romantic region where the lower slopes, clothed with birch and pine forests, are succeeded higher up by "rich open meadows, bright with flowers of every imaginable colour, where sheets of blue iris, great scarlet tiger-lilies, sweet-scented yellow day-lilies, huge orange buttercups or purple monkshood delighted the eye; and beyond were bits of park-like country, with groups of spruce and fir beautifully dotted about, and spangled with great masses of deep-blue gentian, columbines of every shade of mauve or buff, orchids white and red, and many other flowers" (H. E. M. James, *The Long White Mountain*). From the western and eastern slopes flow the Yalu and Tuman, the former to the Bay of Korea, a northern extension of the Yellow Sea, the latter north-east to the Sea of Japan, while several torrents descend northwards to the Sungari.

Korea is said in official language to have thirty-three

cities of the first, twenty-eight of the second, and seventy of the third rank. But the only large centre of population is the capital, Seul or Soul (pronounced Sowl by foreigners, but So-ŭl by the natives), called also Han-yang from the neighbouring river Han, which is the chief watercourse of the interior. After a course of about 150 miles, the Han, which is formed by the junction of the So from the north and of the Yo from the south, develops below the capital an extensive delta in the Yellow Sea, interspersed with many rocks and wooded islets, collectively known as the Prince Imperial Archipelago.

Soul, which contains over 30,000 houses, with a total population (1893) of about 250,000, covers an area within the walls of ten miles. "It is as much the heart of Korea as Paris is the heart of France. It is the object of every Korean gentleman to live in the capital, for there every pleasure and vice is more easy of attainment, the chances of getting favourite posts by judicious flattering and canvassing of superiors are multiplied, while the finest and best of native and foreign produce is to be procured. The darkest side of the picture lies in the crowded collection of hovels, swarming with human (and insect) life, absolutely devoid of even elementary sanitation; where the use of soap and water is confined to a few of the highest classes; where disease and vice have lived in close partnership for several hundred years; where dishonesty and oppression are carried to their utmost limits; where torture and cruelty exercise full sway, and where private and political intrigue hinders any improvement in the condition of the great bulk of the community" (Captain A. E. J. Cavendish, *Korea,* 1894, p. 24).

Of the three treaty ports, Chemulpho lies on the west coast near the capital, with which it communicates

A GATE OF SOUL.

by the River Han. Like many other places, it bears three names, the Korean (Chemulpho), the Chinese (Jen-chuan), and the Japanese (Jinsen), and this superabundant terminology is a source of great confusion in the geographical nomenclature of Korea. At the port of Chemulpho the tides rise no less than 36 feet, and at ebb "very little water remains in the harbour, only a narrow channel in the middle, just wide enough for a steamer to swing in, while the rest of the harbour becomes a vast expanse of mud flat broken in one place by a small rocky island" (Cavendish, p. 16).

Fu-san and Won-san, the other two treaty ports, are situated, the former at the south-east point of the peninsula, the latter about the middle of the east coast, just below Port Lazaref on Broughton Bay. Won-san, a place of about 15,000 inhabitants, is reached by a rough overland route from the capital, occupying in fair weather from six to eight days, and is connected, like Fu-san, by a regular service of steamers with the Japanese port of Nagasaki. A cable has also been laid between Nagasaki and Fu-san, and a telegraph from Won-san across the peninsula to Chemulpho. "It is proposed to extend the Soul line to Ham-heung and Vladivostock, and we saw the poles lying in heaps by the roadside, but many of them have been removed for firewood, and other heaps have been set on fire through wanton mischief. The telegraph at Won-san is worked by a Korean, the office is in the Yamen (court), and the English language is employed, as it is at present impossible to signal Chinese symbols" (Cavendish, p. 92). Won-san is the natural outlet for the produce of the eastern, as Chemulpho is of the western provinces, and here the traveller bound for the interior may now procure all manner of native and foreign wares.

Notwithstanding its great natural resources, the

country is generally described as wretchedly poor, trade and agriculture in a very primitive state, the people of rude and simple manners. A census taken in the year 1793 gave a population of over 7,340,000, but the estimate for 1894 is about 14,000,000, of whom a million are reckoned as fighting men. The broad western valleys sloping seawards seem to be thickly peopled, the east side far less so, and the north very sparingly. Towards the Chinese frontier an artificial wilderness has been created by the Government as a protection against the warlike Manchurians, and for this purpose four large cities, besides many villages, are said to have been razed to the ground. Hence the broad zone of neutral and uninhabited land traced on the maps between Korea and Shing-king (Liao-tung).

Of the products one of the most useful is hemp, of which several varieties are cultivated and manufactured into a strong coarse material for the dress of the lower classes. Many districts are extremely fertile, and all the river valleys, both below and above the capital, are extensively cultivated. Here the chief crops are rice of the wet and dry varieties, wheat, barley, millet, beans, coarse tobacco growing 7 or 8 feet high, while the little garden plots of chillies, cabbages, and turnips are bordered by hedges of the castor-oil plant. "Besides flax, maize, and cotton, here were fields of the small millet (*Setaria italica*), a substitute for porridge, and of the tall millet, Susu or Kaoling (*Holcus sorghum*), with stems 8 to 12 feet high, and as thick as a man's thumb, turning to a golden yellow or bright mahogany colour,—from the latter kind the Koreans make the coarse cloth of which their rough garments are composed, when they do not use Manchester shirtings,—also fields of beans, food for cattle and men, and the foundation of Japanese soy, and of our Worcester sauce. Many of the houses had their roofs covered with gourds, while here and there were patches of melons. In the villages, a little way off the road, we could see fruit-trees, pear, persimmon, but all the fruit except the Ham-heung pears was hard, dry, and tasteless" (Cavendish, p. 54).

Formerly there was little foreign trade, although much ginseng and paper, the only exported articles, were either smuggled across the Chinese frontier or brought to the fairs held at stated times along the border lands under the sanction of the authorities. These commodities were also brought to China in considerable quantities, in the suite of the embassy which proceeds every year to Peking. The Korean paper, made of cotton and the inner bark of a species of mulberry, is very strong, and, like that of Japan, applied to a great variety of purposes.

Since the country has been thrown open to foreign intercourse, its general trade has steadily increased, especially with England, Siberia, and Japan. The imports advanced from £610,000 in 1888 to £945,000 in 1890; the exports from £173,000 to £710,000, and the net revenue from Customs dues from £53,000 to £103,000 for the same years. Of this rising trade England takes by far the largest share, nearly 60 per cent of all the imports being of British origin, while those of Japan and China, Korea's two next best customers, are only about 20 and 12 per cent respectively. The exports comprise a far greater variety of wares than might be supposed, and already include, besides the above-mentioned articles, cereals, cow-hides, and bones, bear, tiger, and leopard skins, and great quantities of fish, salted, dried, and for manure.

"The seas round Korea swarm with fish, and near Won-san I was told that the water was at times literally stiff with the little fish used as manure by the Japanese, and which appears every two years; but the natives are so very lazy about sea-fishing, that they will go out for a few hours, make a good haul, and not go out again until the proceeds are spent, when very likely the shoals of fish have passed on. On the other hand, Japanese fishing-boats, which by a treaty have to pay for a licence to fish within the three-mile limit, make good profits, the average earnings per boat during twelve months being £100. The ming-tai, a kind of haddock, which is caught on the east coast north of Pukchong, and dried without being salted, is a very favourite food, Won-san exporting to Korean ports no less than £72,000 worth" (Cavendish, p. 47).

International relations have increased to such an extent during the last few years, that in 1891 as many as 7000 foreigners were resident in the three treaty ports; of these, however, the great bulk (6435) were Japanese, and not more than thirteen British subjects.

The Korean Archipelagoes.

The south and especially the west coast of Korea are fringed by innumerable insular groups, mostly of small size, and ranging in height from a few hundred to over 2000 feet. Partial surveys have been made of the groups; but the best general description still remains that of Captain Basil Hall, who navigated these waters in all directions early in the nineteenth century. "We threaded our way for upwards of a hundred miles amongst islands, which lie in immense clusters in every direction. At first we thought of counting them, and even attempted to note their places on the charts which we were making of this coast; but their great number completely baffled these endeavours. They vary in size from a few hundred yards in length to five or six miles, and are of all shapes.

"From the mast-head other groups were perceived lying one behind the other to the east and south as far as the eye could reach. Frequently above a hundred islands were in sight from deck at one moment. The sea being quite smooth, the weather fine, and many of the islands wooded and cultivated in the valleys, the scene was at all times lively, and was rendered still more interesting by our rapid passage along the coast, by which the appearances about us were perpetually changing.

"Of this coast we had no charts possessing the slightest pretensions to accuracy, none of the places at which we touched being laid down within sixty miles of their proper position. Only a few islands are noticed in any map, whereas the coast for nearly two hundred miles is completely studded with them, to the distance of fifteen or twenty leagues from the mainland. . . . Farther on we passed for a distance of five miles amongst islands, all except the very smallest inhabited. The villages are built in the valleys, where the houses are nearly hid by trees and hedges. The sides of the hills are cultivated with millet and a species of bean; and in the numerous small gardens near the villages we saw a great variety of plants.

"As the peaked island which we had undertaken to climb was steep, and covered with a long coarse grass, it cost us a tiresome scramble to gain the top, which was about 600 feet above the level of the sea. The mainland of Korea is just discernible in the north-east and east from this elevation. But it commands a splendid view of the islands, lying in thick clusters as far as the eye can reach, from north-west quite round by east to south. We endeavoured to count them. One person, by reckoning only such as were obviously separate islands, made their number one hundred and twenty. Two other gentlemen, by estimating the numbers in each connected cluster, made severally one hundred and thirty-six and one hundred and seventy, a difference which at once shows the difficulty of speaking with precision on this subject. But when it is considered that from one spot which, though considerably elevated, was not concentrical, one hundred and twenty islands could be counted, and that our course for upwards of one hundred miles had been amongst islands no less crowded than these, some idea may be formed of this great archipelago." [1]

Quelpart—Port Hamilton.

Apart and entirely distinct from these clusters is the relatively large Island of Quelpart, which occupies a commanding position at a point where the Eastern and Yellow Seas communicate through Korea Strait with the Sea of Japan. Both politically and geographically Quelpart (Quelpaert) belongs to Korea, from the south-west angle of which it is distant not more than 52 miles. It has an extreme length from east to west of 43 miles, with an average breadth of about 15 miles, a circuit of 114, an area of 750 square miles, and a population of 10,000, all Koreans occupied almost exclusively with agriculture. The fertile soil, apparently of volcanic origin, yields bountiful crops of wheat, barley, turnips, and other vegetables, and seems well adapted for viniculture.

Quelpart, as it was named by the Dutch by whom it was first sighted in the seventeenth century, is the Tsetsin of the Koreans, the Tang-lo of the Chinese, and the Tamura of the Japanese. English names have also been introduced to increase the confusion of this perplexing nomenclature, and the culminating peak Aula (Ra-hansan, Han-ka-san), which rises to a height of 6700 feet near the south coast, figures as "Mount Auckland" on the British Admiralty charts.

[1] Basil Hall, *Voyage to Korea*, etc., p. 42 *et seq.*

Grouped round the main island are a few rocky islets also bearing English names—Eden, Barlow, Gifford, Beaufort,—which collectively form a little archipelago disposed in the direction from south-west to north-east, and which appear to be connected geologically with the Japanese volcanic system. Auckland itself looks like the northern flank of an igneous cone, the other three sides of which have been submerged, and scattered over the hilly surface of Quelpart are the remains of several other more or less obliterated craters now densely wooded. Hence the diversified character of the picturesque scenery as viewed from the sea. The succession of thickly wooded heights, rich valleys, and groups of rustic habitations excited the admiration of La Perouse, who declared that it would be scarcely possible anywhere to find an island presenting a more delightful aspect.

Between Quelpart and the south coast of Korea lies the rocky island which the English surveyors have named Port Hamilton, and which was occupied in 1885 as a British naval station with the consent of the Imperial Government. The object of the occupation was to secure a strong position in these waters, in the event of Russia seizing any of the Korean seaports. But the station, described by some as valueless, by others as a well-sheltered harbour, was afterwards abandoned on the assurance of the Tsar's government that there was no intention of occupying any part of the peninsula. But after the outbreak of the war between China and Japan (1894) it was stated that the conditions were altered, and while Russia reserved to itself complete freedom of action, it would not tolerate the reoccupation of Port Hamilton by the English.

The peninsula abounds in minerals, such as gold, silver, iron, copper, lead, and coal. But the State reserves to itself the exclusive monopoly of the mines.

The presence of the precious metals has long been known, and the annual tribute formerly paid to China included 100 ounces of gold and 1000 ounces of silver, besides a gold casket ornamented with pearls and other gems. At present considerable quantities of gold are obtained by primitive processes from the diggings and washings of the Keum-ha-wön, Yeung-heung, and other auriferous districts, especially in the central provinces north and north-east of the capital. Silver and copper occur in close proximity on the Chang-jin plateau (3500 feet), east of the river of like name, south of the White Mountain, and systematic surveys will no doubt show that these and other minerals are widely diffused throughout the main range and its lateral spurs. Meantime considerably over £100,000 worth of gold is annually exported from the three treaty ports, though the amount fluctuates, having fallen from £157,000 in 1889 to £120,000 in 1890, and £110,000 in 1891. Nearly as much more is supposed to leave the country undeclared or smuggled across the borders. Coal is said to occur chiefly in the north, but it appears to be inferior to that of China and Japan.

The natives are quite as skilful as the Japanese in the working of metals, often betraying much artistic taste in the designs. But the more useful arts are in a very backward state. Navigation is conducted on the rivers with flat-bottomed boats, on the coast with small and crazy-looking junks.

Regarding the political institutions and the details of the administration, our knowledge has lately been much enlarged. The monarchy is known to be of an absolute type, modelled on that of China. But besides the royal family there are privileged classes and a hereditary aristocracy, an institution unknown in China. Amongst the nobles several parties have been formed, of which the

State is compelled to take account. Although the crown is hereditary, the succession often gives rise to contentions, in which the magnates play an important part.

This subject of social classes, hitherto so little understood, has been specially studied by Mr. C. W. Campbell of the British Consulate-General at Seul. It appears that although in theory Korean society is broadly divided into an upper, a middle, and a lower class, the social

GROUP OF KOREANS.

grades are as endless as the Hindu castes themselves, all being distinguished by certain honorific forms of speech, dictated by long-established usage, and insisted upon not merely as a matter of courtesy, but of right and privilege.

But the two most clearly recognised social groups are the *Nyang-pan*, or highest privileged class, and the *Ha-in*, literally the "low" or "inferior people." In the former are included the "upper ten thousand," the term being now applied even to the descendants of all high functionaries. Thus in Korea a sort of aristocracy of blood is in process of development, and ancestry has already become almost the chief factor in determining the highest rank. At one time, in fact, those descended from officials of the early kings of the present dynasty were alone regarded as genuine members of the Nyang-pan circle. Although others were later included, the old families have always looked on them as intruders, treating them with scant courtesy unless they are able to command respect by their personal wealth and influence.

Thus the highest members of this bureaucracy correspond somewhat to the peers of the realm in England, and like them enjoy very important privileges at court and before the law. They are free from arrest except by order of the king or of the provincial governor, and even then only for the gravest crimes, such as treason or extortion. Their autocratic power is such that they may resent at pleasure any real or fancied insults by the lower classes. On the other hand, they lose caste should they engage in any occupation except the public service and teaching, which, as in China, is held in high honour.

Such a system is necessarily liable to abuse, and one of the chief difficulties of the Korean administration arises from the attitude of the Nyang-pan towards the executive and the mass of the people. While the lack-land live from hand to mouth, the income of those possessing estates is mainly derived from rent paid in money or in kind, and there can be no doubt that in exacting this income much injustice and cruelty is inflicted on the rural classes, that is, on the great bulk of the Korean population. Hence the constant troubles, the disorder and revolts against intolerable oppression which have recently led to the active intervention of Japan, whose obvious interest it is that an efficient administration should be established and maintained in the peninsula, if only to deprive the Russians of any pretext for interfering in the internal affairs of the kingdom.

But the Nyang-pan class is itself divided into factions, some of which are of long standing, and all of which tend to aggravate the general disorder. The four chief rival parties had their origin some three centuries ago in a dispute over some frivolous points of etiquette, and ever since that time each has in its turn enjoyed court favour through intrigue and deeds of violence. The first cause of these palace broils has long been forgotten; but the strife continues, fostered by hereditary hatreds of a virulent character. Even before his arrest by the Japanese, the present king, who although weak has the reputation of being possessed of some common sense and love of justice, never encouraged these aimless feuds, and endeavoured to treat all with perfect impartiality. Hence under this ruler all the four factions have enjoyed a share in the government of the State; but their mutual animosities continue unabated, and at the outbreak of hostilities between China and Japan over the "Korean question," in 1894, the extraordinary spectacle was witnessed of the native officials at daggers drawn, while the kingdom was made the battlefield of its powerful neighbours.

Nor was much patriotism shown by the Ha-in class, which probably comprises four-fifths of the whole nation. In it is included all the commercial element, trade being regarded as an ignoble pursuit. The prejudice, however, entertained by the aristocrats towards merchants and dealers has been considerably mitigated in Korea, as well as in Japan, since the country has been thrown open to foreign intercourse. Unfortunately the antiquated system of monopolies still flourishes, and it is stated that the trade of the capital and surrounding district in silk, cotton goods, hemp and grass cloth, native silk and paper, is entirely controlled by six powerful guilds with their headquarters in Seul. Both the producers and the retail traders are at the mercy of these guilds, which in their turn are "squeezed" by the government, and which are consequently regarded as a convenient means of collecting revenue.

The war between China and Japan is directly traceable to one of those periodical risings which occurred early in the year 1894, and which led to the occupation of the country by the two rival powers. Like all preceding outbreaks, the rising itself was entirely due to the hopeless maladministration, as was pointed out by an intelligent Korean at the time. "If you tread on a cuttle-fish," he remarked, "it will wriggle. The people of Korea have arrived at the point where even their patience has a limit. Official squeezing has become so merciless that productive farms are everywhere deserted, the oppressed farmers either emigrating to China or Russia, or joining the bands of robbers that obstruct what little traffic there is in the country. Let the oppression proceed a little farther, or a famine take place, and the people must rise to break off the iron fetters which drag heavier at every step."

On this occasion the insurrection assumed alarming proportions, and its movements were said to be controlled by the *Tong-Hak* or "Korean School," that is, the national party, which has lately been developed with surprising rapidity. But the growth of a healthy national party suits none of the neighbouring states, and the outbreak served as a convenient pretext for the joint occupation of the peninsula by China as the suzerain power, and by Japan in virtue of certain international conventions.[1] They came, however, with different in-

[1] More particularly the Treaty of 18th April 1885, which established the

tentions, and while the Japanese expressed their determination not only to restore order, but to establish a strong and just administration, the Chinese deprecated all interference in the internal affairs of the kingdom, and limited their action to the maintenance of the *status quo*. All concerted action was thus rendered impossible, and the usual diplomatic negotiations were followed by the Chino-Japanese war of 1894-95, which resulted in the expulsion of the Chinese and their final abandonment of all claims to the over-lordship of Korea. The country was thus left at the mercy of the Japanese, who, however, through fear of Russian opposition, have hitherto refrained from declaring a protectorate, and have even expressed their willingness to withdraw their forces as soon as order is restored and a strong government established.

King Li-Hi, who had already been arrested by the Japanese on 23rd July, ascended to the throne in 1864, and is the twenty-ninth in succession since the founding of the present dynasty in 1392. The Hon. G. Curzon, who was accorded an interview with the king in 1892, describes him as "a man of small stature and sallow complexion, with a thin black moustache and tuft below the chin. His countenance wears a singularly gentle and pleasing expression, indicative of much amiability of character; and many instances are related of his personal charm of disposition and bearing."

Officials are said to be mostly appointed, as in China, after a searching investigation, although they occasionally

right of Japan to send troops to Korea in case of emergency. The last article of that Treaty stipulates that "in case of any disturbance of a serious character occurring in Korea rendering it necessary for the respective countries (Japan and China), or either of them, to send troops to Korea, it is hereby understood that they shall give, each to the other, previous notice in writing of their intention so to do, and after the matter is settled they shall withdraw their troops immediately and not further station them there."

acquire office by purchase or the royal favour. The higher functionaries have almost unlimited control over the lives and property of those within their jurisdiction. The penal code is atrociously cruel, and the application of the bamboo of daily occurrence for the most trivial offences.

All are liable to military service, yet there is no standing army, and the bodyguard of the king are alone entitled to be regarded as soldiers in the ordinary sense of the term. Discipline and tactics are of course unknown; but the rural population, who are bound by a sort of villanage to the crown, are summoned at stated periods to the chief towns of the circles, where they serve as soldiers or armed police. Their weapons are the spear, bow and arrow, and matchlocks, although some really good firearms are manufactured in the capital. The guards wear helmets and breastplates, and in war long overcoats, so thickly padded with cotton as to be proof against sword-cuts and musket-shots, but not against the rifle. But this uniform is so heavy that it prevents all free and rapid movements of the troops. A large army of Koreans would be almost helpless in the presence of a well-appointed company of Europeans.

The State religion, like so many other social features, resembles that of China. Both Buddhism and the Lao-tse doctrines are widely spread amongst the people, while the upper classes rest satisfied with the colourless moral teachings of Confucius. The Korean language, probably intermediate between the Mongolo-Tatar and Japanese, is written with a true alphabet of twenty-seven letters, and of uncertain origin. But this alphabet is held in slight esteem, being employed chiefly by women and children, while all the lettered classes are familiar with the Chinese ideographic system. By this means the Chinese and Koreans, although speaking totally distinct languages, are able to communicate their thoughts in writing. Thus the sign for *man*, read off as *yen* by the former, and *saram* by the latter, will convey to both alike the same concept of *man*, just as all Europeans, however they may pronounce them, attach the same value to the Arabic ciphers.

China Proper.

Comprising most of the Hoang-ho and Yang-tse drainage, besides the Pei-ho, Kiang-si, and other smaller river basins, China proper occupies altogether rather more than one-third of the whole empire. But this smaller section is immeasurably the most important in respect of population, products, trade, industries, and material resources of every kind. To such an extent is this the case, that the loss of all the other great divisions would not appreciably diminish the status of China as one of the great powers of the world. Indeed it may be doubted whether its position would not be thereby strengthened. For ages the resources of the Central Government have been strained to the utmost in the endeavour to keep together an unwieldy and, to some extent, an incongruous political system, the vast frontier line of which it is almost impossible to defend at all points. Were this frontier line contracted to the still broad limits of China proper, the gain in greater concentration alone would probably more than balance the loss of the vast sandy wastes of Mongolia and the bleak upland Tibetan plateau.

Within its natural limits China presents the form of an irregular circle, the landward and seaward semicircles of which are about equal in extent. The inner curve sweeping round from the head of the Yellow Sea to the Gulf of Tonking, runs successively along the borders of the conterminous regions of Korea, Manchuria, Mongolia,

Tibet, Burma, Siam, and Annam. The outer curve or coast line follows the Pacific seaboard through its entire course along the Gulf of Tonking, the China Sea, Fukien Strait, the Eastern and Yellow Seas, the Gulfs of Pe-chi-li and Liao-tung, and Korea Bay.

Excluding the great islands of Hainan and Formosa, this compact mass of land stretches from 20° to 42° N. and from 98° to 122° E., or very nearly 1400 miles both ways, with a total area of about 1,556,000 square miles. From the Tibetan plateau the surface slopes uniformly eastwards, in which direction all the main streams drain to the Pacific. Of these the Yang-tse-kiang, traversing all the central provinces, and dividing the country into two nearly equal portions, is by far the most important artery of trade, and affords the easiest means of access to the interior. Hence this line has been followed by Cooper, Margary, Gill, M'Carthy, and most of the intrepid explorers who have in recent years traversed the land from the Pacific coast to Burma and British India. A good idea of its general features may be had by following in the footsteps of any of these travellers. One of the most instructive journeys, notwithstanding its tragic termination, was that undertaken in 1874 by the unfortunate Augustus Margary, who perished almost at the very goal of his ambition.

Lieutenant Margary followed the course of the Yang-tse to Hankow, whence he started in a native junk on 4th September 1874. In the province of Hu-pe, in recent times more than once wasted by marauders and the Taiping rebels, the level tracts along both sides of the river were planted with cotton and sesame. Here numerous rafts were met, and in Losfan an inquisitive and importunate crowd flocked round the stranger, who was the first European visitor to that place. Round about Losfan runs a crescent of dunes, beyond which stretches an interminable level plain, forming one of the great rice-fields of the empire.

On 20th September the expedition reached the island of Chun-shan at the entrance of the great Lake Tung-ting. Here the turbid Yang-tse was succeeded by the clear water of the shallow lagoon, across which a favourable breeze wafted the vessel to the River Yu-an. At Ni-hsien-tang, 36 miles from the mouth of this river, the hitherto bare or slimy banks gave place to fine grassy tracts and meadow-lands, carefully-cultivated fields mostly planted with cotton, stately residences, and a general appearance of prosperity. Landing at Chang-te, the traveller crossed over to Tao-yuan-hsien, a large and flourishing city, formerly a chief centre of the earthenware industry. From this town the pottery is sent in large quantities to Chen-chu-fu, which has of late years been the most unruly and turbulent place in the province. Beyond Tao-yuan-hsien a picturesque highland region was reached, where the crests and slopes of the hills were clothed with pines, aged oaks, and palms. The stream now flowed in a narrow bed between bold and sheer rocky walls, while higher up the channel was obstructed by ledges, forming dangerous rapids.

At Chen-chi-hsien the river describes a great curve, sweeping for 60 miles southwards and the same distance northwards, after which it again resumes its normal westerly course. On 27th October Margary reached Chen-yuan-fu, a city enclosed by grand rocky heights. Here ended the wearisome journey by water, during which he was so ill that he lay nearly the whole time prostrate on his couch. Nevertheless he at once resumed the land journey in a palanquin, and soon after recovered his wonted health. At Ching-ping-hsien coal from the neighbouring pits was seen exposed for sale. As the capital of the province was neared, the towns became larger and the villages more numerous, while the cultivated tracts, chiefly under rice and tobacco, covered continually wider spaces. Beyond Kwei-ting-hsien the route lay mostly through narrow gorges, the grassy hills approaching so near that little room was left for tillage. Yet the road was here lined on both sides by close quickset hedges. On 5th November the traveller entered Kwei-yang, capital of Kwei-chow, which lies in a rolling plain planted with trees and encircled by hills. These hills are mostly isolated and clothed with a rich vegetation to their summits, which consist of level black rocks mostly crowned with splendid temples. At the end of the highway, and facing the city, stand a number of remarkable white marble triumphal arches, erected to the memory of devoted women.

Beyond this place stretched spacious arable plains, now overgrown with tall grasses, but still betraying many signs of former cultivation. The valleys extended mostly in the direction from west to east, wild-flowers fringed the roadside, and in the hedges the wild tea-plant was in full bloom. Very remarkable are the numerous isolated hills which rise to an average height of about 300 feet, in the Kwei-yang plain, in the districts beyond Ching-chi-hsien and away to the south. Kwei-chow, usually described as an inaccessible highland region, Margary was agreeably surprised to find diversified with many smiling plains, while the route lay chiefly over grassy hills and tracts of rolling steppe all the way to Hang-tai.

Here he at last entered a really highland region, where the road was often extremely precipitous, winding its way over hills rising to a height of 4000 feet and upwards. In this neighbourhood the River Metou marks the limits of the domain of the wild Miao-tse and Chung-chia tribes. Farther west the land under tillage becomes more extensive, the villages grow more numerous, and a limited trade is done in oranges and other local produce. The climate of Yun-nan is colder than that of Kwei-chow, where the dwellings are not calculated to resist the effects of severe winters.

On 20th November the Yun-nan Pass was crossed, and the expedition reached the first frontier town of Yun-nan. Between Yun-nan-fu, capital of the province, and the Irawadi valley, North Burma, the natural difficulties of the route offer very serious obstacles to the local trade. The whole of Yun-nan is covered with mountains of an extremely rugged character. From the capital to Bhamo, in Burma, the track lies over ranges rising to 3500 or 4000 feet above the plain, which is itself about the same height above sea-level. The country is thinly peopled, and even the valleys are but partially cultivated. Here opium is produced in considerable quantities, one-third of all the crops apparently consisting of poppies. The natives are generally poor and wretchedly housed. The so-called "lekin," or stations, seem to be the most flourishing establishments. They are met everywhere, and are constantly used as residences, being usually the best and cleanest places in the towns.

As was the region traversed by Margary, so is China generally—a land almost everywhere presenting the most violent contrasts, both physical and social, amidst an all-prevailing and undefinable sense of monotonous uniformity. We feel that this is still the home of the children of Han, a land of wealth and want, exposed to the most violent vicissitudes in the midst of an eternal stability, devastated by sudden outbursts of nature's pent-up forces or by the still more terrible display of human passion, but ever rising with fresh youth from the desolation of its ruined cities and wasted plains.

This surprising vitality is due partly to the inexhaustible fertility of the soil, partly to the character of the people. Not only is the "Chinese Mesopotamia" the richest granary in the world, but its productiveness is enhanced by the ingenuity of a laborious and patient race. Wheat, barley, tobacco, pulse, are the chief products of the northern provinces, where cold winters are succeeded by hot summers. Cotton, the sugar-cane, pepper, betel, spices, tropical fruits, and especially rice, which is the staple of food, are mainly cultivated in the central provinces, where the heats are excessive and the cold seasons accompanied by storms and a heavy rainfall. The chief rice-growing tracts form the heart of the country, the seat of the densest population, and the focus of commercial life.

The western hills, or rather their western slopes, are one of the few regions where the valuable medicinal rhubarb plant is indigenous, and here also the opium-producing poppy is largely grown. The taste for this narcotic is said to have been unknown in China during former times, although doubtless some other drugs must have been used. During recent times the consumption has increased to a large extent among the inhabitants of whole provinces. One-fourth of the people are now said to be opium-smokers, and the cultivation of the plant is now monopolised by the Chinese Government, which even appears to encourage its growth. The home-grown article comes chiefly from Mongolia and North Manchuria, and although some more southern provinces have of late years begun to take part in its cultivation, still the local supply is insufficient for the demand. Large quantities are imported from India for the use of the upper classes. The Indian Government levies high taxes on the exportation to the amount of £6,000,000 to £7,000,000 annually (7,966,000 Rx. in 1893). This taxation has been much discussed lately, but it rests on the same foundation as the taxation levied by other Governments on spirits, and in principle it is defended as rendering dear an article which, if consumed temperately, may be harmless, but if consumed to excess will be injurious.

It is very remarkable that, notwithstanding the general deficiency of timber, the Chinese peasantry carefully clear the land of all its bush and forest growths. Although the great demand for wood and the want of more space for tillage may elsewhere account for this practice, such motives cannot explain the systematic burning of the hilly woodlands, more especially that the ground is not needed for stock-breeding, an occupation very little pursued in China. The Abbé David attributes this reckless destruction to the desire to get rid of wild beasts, such as the tiger and panther, which still infest many provinces. But whatever the cause, the consequences are most disastrous, leading to denudation of the surface soil in the uplands, and to constantly-increasing inundations in the lowlands. Unless the process be checked, it must ultimately reduce some of the richest lands to deserts, and bring about a state of things similar to that of Persia and parts of Central Italy, where the disappearance of forests on the uplands has been followed by the disappearance of agriculture on the lowlands.

A chief source of national wealth is the production of raw silk, which forms a staple of trade in a country where sericulture has from time immemorial been indigenous. At Shanghai silk ranks with tea and cotton as a staple export, while in Canton it takes the first place.

But next to agriculture the main resource of China lies in the ground itself, which harbours supplies of ores and coal sufficient some day to revolutionise the trade of the world. Although coal has long been used as an article of fuel, the attention of Europeans has but recently been directed to the vast coal-measures of the great river basins. Some of the more important mines had already been visited by Kingsmill, Pumpelly, David, and other intelligent explorers. But for a more accurate knowledge of China's carboniferous deposits we are indebted mainly to Baron von Richthofen. South of the crystalline Peling (Tsin-ling-shan), over 11,000 feet high, which with the Funiu-shan forms the eastern extension of the Kuen-lun between the Hoang-ho and Yang-tse valleys, the measures, although less important than on the northern side, are still far from inconsiderable. The superficial area of the vast coal-basin in Se-chuen alone is estimated by Richthofen at 100,000 square miles. This basin is enclosed on all sides by lofty ranges, and the deep channels excavated by the affluents of the Yang-tse everywhere show the fossil cropping out, while the working of the mines is facilitated by the streams themselves, which are navigable to the limits of the basin. Here the quality varies, the good bituminous beds of the north and west being elsewhere replaced by moderately good anthracites. Large deposits of excellent anthracite also occur in Yun-nan, here associated with rich copper, tin, zinc, and lead ores.

But the old and true carboniferous formations lie in the more central province of Hunan, which basin may be compared with the rich Pennsylvanian deposits. It has an area of 16,000 square miles, stretching along both banks of the Tse-kiang (Lo-kiang) as far as Siang-Tang, and contains excellent anthracite in the south, and bituminous coal in the north.

In the northern provinces, however, are found the most extensive and richest measures, spreading over 25 degrees of the meridian from the western deserts to the Yellow Sea. They belong mostly to the old carboniferous formations, and iron ores also abound in Shan-si, a chief centre of these deposits. Not only are the coal-fields extremely rich, but they might be worked under more favourable conditions than perhaps any others in the world. The exceptionally thick seams are mostly horizontal, spreading over a vast plateau 32,000 square miles in extent, and resting on a horizontal limestone foundation. This plateau of Tal-hal-shan rises from 2000 to 3000 feet above the surrounding plains, and in many places the seams have been denuded by the water-courses right down to the lower limestones. Here the anthracite is of the best quality, very thick and pure. Honan and Kan-su bordering on Shen-si are less rich in coal-fields, which, however, are also found in Pe-chi-li. Here the Government has already begun to utilise the vast coal, copper, and iron deposits north and west of Tientsin, which were surveyed by Henderson a few years ago. The coal of this district is of prime quality.

Islands—Hainan.

The Gulf of Tonking is enclosed on the east by a peninsula projecting southwards, and separated only by a narrow channel from the mountainous but unhealthy island of Hainan. The interior, which contains productive gold-mines and valuable timbers, has been explored chiefly by Mr. B. C. Henry, who has made extensive journeys to the mountainous inland districts. In 1886 he penetrated to the geographical centre of the Li territory, and showed that this region, previously supposed to be impracticable, may be traversed in all directions with comparative ease. The interior is still mostly held by the independent Li tribe, who appear to be more akin to the Shans than to the Chinese. The Chinese hold possession of the coast only, on the north side of which is the large seaport of Kien-chow. The island, which seems to culminate with the snowy Ta-uchi-shan, has an area of over 16,000 square miles, with a population estimated at 1,500,000.

The whole eastern seaboard of Asia is fringed by a rampart of islands, and insular groups stretching from Kamchatka in a series of graceful curves southwards to the Philippines. The Kuriles and Japan are followed by Linshoten and Liu-kiu, reaching to Formosa, beyond which the Bashi and Babuyan groups complete the chain to Luzon. These islands have been compared to a line of fortifications, and the comparison is justified by the fact that from the Yellow Sea to Amoy the Chinese sea-ports are completely sheltered by them from the typhoons which elsewhere sweep the Eastern seas. Arrested by the rocky barrier of Formosa, these terrific tornadoes are diverted to the southern seaboard, where they often spend their fury on the exposed rocks of Hong-Kong and Macao.

Macao—Hong-Kong.

These two islets, at the eastern and western entrance of the Canton River, symbolise the setting star of Lusitania and the rising sun of England in the Eastern waters. Macao, for ever associated with the names of Camoens and Xavier, is all that now remains in these regions to the Portuguese, who first revealed to the West the water highway of the Eastern world, and whose influence was at one time paramount from Mozambique to Japan. And even for this rock and decayed seaport with the high-sounding name of Cidade do Santo Nome de Dios de Macao, they pay an annual tribute of £150 to the imperial exchequer; for the Chinese Government has never acknowledged the absolute right of Portugal to the possession of this little tongue of land at the southern extremity of the Si-kiang delta. In a space of about 12

HONG-KONG.

square miles there is here massed a population (1894) of about 70,000, including 5000 Eurasians. But the monopoly of the trade with Europe enjoyed by Macao for three centuries disappeared with the opening of the Chinese Free Ports, and the traffic in slaves, disguised under the name of Coolie emigration to Peru and the West Indies, was finally suppressed by the Imperial Government in 1874. Since then Macao has continued to decline, and it is now chiefly noted for its gambling-houses. The sea-going vessels that formerly crowded its spacious harbour now either pass up the river to Whampoa for Canton, or else cross over to Victoria on the north side of Hong-Kong.

This barren granite and basalt mass, some 36 square miles in extent, had a population of about 2000 when it was occupied by the British in 1841. Now it is one of the great marts of Eastern trade, with magnificent quays, dockyards, arsenals, a yearly tonnage of over 5,000,000, and exchanges exceeding £6,000,000! The whole island is dotted over with large villages, suburban villas, public buildings, frowning battlements, and is overcrowded with a motley population (1891) of 221,000 Chinese, Hindus, Burmese, Malays, Polynesians, Europeans, and Americans. Formerly a very unhealthy place, Victoria has by a proper system of drainage become a sanitarium for the English residents in China. But it lies unfortunately within the limits of the cyclones, during one of which in 1874 over 1000 houses were blown down, 33 large vessels and several hundred junks wrecked, and many thousand lives destroyed. It was also ravaged in 1894 by the plague, introduced from Canton.

Hong-Kong is a Crown colony, administered by a governor and legislative council. In 1881 this body voted 30,000 dollars for an observatory and time-ball to warn shipping against approaching storms.

5. *Climate: Prevailing Dryness—Steppe Storms—Typhoons.*

The climatic conditions are essentially different in Central and East Asia. Considerable uniformity prevails on the lofty Tibetan plateau and the less elevated plains of Mongolia, where the prevailing features are great extremes of heat and cold combined with excessive dryness. But the Hoang-ho and Yang-tse basins, being exposed to the soft moisture-bearing winds from the Pacific, enjoy a far more equable temperature. Here the winters are doubtless severe in the north, the summers hot in the south; but there are nowhere such extreme vicissitudes as are characteristic of strictly continental climates, while there is nearly everywhere a sufficient rainfall.

At the same time the copious rains would appear to extend much farther inland than is commonly supposed, embracing not only the region of the "Cross Ridges" on the Tibeto-Chinese frontier, but even the Ala-shan highlands of South-East Mongolia. Here Prjevalsky was more than once overtaken by tremendous downpours of quite a tropical character. On one occasion he tells us "the rainfall was so great that streams formed in every cleft and gorge, even falling from the precipitous cliffs, and uniting in the principal ravine, where our tent happened to be pitched, descended in an impetuous torrent with terrific roar and speed. Dull echoes high up in the mountains warned us of its approach, and in a few minutes the deep bed of our ravine was inundated with a turbid coffee-coloured stream carrying with it rocks and heaps of smaller fragments, while it dashed with such violence against the sides that the very ground trembled as though with the shock of an earthquake. Above the roar of the waters we could hear the clash of great boulders as they met in their headlong course."[1]

This was in the month of July, and in a district usually included within the rainless zone of the Gobi. But the explanation lies in the direction of the Indo-Chinese ranges and East Tibetan "Cross Ridges," which run south and north, or south-east and north-west, thus giving free access to the rain-clouds from the Indian Ocean and China Sea far into the interior of the continent. This orographic disposition also serves to account for the extraordinary number of large streams which here take their rise, and which have in the course of ages cut up the East Tibetan plateau into so many independent river valleys. The mountains, which are not upheavals but the result of fluvial action, grew, so to say, hand in hand with the rainfall, resulting in an intricate highland river system elsewhere unparalleled.

Farther west, such is the extreme dryness that for months together not a single snowflake will fall on the elevated Tibetan plateau. It is a remarkable fact that the snow-line descends much lower on the southern than on the northern slopes of the Himalayas, where Forsyth found the Cayley Pass quite free at an elevation of over 19,000 feet. Owing to this absence of moisture the passes between Kashmir and Yarkand are open throughout the year, and some of the more difficult passes are elsewhere lined with the withered or mummified bodies of yaks, horses, or sheep, which dry up where they fall without passing through the process of putrefaction. But notwithstanding the absence of snow the cold is not the less intense, and this, combined with the "mountain sickness," produced by the extreme rarefaction of the air, causes great sufferings to travellers and animals in winter. Even in summer the streams often freeze.

[1] *Mongolia*, ii. p. 264.

In the Tarim basin and West Mongolia the air is also extremely dry. Here there is scarcely any spring, intensely cold and late winters being followed almost immediately by equally intense heat, when the glass rises even in April to 93° F. in the shade (Prjevalsky).

In East Mongolia the spring is also cut up by late frosts, lasting even into May, when still water sometimes freezes an inch thick during the night. Here the sudden changes of temperature are very trying, and north-westerly gales prevail for weeks together, obscuring the sun's rays and filling the air with clouds of sand mixed with fine particles of salt from the saline marshes. Steppe storms rage at times with great violence, "during which even the camels accustomed to the desert would turn their backs to the storm and wait till its fury had abated."[1]

The climate of Korea is healthy, although severe in the north, while resembling that of Japan in the south. In winter, however, snow and ice are common everywhere. Korea being washed on three sides by the sea, the rainfall is very heavy and the vegetation correspondingly vigorous.

Owing to the steady trade-winds from the Bay of Bengal and the monsoons from the Pacific, China proper enjoys both a more copious rainfall[2] and more regular seasons than Central Europe. Although it lies much farther south, reaching from about 40° N. into the tropical zone, the mean temperature of both regions differs but slightly. China, however, is at once colder and hotter than Europe. The dreaded typhoons visiting the Eastern seaboard are produced by the conflict of the south-westerly and south-easterly trade-winds which

[1] Prjevalsky, *Mongolia*, i. p. 119.

[2] Mean rainfall on east coast, 40 inches; at Canton, 50; at Shanghai, 42; at Peking, 25.

meet in the Pacific not far from the coast. The word typhoon, which has nothing more than a curious coincidence in sound and meaning with the Greek **τυφῶν**, is derived either from the Chinese *ta-fung* = great wind, or much more probably from the Formosan *tai-fung*, thus quaintly described in the *Tai-wan-fu-chih*, or native annals of Formosa—"The winds of our sea are very different from those of other seas. A fierce storm, which here blows, is called *keu*; but greater strength is possessed by the *tai*. The keu rises suddenly, as it also ceases suddenly, while the tai rages ceaselessly day and night. The keu blows in the season between February and May, the tai from June to September; in September follows the north wind (north-east monsoon). . . . The tai is accompanied by strong rain, uproots trees, blows down brick walls, unroofs houses, and rends rocks. Ships at anchor are dashed to pieces; but as soon as thunder is heard the storm is over. When the north wind rises in the seventh month, a *tai-fung* is very likely at hand. To recognise a true tai we must observe its course. For the tai is a storm which blows from every quarter, and all the tai winds follow this law; but an ordinary storm rages in one and the same direction."

The form "tai-fung" and the description here given leave little doubt as to the origin of the term typhoon, which was first used by Pinto (1560), who derives it not from the Greek but from the Chinese. The old English spelling "tuffoon," and the Arabic "tufán," are probably due to Greek influence.

6. *Flora and Fauna: Rhubarb and Ginseng—The Yak, Wild Camel, Birds of Passage.*

The great elevation of Central and West Tibet is unfavourable to the growth of trees, which are here represented chiefly by a few poplars and hardy fruit trees, in some places found at elevations of 12,000 and 13,000 feet. At these heights the breezy plateaux are often covered with coarse grasses, strong and sharp enough to pierce the boots of travellers and the hoofs of pack animals. Barley grows in sheltered spots as high as 15,000 feet, and many of the less exposed plains are covered with rich pasturage. Forests abound in the well-watered valleys of South-East Tibet, where a variety of the holly rivals the conifers in height, while far surpassing them in the richness of its foliage.

Equally rich in timber is the highland region between Kulja and the Tarim basin, where the spruce, larch, poplar, birch, apple, and apricot are varied with an undergrowth of hawthorn, woodbine, brier, and wild hop. The poplar, aspen, peach, willow, birch, and some other trees also flourish on the uplands of Eastern Mongolia. But in the Ala-shan and Kuku-nor districts the scanty vegetation is represented chiefly by a few dwarf elms, clumps of acacia, two or three species of shrub, and some flowering grasses. Even in the neighbouring province of Kansu, notwithstanding its rich and varied flora, forests occur only on the northern slopes of the southern ranges. Here the red birch, mountain ash, spruce, poplar, willow, and arboreous juniper, growing to a height of 20 feet, are interspersed with the wild rose, barberry, gooseberry, raspberry, currant, and other shrubs.

Indigenous in these regions is the medicinal rhubarb, the Shara-moto or "yellow tree" of the Mongolians, which grows 8 or 10 feet high, with a stalk nearly 2 inches thick, and leaves 3 feet by 2. This useful plant, which is found as high as 10,000 feet above the sea, is sent down the Hoang-ho for Peking and Tien-tsin, where it is shipped for Europe.[1]

[1] During his last expedition (1880) to Tibet, Prjevalsky found the

Owing to its copious rainfall, Korea possesses an abundant vegetation, including rice, millet, maize, and other cereals, which are largely cultivated. Tobacco, apples, pears, apricots, pomegranates, hemp, and cotton also flourish here. Of great commercial importance is ginseng, a species of *pannax*, which is raised from the seed sown under sheds covered with pine bark. The roots arrive at maturity in five years, and are then collected and dried for the China market. Although less esteemed than that of Manchuria, the Korean variety fetches in Peking from £3 to £4 per pound.

The Chinese flora is extremely rich. Forests, in the European sense, are rare; but evergreens, flowering shrubs, and especially resinous plants, are found in great variety. Proceeding southwards, the transition is very gradual from the Manchurian to the tropical flora of Indo-China. Hence in some of the central districts there is a remarkable intermingling of species belonging to different zones, the bamboo flourishing by the side of the oak, while wheat and maize crops are interspersed with paddy-fields, sugar and cotton plantations. In general the cultivated species are everywhere encroaching on the wild flora, and in many districts little is seen except vegetables, fruit trees, sugar-cane, cotton, the poppy, mulberry, rice, the bamboo, and tea plant. The three last named are of vast economic importance, rice supplying the staple food of hundreds of millions, the bamboo yielding the chief material for the construction of their houses and furniture, while tea, forming their national beverage, is still largely exported to England, Russia, Australia, and America.

Notwithstanding its scanty vegetation, Tibet, which

rhubarb in the district beyond Gomi, on the Upper Hoang-ho, growing to a colossal size. One root taken by him "measured 16 inches in length, 12 in breadth, and 7 in thickness, and weighed 26 lbs.'

by some naturalists is regarded as a chief centre of evolution for animal life, possesses a fauna of extraordinary richness. In the west, Nain Sing met the antelope in herds of as many as two thousand bounding over the plains. Here the yak, gazelle, wild goat, various species of sheep, wild ass, fox, jackal, wild dog, white wolf, and even a white bear, resembling the polar bear in appearance, are amongst the most characteristic animals. But

MUSK DEER.

the avifauna is poor, chiefly comprising the eagle, vulture, raven, but no songsters except birds of passage. In East Tibet numerous herds of buffaloes are preyed upon by the wolf and panther. The musk deer is found as high as 8500 feet, while the monkey and squirrel here form the transition to the Indian fauna. Amongst the birds are the pheasant, cuckoo, and lark, which latter is said to soar to a height of 15,000 feet. The statement may be credited in a region where yaks range to nearly 20,000 feet (Schlagintweit), and where marmots are found burrowing at altitudes of 17,000 or 18,000 feet.

The Tibetans have domesticated the yak, sheep, and horse, as beasts of burden. The sheep carries loads of 20 to 30 lbs. over the highest passes, and from the yak and Indian zebu, the dzo, a useful cross-breed, has been obtained, which however reverts after the fourth generation to the original types. There is a formidable species of watch-dog, which degenerates in India, but has been acclimatised in England. But of all the animals the most valuable is the goat, whose soft down (pashm), growing under the outer coat, supplies the material for the finest Kashmir shawls.

Amongst the wild animals of the Tarim basin and South Mongolia are the wild boar, tiger, antelope, and hare. The wild camel was formerly common in the Lob district and along the Altyn-tagh range, but is now almost entirely confined to the desert tracts east of the Tarim. Unlike the domestic species, the wild variety is remarkable for its sagacity and highly-developed sense of smell, sight, and hearing. It will climb the most inaccessible places, and when once it takes to flight will run for 20 or 30 miles at a stretch.

A notable feature of the Lob, Dalai, and other great steppe lakes, are the birds of passage which alight at these halting-places on their long journeys north and south in the spring and autumn.

Old writers speak of the rhinoceros, tapir, and elephant as formerly roaming over the plains of China. But the only large wild animals now found in the Yang-tse and Hoang-ho basins are the tiger and panther, and even these seem to be disappearing. Nevertheless the fauna, especially of the wooded highlands towards the west, is very rich, especially in animals of the snake, salamander, and lizard orders. On the whole, the species common to Europe are few, including 146 out of 764 birds, 10 out of 200 mammals, and no fresh-water fish except the eel. There are several species of the monkey, some of which are met as far north as the neighbourhood of Peking.

Stockbreeding is not a favourite pursuit in China; hence cattle, sheep, and even horses are comparatively rare, and of indifferent breeds. But buffaloes and swine are very common, as well as ducks, which, with fish, vegetables, and rice, form the chief articles of food. In many places the cormorant is trained to fish.

The crack of the sportsman's gun has scarcely yet been heard in Korea, which region still abounds in game of all kinds. In his official report for 1891 Mr. W. C. Hillier states that tigers, leopards, bears, and deer appear to be plentiful, especially in the northern districts, which are now accessible from the east coast. Pheasants and wild-fowl also abound, some of the coast-lands being frequented by millions of wild swans, geese, duck, and teal during the season. All the streams on the north-east coast are also well stocked with trout and salmon. But the true salmon does not frequent the west coast, where, however, it appears to be represented by a species of salmon-trout.

7. *Inhabitants: Table of Races in the Chinese Empire—The Chinese; Jews, Muhammadans, and Christians, in China—The Tibetans—Buddhism—The Mongolians.*

Within the limits of the Chinese Empire are probably comprised one-fourth of mankind, and of these at least nine-tenths are concentrated in China proper. Korea and parts of Manchuria seem to be fairly well peopled. But most of the other outlying regions are, to a large extent, uninhabitable. In Tibet the population is mainly confined to the San-po basin, in Mongolia to the northern and eastern edges of the Gobi desert, in the Tarim basin to the middle courses of the Kashgar, Yarkand, and Aksu rivers. In China the great river valleys are amongst the most densely-peopled regions on the globe, and even the most inaccessible highlands on the Indo-Chinese and Tibetan frontiers are occupied by numerous hill tribes.

Apart from these hill tribes, whose affinities are still largely undetermined, all the inhabitants of the empire, except the few Iranians of Kashgaria and Zungaria, and Malays of Formosa, belong physically to various branches of the great Mongolo-Tatar family. But there are at least six fundamentally distinct linguistic groups, as shown in the subjoined table of all the races of the empire.

Most of the hill tribes are still nature-worshippers. Muhammadanism has been largely diffused throughout Kashgaria, Zungaria, West and North-West China. Christianity has secured a footing in various parts of China proper, and even in Mongolia and Manchuria. The Jews, or "Blue Muhammadans," as the Chinese call them, were formerly numerous, but are now reduced to a few hundreds, mostly centred in Kai-fung-fu, capital of Honan. They claim descent from the tribe of Asser, and say they reached China in the Han dynasty (202 B.C.—264 A.D.). But they have forgotten their language, and the few "Aronists," or rabbis, who can still decipher the Pentateuch, pronounce the Hebrew words Chinese fashion, so that *Israel* becomes *Ye-se-lo-ni.* But nobody understands the text, and their traditions have got so confused that they believe Mecca and Medina to be their holy cities. But the great bulk of the inhabitants have long adopted the Buddhist tenets, variously modified according to the national temperament, usages, and traditions.

I. Mongoloid Races of Mongolo-Tatar Polysyllabic Speech.

Group	Subgroup	Tribe	Region
Sharra or Eastern Mongolians	Khalka	Tushetu	N. Mongolia mainly.
		Tsi-tseng	
		Jasaktu	
		Sain-noin	
	Uchumsin ; Chakar		S., E., and S.E. Mongolia.
	Genshikten ; Barin		
	Kartsin ; Jarot		
	Uniot ; Sunni		
	Tumet ; Kortsin		
	Durban ; Urut		
	Naiman ; Ahkhanar		
	Ordos		N. bend of the Hoang-ho.
Eleuts (Kalmuks) or Western Mongolians	Chorass		Zungaria, Kulja, N.W. Mongolia.
	Turgut		
	Khoshot		
	Durbat		
Urianhai			Upper Yenisei basin.
Sok-pa			N.E. Kachi (N.E. Tibet).
Taldi (?)			W. Kansu.
Tungus Family	Manchus		Manchuria.
	Tungus		
	Solons		Upper Ili valley, Kulja.
	Sibos		
Turki Family	Taranchi		Kulja.
	Kirghiz-Kazaks		
	Kara-Kirghiz		Central Tian-shan.
	Kashgarians		Tarim basin, Kulja.
	Dolans		Kashgaria.
	Salars (Kara-Tanguts)		? About source of Yang-tse.
	Hor-pa		W. Kachi (N.W. Tibet).

II. Mongoloid Race of Korean Polysyllabic Speech.

Koreans	Korea.

III. Mongoloid Races of Tibetan Intermediate Speech.

Bod-pa (Tibetans proper)		San-po basin mainly.
Tanguts (Northern Tibetans)		Kansu, Kuku-nor, Tsaidam.
Drok-pa		Central Kachi, between Sok-pa and Hor-pa.
Chak-pa		
Cham-pa		East of Noh, Tibet.
Kham-pa		? Central lake region, Tibet.
Chang-pa		East of the Kham-pa.
Si-fan	Andoan, Tochu, Arru, Gyarung, Telu, Manyak, Melam	Tibeto-Chinese frontier from Kuku-nor to Yun-nan.

IV. Mongoloid Races of Chinese Isolating Speech.

Chinese proper	N. and Central China.
Punti Hwui-chan	Kwang-tung.
Hakka	Kwang-tung, Fo-kien.
Hok-lo	Swatow district (Fo-kien).
Tungans	Kansu, Zungaria, Kulja.
Khambing Chimpan, Khatozun (extinct ?)	Kulja.

V. Highland Races of Undetermined Ethnical and Linguistic Affinities.

Group	Race	Region
	Man-tse (I-jeu) Sumu	W. Se-chuen.
Miao-tse, or Nan-man group.	Pe-Lolo Shu-Lolo He-Lolo Sen-Lolo	S. bend of the Yang-tse. S. Se-chuen, N. Yun-nan.
Miao-tse, or Nan-man group.	Chung, Nguchung, Tu-man, Kilao, Kitao	Kwei-chow uplands.
Miao-tse, or Nan-man group.	Yao	Lipo district, S. side Nan-ling mts.
Miao-tse, or Nan-man group.	Seng	Nan-ling mts.
Miao-tse, or Nan-man group.	Tung	N. Kwang-tung.
	Lyssu	S.E. Tibet, between Lu-tse-kiang and Lan-tsan-kiang.
	Mosso (civilised Lyssu)	N.W. Yun-nan, S. of the Lyssu.
	Lu-tse (Anong)	N. of the Lyssu.
	Remepang	E. of the Lyssu.
	Pagni (Pai, Terong, or Ba-yul)	W. of the Lu-tse.
	Tsarong	N. of the Lu-tse.
	Ku-tse	N. of the Remepang.
	Diju	N. of the Ku-tse.
	Jrupa	N. of the Diju.
	Mu-ua (Anampel)	Upper Irawadi, Burmese frontier.
	Shutung	W. Yun-nan.
	Shang-lai Shuk-lai	Island of Hainan.

VI. Aryan Stock and Speech.

Tajiks	Kashgaria, Kulja.
Kara-Kultsi (?)	Lower Tarim River.
Lobnorski or Kara-kurchin	Lob-nor district.

Of the peoples comprised in this table scarcely any can pretend to claim unsullied lineage. In the Ili basin, where whole populations have been more than once extirpated by fire and sword, the Taranchi, Tungans, Solons, and other immigrants from Kashgaria, Kansu, and Manchuria, present the most varied types with almost every shade of transition between the fair and yellow stocks. The Taldi seem to be half-caste Chinese and Mongolians, the Salars a mixed Tibetan and Turki people, the Kashgarians a curious blending of Turki and Iranian, the Tanguts a still more remarkable fusion of Tibetan, Mongol, and Chinese elements. Such at least is the inference to be drawn from the remarks of Prjevalsky, who first studied them and gave them the Mongolic name, Tangut, by which they are now best known. But Mr. Rockhill, who had the advantage of being a Chinese and Tibetan scholar, points out that all these Si-fan "call themselves Bo-pa (written Bod-pa), the name used by all Tibetans save those of East Tibet, in speaking of themselves; hence it would be infinitely better to call these Si-fan Tibetans, or Northern Tibetans" (*Geo. Proc.*, 1889, p. 731).

Of the original stock of the aborigines in the southern and south-western Chinese uplands, we know next to nothing; but we do know that these aborigines were already in possession of the Yang-tse basin long before the arrival of the children of Han. Here they were partly extirpated, partly absorbed, or driven into their present inaccessible fastnesses by the yellow intruders from the Tibetan plateau, who gradually re-settled the land. Hence the Chinese themselves can in no sense be regarded as a pure race, and the many diversely-modified forms of the Mongol type which they present must be attributed to the various interminglings that took place between them and the aborigines during the conquest and settlement of the "Middle Kingdom."

The Chinese.

Nevertheless the Chinese are the most important branch of the Mongolian family of mankind, far surpassing all the rest combined in numbers, wealth, and power. The great antiquity of their culture is reflected in the present state of their isolating speech, which in the course of ages has lost all trace of grammatical inflection. The same root is capable of representing all the parts of speech without further change, and in virtue of its position alone. Yet so subtle are the laws regulating the place of the word in the sentence, that the language not only amply suffices for ordinary intercourse, but has become an adequate instrument in the hands of the lawgiver, philosopher, historian, and poet. In their simple speech the Chinese have shown what great things may be accomplished by small means (R. K. Douglas).

Equally original and characteristic is the writing system, which is still mainly ideographic—that is, expresses not sound but thought. There is consequently no such thing as an alphabet, and the only approach to true phonetics are the 214 so-called "keys" or "tribunals" used in combination to indicate the pronunciation of unknown characters. There are practically as many symbols as there are words in the language, or 43,496 altogether; but of this number 13,000 are totally irrelevant, and for the expressions in ordinary literature about 4000 signs appear to suffice. The writings of Confucius and his disciples can even be read by the help of only 2500, and a knowledge of these will enable the student tolerably to understand all Chinese works on history and philosophy.[1]

Owing to their tenacious adherence to these and

[1] F. Ballhorn, *Grammatography*, p. 32.

other primitive methods and traditions, the Chinese now find themselves far behind other nations in scientific attainments. Nevertheless, the patriarchal conception of the State as an enlarged family, and of the family as the State in miniature, has given rise to many excellent institutions and some charming features in the social and domestic life. Hence it would be a manifest mistake to regard the Chinese as a decrepit or hopelessly corrupt people on account of the present low state of their culture. Their standard of morality certainly differs widely from ours; but in the midst of much vice the nation has, so to say, been safeguarded by its extreme frugality and thrift. Drunkenness has hitherto been a rare phenomenon amongst them; but they have during recent times largely extended their habit of opium-smoking. They not only consume all the opium that can be obtained from India, but they grow it largely for themselves. The Indian opium is consumed by the upper classes, the indigenous or Chinese opium by the humbler classes. Whether opium-smoking is deleterious, even in moderation, is a question much discussed at present. Many authorities believe that it is not more harmful than spirituous liquors in European countries. Some even contend that it is less injurious than the other stimulants which, in some form or other, are used by nearly all nations.

The status of woman is not so bad as is often supposed. Those of the lower classes have doubtless to work hard for the support of the family, whose ordinary diet is rice and cabbage. Still, their lot is not, perhaps, any harder than that of the same class elsewhere, more especially as with the Chinese hard work has become a second nature. The husband has, in certain cases, the power of life and death over his helpmate, yet he seldom strikes her, although the reverse would seem to be far from rare.

A special occasion for rejoicing is the birth of a son; but in the case of a daughter the midwife would find it as difficult to recover her fee as is so often the case in India.[1] That the birth of a son should be regarded as a propitious event is all the more natural that, according to universal usage, the son remains in the house and becomes the support of his parents in their old age, while the daughter either founds a new home or becomes a burden to the family. Old age is held in special veneration, and a man advanced in years is highly flattered by the inquiry after his "honourable teeth," the conventional phrase employed in asking people their age.

At the death of the Emperor the whole nation is thrown into mourning, the rites attending which are of a very stringent character. For a hundred days the court and people of rank wear white trimmed with white fur, this being the mourning colour in China. For the same period the men abstain from shaving, while the women lay aside the favourite ornamental head-dress. After this first term the garments assume a black or dark hue for the following twelve months. During this period no betrothals take place amongst the better classes, while for others the corresponding term is limited to a hundred days. All entertainments and public rejoicings are strictly suspended for a year.

The Chinese, if not the greatest, are among the oldest traders in the world, and during the course of ages all the complicated relations of buyer and seller have been regulated by prescriptive usage. The legal rate of monthly interest on advances is fixed at 3 per cent, this high rate being partly explained by the great risk incurred by the lender.

1 "In midwifery cases a large fee will sometimes be paid spontaneously if the child be a boy, and no fee whatever if it be a girl" (Mrs. S. Heckford, *The Queen*, Nov. 19, 1881).

religion amongst the natives. Of the 500 European missionaries three-fourths are Frenchmen, and to the same nationality belong nearly all the "sisters" engaged at the stations. These stations stretch in an unbroken chain from the coast to the western frontiers. The French priests display remarkable zeal and self-devotion; they are compelled to assume the costume and conform to the customs of the people, and were formerly even obliged to renounce their native land for ever. They were not allowed to leave the country or give strangers any information regarding the interior. Thus alone was it found possible to allay the ever-watchful suspicion of the authorities.[1] These missions are said to be at present in a flourishing state, with native congregations of perhaps not less than 500,000 altogether.

In recent years the question of Chinese emigration has been much discussed by political economists, and fears have been expressed that the white races may yet be swamped by the surging tide of Mongol invasion. But those who have struck this note of alarm do not understand the very elements of the question. It is assumed that the population of China is so vast, or rather that the land is so over-peopled, that once the stream of emigration sets in the direction of any given region, the inhabitants of that region must sooner or later inevitably become absorbed or assimilated to the yellow type. But there is good reason to believe that the current notions regarding the excessive population of China are greatly exaggerated, and that the total falls short of 350 millions. In any case it is certain that the country is far from being over-peopled, except perhaps in a few highly-favoured districts, such as occur in India, Belgium, England, and other long-settled regions.

But a more important consideration is the fact that,

[1] T. Cooper's *Travels*. London, 1871.

beyond the limits of the empire, there is, strictly speaking, no Chinese emigration at all, that is, no movement by which the Chinese race could perpetuate itself in any foreign country. Emigration in the strict sense takes place to Mongolia, where large tracts of steppe or pasturage have been occupied by Chinese agricultural settlers, and especially to Manchuria, which has practically become a northern province of China proper. But elsewhere emigration is almost entirely restricted to the men, the women being prohibited by a fundamental law of the empire from quitting the country even temporarily. These emigrants—mostly peasants, miners, traders, day labourers, and menials of all kinds—often settle down in those lands, such as Siam, Java, and other parts of Malaysia, where the physical and ethnical surroundings differ little from their own, and where they become themselves absorbed by intermarriage with the natives. For countless generations this movement has been going on without to any great extent affecting the relations in the Eastern Archipelago, which still remains essentially a Malay land. In other countries, such as Peru, Brazil, and especially the United States and the British colonies, scarcely any remain, the vast majority returning to China, literally "dead or alive," that is, either in their coffins for burial in their birthplace, or else to spend the rest of their days amongst their kindred on the fortunes amassed abroad. Hence in such countries there can be no question of any ascendancy of the yellow race.

This subject has lately been studied by M. J. de Groot of the Dutch East India service, who shows that Chinese emigration is due, not to overcrowding, but to the poverty of the soil in the provinces whence the emigrants come. It is the bare upland valleys of the eastern provinces that furnish the emigrants to the Spanish, Dutch, and English colonies, to California,

Australia, and especially Indo-China and Cochin-China. The prevailing formation is granitic, the rainfall is slight, timber and fuel have become scarce, and the soil yields little except a little rice, potatoes, and other vegetables of poor quality. Hence the people are compelled to seek subsistence in other countries. But whenever China undertakes the construction of railway and other great works, the stream of emigration, which is causing so much anxiety in many parts of the world, will cease, as the people will find at home the employment they now seek elsewhere.

It is also to be noticed that the Chinese themselves are far from possessing the homogeneous character with which they are popularly credited. Revolts, massacres, famines, inundations, by which whole provinces have been wasted, have at all times caused great dislocations and re-settlements, tending to bring about a certain uniformity. Nevertheless wide differences in speech, usages, traditions, and temperament still prevail, and the inhabitants of the central and northern provinces scarcely regard those of the extreme south-east districts as fellow-countrymen at all. "The province of Kwang-Tung is a peculiar one, and the Chinese of the north hardly look upon the Cantonese as fellow nationals. I have heard a Shanghai native remark—'There were seven Chinamen and two Cantonese.' Those in the north call themselves the descendants of Han, those in the south call themselves the descendants of T'ang, Han and T'ang being the names of Chinese dynasties" (C. F. R. Allen, *Proc. Geo. Soc.*, 1891, p. 264). Hence the lack of solidarity, the rival secret societies, and hostile factions, which have so often led to murderous conflicts between the Chinese coolies from different provinces thrown together on the plantations in Malaysia and elsewhere. To the same cause is also largely due the absence of a common national sentiment, which has enabled a handful of Manchus, a mere military caste, to hold sway for more than three centuries over hundreds of millions of Chinese not deficient in personal courage. Until these multitudes are merged in a more uniform nationality, the world runs little danger of being overrun by the yellow race.

The Tibetans—Buddhism.

According to a Chinese official estimate quoted by Desgodins, the inhabitants of Tibet number about 4,000,000, and Klaproth estimated them at not more than 5,000,000. Such a sparse population in such a vast area is doubtless mainly due to the sterility and bleakness of the land. But the great number of celibate monks, combined with the custom of polyandry and the low tone of morality, are also contributing causes.

The Tibetans constitute a very distinct branch of the Mongol family, and are described by Huc as a people with small, contracted, black eyes, thin beard, high cheek-bones, flat nose, wide mouth, and thin lips. The skin of the upper classes is as white as that of Europeans, but the ordinary complexion is tawny. They are of middle height, and combine agility and suppleness with strength and energy. They have the reputation of being frank and generous, brave in war, extremely superstitious, and fond of display. They have domesticated the yak, they breed ponies, sheep, and goats in large numbers, cultivate such cereals as will ripen in their climate, work the precious metals, and are skilful weavers and potters.

Although Buddhism was not introduced till the seventh century A.D., Tibet has become the centre of the Buddhist world. A native king, founder of Lassa in 617, having married a Chinese princess, is said to have sent to India for the Buddhist Scripture, causing it to be

TIBETANS.

translated into Tibetan with an alphabet derived from the Devanagari. In the fourteenth century the great reforming lama Tsong-kaba introduced many changes, prohibiting clerical marriages and necromancy, and organising frequent conferences of the priesthood. His followers were distinguished by a yellow dress and cap, and called Duk-pa, while the old unreformed party were called the Red Sect or Gelukpa. At present the Red Sect prevails in Ladak, Bhutan, and Sikkim, the Yellow in Tibet proper (Colonel Yule).

Gedun-tubpa, another reformer, arose soon after, and built the monastery at Teshu-lumbo in 1445, and it was in his person that the system of perpetual incarnation was supposed to begin. The sixth in succession of those incarnations, called Navang Lotsang, made himself master of all Tibet in the middle of the seventeenth century, and founded the Dalai and Teshu Lamaships as they now exist at Potala near Lassa and Teshu-lumbo respectively. There is a third incarnation called the Khutuktu, resident at Urga in Mongolia, and probably a fourth in the person of the Changay Lama or High Priest of Peking, besides a female incarnation—an abbess of a convent on the island in Lake Palti, referred to by Bogle and by Giorgi.[1]

The professed monks or clergy, subordinate to the holy lamas, are also called lamas, and are very numerous. They live in monasteries, some of vast extent, scattered not only over the inhabited valleys, but even in some of the wildest parts of Tibet. Their ritual consists mainly in the recitation and chanting of the *sutras*, or precepts and rules of discipline, to the sound of musical instruments. A characteristic feature is the prayer-wheels—metal cylinders charged with rolls of prayers—which are

[1] In his great work, the *Alphabetum Tibetanum*, issued by the press of the Propaganda in the last century.

kept revolving during the service and placed over streams to be turned by the current. They have been in use over 1000 years, being mentioned by the pilgrim Fa-Hian. Another peculiarity is the votive pile of stones frequently met with along the road-side, from a few feet to half a mile in length, and stuck over with flags

PRAYER-WHEEL AT DOTON.

inscribed with the mystic formula, *om mani padme hum*, *i.e.* "O the Jewel in the Lotus! Amen!" These primeval six syllables, "among all prayers on earth, form that which is most abundantly recited, written, printed, and even spun by machines, for the good of the faithful. They are the only prayers known to the ordinary Tibetans and Mongols; the first words the child learns to stammer, the last gasping utterance of the dying. The wanderer murmurs them on his way, the herdsman beside his cattle, the matron at her household tasks, the monk in all the stages of contemplation; they form at once a cry of battle and a shout of victory! They are to be read wherever the Lama Church has spread, upon banners, upon rocks, upon trees, upon walls, upon monuments of stone, upon household utensils, upon strips of paper, upon human skulls and skeletons! They form the utmost conception of all revelation, the path of rescue, and the gate of salvation!" (Colonel Yule).

But the most extraordinary feature in the Tibetan Buddhist system is undoubtedly the external resemblance between its ritual and that of the Roman Catholic Church, a resemblance often extending to the minutest details. The priests of both hierarchies wear the tonsure together with flowing robes covered with gold embroidery. They fast and mortify the flesh, observe spiritual retreats, confess the faithful, intercede for them with the saints of heaven, make long pilgrimages to shrines where relics are devoutly preserved. "Celibacy is common to both, and, under the shadow of church and temple alike, communities of men and women devote themselves entirely to a life of contemplation. Church and temple are in the same way furnished with high altar, candlesticks, reliquaries, holy water fonts, and belfries. The lama, like the priest and bishop, officiates with mitre and crozier, cope and dalmatica, salutes the altar, bends the knee before the relics, intones the service, recites the litanies, utters prayers in a language unknown to the congregation, solicits offerings for the repose of the faithful departed, heads the processions, pronounces blessings and exorcisms. Around him the choristers sway the incense-burner, and the devout tell their beads."[1]

The early missionaries were struck with the outward

[1] Reclus, vii. p. 80.

identity of the two rituals. Some have endeavoured to trace it to the early Christian Church of India, with which country Tibet has had direct relations since the seventh century. But from India Tibet derived not Christianity but Buddhism, and a more probable solution may perhaps be found in the pre-Christian Zoroastrian rites, spreading east and west from Irania, and influencing the religious thought of both regions during restless periods of transition. It is curious that the name of the Persian *mitre* should survive in Roman ecclesiastical nomenclature, while the object, variously modified, is still in use in the Latin and Greek churches as well as in the Tibetan and Mongolian Buddhist temples.

The salient features of Buddhism as originally constituted are threefold :—

1. *Socially*, Buddhism claims for itself a superiority over worldly power, holds that religion has a first claim upon all property, and forbids caste distinctions.

2. *Dogmatically*, it cannot be designated as theistic, as it deifies humanity and moral ideas.

3. *Ethically*, it teaches the vanity and emptiness of all mundane things, the transmigration of the soul and its ultimate absorption in *Nirvâna.*

But Buddha himself, like Confucius, was personally a philosopher, or expounder of an ethical code and a mirror of virtue, not professing to be a redeemer of fallen humanity, but declaring that man can work out his own redemption.

The Mongolians.

The Mongolians, once the terror of the world, and founders of ephemeral empires stretching from the Pacific seaboard to Central Europe, have ceased to possess any political cohesion since the destruction of the Zungar power in the second half of the last century.

Their primeval home is undoubtedly the region still known as Mongolia, where every prominent feature of the land is associated with some national legend, where every mountain is a king, every lake or stream a divinity. During centuries of migrations and ceaseless military expeditions the race has become largely affected by Chinese, Turki, Tibetan, Iranian, and other foreign elements, and the original type seems now best preserved in the Khalka branch occupying the whole of North Mongolia, besides the Ala-shan district in the south. Capt. Younghusband noticed a distinct difference between these eastern Mongols and those of Western Mongolia, "the features of the former being rounder and fuller, while the latter had rather more elongated faces with noses slightly more prominent and less equal than the eastern Mongols" (*loc. cit.*, p. 497).

The average Mongolian is perhaps slightly under the middle height, robust and capable of enduring hardships that would kill an ordinary European. But these must be such as he is accustomed to; for although he will keep his seat on the camel for 15 hours at a stretch with the glass marking $-20°$ F., a short walk across the steppe will completely subdue him with fatigue. He is always mounted, a skilful horseman, and extremely fond of racing, in which the whole encampment takes part. But in other respects the race has greatly degenerated, and under 200 years of Chinese government and lama influences it has even lost the personal courage which formerly enabled its warlike hosts to overrun continents. Vanquished, disorganised, and broken up into hostile factions, the nation has even acquired a sense of its helplessness. The independence, love of freedom, equity, and tolerance which, fully as much as their martial spirit, formerly proclaimed their immense superiority over the surrounding Asiatic races, are now far less conspicuous than the degrading superstition, gluttony, indolence, filthy habits, and other vices attributed to them by all recent observers. "The gluttony of this people exceeds all description. A Mongol will eat several pounds of meat at one sitting, and some have been even said to devour an average-sized sheep in the course of twenty-four hours. But the most striking trait in their character

A MONGOLIAN TENT.

is sloth. Their whole lives are passed in holiday-making, which harmonises with their pastoral pursuits. Their cattle are their only care, and even they do not cause them much trouble. The Mongol is so indolent that he will never walk any distance, no matter how short, if he can ride. His legs are bowed by constant equestrianism, and he grasps the saddle like a centaur. The wildest steppe-horse cannot unseat its Mongol rider.

"But the first thing which strikes the traveller in the

life of the Mongol is his excessive dirtiness. He never washes his body, and very seldom his face and hands. His clothing swarms with parasites, which he amuses himself in killing in the most unceremonious way. The uncleanliness and dirt in which they live is partly attributable to their dislike, almost amounting to dread, of water or damp. Nothing will induce a Mongol to cross the smallest marsh where he might possibly wet his feet, and he carefully avoids pitching his yurta anywhere near damp ground.

"Lamaism, which has struck deep root in their midst, is represented by the *Kutukhtu* of Urga, ranking next to the Dalai Lama and Pan-tsin-Erdeni of Tibet. Besides the Kutukhtu there are over one hundred 'Gigens' or minor saints, who never die, but pass from one body to another. Their personal influence is unlimited. A prayer offered up to one of them, the touch of his garments, his benediction, are regarded in the light of the greatest blessings humanity can enjoy. But they are not to be had gratis. Every believer must bring his offering, which in some cases is very large. Lamaism is open to objection as attracting an undue part of the male population, and, by its unbounded influence, deprives the people of the power of rising in the intellectual scale.

"But although this religion has taken so strong a hold on them, superstitions are equally prevalent, and evil spirits and witchcraft beset the Mongol's path.

"The Mongols expose the bodies of their dead to be devoured by birds and beasts of prey, their lamas deciding in which direction the head should lie. Princes, gigens, and lamas of importance are interred or burned after death. Masses are said for the departed for forty days on payment of a sum of money."[1]

[1] Prjevalsky, *Mongolia*, *passim*.

8. *Topography: Lassa.*

Outside of China proper, large centres of population are extremely rare. In Tibet the only real town is the capital, Lassa (Lhasa)—that is, "The Place of God"—which lies towards the south-east, on a plain surrounded by mountains and dotted over with large monasteries. It has a circumference of 2½ miles, in the centre of which stands a large temple containing images richly inlaid with gold and precious stones, and surrounded by bazaars with shops kept by Tibetan, Kashmiri, Nepalese, Chinese, and Muhammadan traders. Close by is the palace and monastery of Potala, residence of the Dalai Lama, encircled by eleven other monasteries, with an aggregate of 20,400 monks. The city itself has a population of scarcely more that 15,000, besides a garrison of 1000 Tibetans and 500 Chinese armed with flint guns.

Yarkand—Kashgar

When visited by the British Mission to Yakub Beg in 1874, Yarkand, the ancient capital of Kashgaria, was still the largest and wealthiest place in the Tarim basin. It stands on an open plain on the left bank of the Yarkand River, and is surrounded by strong walls 4 miles in circuit, beyond which lie several populous suburbs. Close by is the Chinese fort of Yang-shahr, or "New Town," residence of the governor, with his officials and garrison troops. The city had a population of some 20,000 in 1874, and was a busy place, with a staple industry of leather ware. Before Yakub Beg's revolt, Yarkand was a flourishing emporium, and apparently a much pleasanter place to live in than under the Muhammadan rule. "People bought and sold every day," said a citizen to Dr. Bellew, "and the weekly markets were much jollier affairs. There was no kazi, with his six satellites, to flog the people off to prayers and drive the women out of the streets, and nobody was bastinadoed for drinking spirits and eating forbidden meats. There were also musicians, acrobats, fortune-tellers, and story-tellers, who moved about and diverted the people."

Kashgar, the present seat of government, whence the province takes its European name, consists of the Kuhna-

KASHGAR.

shahr, "or Old Town," on the Tumen River, and Yangi-shahr, "or New Town," on the plain, 5 miles farther south, with the Kizil River flowing between. The Yangi-shahr, built in 1838, is surrounded by lofty massive walls and a wide ditch, and in 1874 contained the Amir's residence with the garrison and military bazaars. The Earl of Dunmore, who visited the place in 1892, gives to Kuhna-shahr and its suburbs a present population of 60,000, and to Yangi-shahr not more than 2000. Mangohin, as the Chinese call it, "is simply a large fort entered from the east by a double gate, over a draw-bridge, which spans a ditch 30 feet wide and 30 feet deep. Within this fort are 400 houses and a population of 2000 souls" (*Geograph. Jour.*, Nov. 1893, p. 389).

Karashahr—Turfan—Hami—Urumtsi—Kulja.

East of Kashgar follow successively Aksu, Karashahr, Turfan, and Hami (Khami), all more or less important towns and stations on the old historical Tian-shan Nanlu, or southern Tian-shan route between China and Samarkand. Aksu, the old Arpadil (Ardabil), is a very ancient place, and was formerly the centre of the Chinese trade and the limit of the trade privileges granted by the government to the Khokand Khan over the cities of the western division.

Karashahr, which stands near the north-west shore of Lake Baghrach, about midway between Aksu and Hami, is appropriately named the "Black Town."[1] No more unpleasant place of residence could well be imagined, and Mr. Carey, who visited it in 1885, describes it as "a poor, dirty town, inhabited by Tunganis (Dungans) and Chinese, with numerous encampments of Kalmaks in the vicinity. The Kalmaks expose their dead to be eaten by the ownerless dogs which swarm in the town; and I was told that it was no uncommon thing for living persons lying drunk on the ground to be killed and eaten by the dogs. There is no improbability in the statement, as the Kalmaks are much addicted to drinking. I found Karashahr so unpleasant a place of residence that, as soon as I had satisfied the requirements of politeness by exchanging visits with the local officials, I retraced my steps to Kurla, a much larger and cleaner town, peopled by Turks, who are

[1] From Turki *kara* = black, and Persian *shahr* = town. Such Turko-Iranian compound terms are common enough in Central Asia, where the Mongolo-Tatar and Aryan populations have for ages been intermingled. Cf. *Surma-tash* (commonly but wrongly written *Soma-tash*), the "Black Stone," from Persian *surma* = black, and Turki *tash* = stone. This is the famous inscribed stone lately removed by the Russians from Lake Yashil-kul on the Pamir to the museum at Tashkend.

preferable as neighbours to the forward and inquisitive Chinese and Tunganis" (*loc. cit.*, p. 737).

Turfan, on the north-eastern verge of the Tarim basin, is sometimes called Kuhna-Turfan, in distinction from Ush-Turfan, lying farther west. In the district are the extensive ruins of an older Turfan, capital of the Uighur empire, visited in 1880 by Dr. Regel, and in 1889 by the brothers Grum Grijimailo, who describe the Turfan district, or at least its northern section between the Nameless Range and the Tuz-tau Mountains, as a relatively thickly-peopled country with a population (1888) of about 65,000. The enormous supply of water required for irrigation purposes in this rainless region is obtained partly from springs, partly from the so-called *karys*, or underground conduits like those of Persia and Afghanistan; the quantity might be greatly increased by developing this system in connection with the subterranean reservoirs in the Lukchin and Khandui depressions east of the town of Turfan. These reservoirs are replenished by the Utyn-Auza, Kok-yar, and other streams descending from the southern slopes of the Nameless Range.

But any great increase of irrigation would probably be attended by disastrous consequences, at least in the salt marshy district south of Turfan, between the Tuz-tau and Chol-tau ranges, where the drift sands are already encroaching steadily on the cultivated tracts. Here the Russian explorers were witnesses of a marvellous struggle between nature and man. The native of Turfan, whose energy is astonishing, exploits his land so thoroughly as not to leave a drop of moisture in it—he takes it all. Nature, deprived of her portion, dies, but in dying avenges herself on those who have robbed her by laying waste their settlements and destroying their fields and gardens. In this way the native of Turfan digs his own grave;

there are but few spots left where the reeds can grow, and the wild boar and other animals find shelter; doubtless these too will disappear with the increase of population and the digging of more water-channels" (*Proc. Geo. Soc.*, 1891, p. 220). Thus also perished the former flourishing cities and culture of the Uighur nation in this very region.

Just south of Turfan occurs the remarkable depression known as the "Assa Hollow," which lies no less than 160 feet below sea-level. It was discovered by the Pevtsof expedition of 1889-90, but had already been seen at its eastern margin by the brothers Grum Grijimailo in 1888.

The highway leads east of Turfan to Hami, at the junction of the route over the Tian-shan from Barkul. Lying in a fertile oasis on the skirt of the desert, and at the converging points of the Moslem and Buddhist worlds, Hami presents a curious mixture of Buddhist and Muhammadan monuments, including a fine temple and a magnificent mosque, dating from Shah Rukh's embassy to China (1420 A.D.).

In the Tian-shan Pelu, or northern Tian-shan route (Zungaria), Barkul and Urumtsi correspond to Hami and Turfan of the southern route respectively. Their importance is due mainly to the fact that between them the Tian-shan range is crossed by only one difficult pass, so that they form vital points on the great trade and military route from Peking *viâ* Bautu and Hami to Kulja. Urumtsi, the Bish-Bakil of mediæval times, occupies a prominent place in the history of Central Asia during the last 600 years. It lies near the foot of the triple-crested Bogdo-ola (14,000 feet), and consists of an old town on the slope of the mountain, and a new or Manchu quarter lower down. In the neighbourhood are some hot sulphur springs, but Humboldt's view of the volcanic nature of the district has been denied by Severtzoff. Like Hami, Urumtsi is a great centre of the trade between China and the West, and has a population of about 15,000, according to Regel. Urumtsi, the Hung-Miotza of the Chinese, is now the

TARANCHI MARKET AT KULJA.

administrative capital and military headquarters of the province of Sin-Kiang.

In the Ili valley the chief place is Kulja, a name which has recently been extended to the whole province. There are or were two towns of the name—the old or Taranchi town, lately the headquarters of the Russian administration, and the new or Manchu Kulja, 25 miles lower down, which was a thriving city of 75,000 inhabitants before the revolt of the Tungans, who exterminated its Chinese population, and laid the place in ashes.

Saitu—Su-chau.

Saitu or Sachu, about 220 miles south by east of Hami, although a small place with scarcely 3000 permanent residents, is one of the most important stations in the whole of Mongolia. It stands at an altitude of 5225 feet, near the southern verge of the Gobi wilderness, at the converging point of some of the most frequented caravan routes between China and her Turkestan possessions. All travellers proceeding from Lob-nor eastwards to Kansu, or from Hami southwards to Tsaidam and Tibet, necessarily pass by Saitu, which has consequently been visited in recent years by several Central Asian explorers, such as Prjevalsky, Carey, and Littledale. "The town of Saitu is situated in a small but fertile oasis. It is on the right bank of the Danga Gol River, which is crossed by a wooden bridge about 70 yards in length. Each side of the town measures about half a mile. It is surrounded by a mud wall in fairly good repair, with several gates surmounted by guard-houses of the usual Chinese junk pattern. The interior of the town is uninteresting; the houses are poor, and in many cases dilapidated, and there are no large buildings. It contains a bazaar with several good shops, but the trade is confined to a retail traffic for the supply of the wants of the residents and of the Mongol nomads south of the mountains. Farmhouses are scattered along the banks of the river both above and below the town, and the land appears to be very carefully tilled. The river is the only source of water supply, and outside the strip of irrigated land the country is a sandy desert. On the left bank is the old city of Sachu, the limits of which can easily be traced by the ruined walls still standing. The ground inside the wall has been ploughed up and cultivated" (Carey, *loc. cit.*, p. 747).

Old Saitu was the city known to Marco Polo, who describes it as a month's journey from Lob; although this statement has been questioned by Colonel Yule, its accuracy has been fully confirmed by Littledale, who took exactly thirty days to cover the distance. Shortly before reaching Saitu this traveller came upon an artificial embankment 4 or 5 feet high and about 10 yards thick. The earthworks were followed for seven or eight miles, and where the track branched off the embankment continued as far as could be seen. "I have heard that the Great Wall of China extended beyond Sachu, but this certainly much resembled parts of the wall that we afterwards saw, and if it was not the wall I am quite at a loss to say what its use could have been" (*loc. cit.*, p. 458).

East of Saitu follows Nainshe (An-si), where the route bifurcates, one track striking north across the Gobi for Hami, the other running east by south to Su-chau, just within the Great Wall, about 30 miles from its western extremity. Here is the last gate which gives access from Mongolia to China proper, and which is known in history as the "Jade Gate," because through it ran the caravan route to the jade mines in the far west. Su-chau suffered much during the Dungan insurrection, by which a great part of the surrounding province of Kansu was devastated. But since then it has been rebuilt, and is now gradually recovering its former prosperity as one of the chief stations on the great historical highway between China and Central Asia. Since 1881, when it was thrown open to Russian trade, it is much frequented by dealers from Turkestan.

Urga, capital of Mongolia, and in Buddhist eyes second in sanctity only to Lassa, lies due south of Kiakhta, on the high road to Peking, in 47° 55′ N., 106° 41′ E. The large monastery of this city, visited by countless thousands from every part of the empire, is the residence of the Kutukhtu, Lama king, or High Priest of the Mongolians, who, like the Dalai Lama, is worshipped as a god incarnate. But the city itself, with its irregular and straggling houses and busy tent bazaar, has more the appearance of a nomad encampment than of a town, while the prevalent practice of exposing the dead converts the neighbourhood into a haceldama. The Kutukhtu is very wealthy, owning 150,000 slaves, and receiving a constant flow of rich offerings from his pious votaries.

Not far from Urga stood Karakorum, the ancient capital of the Mongolian Khans, which was visited by Marco Polo, but the very site of which had long been forgotten till its position was found by M. Yadrintzef in 1889. Following the course of the Selenga to the Orkhon confluence, this explorer came first upon some fairly preserved ruins on the Tula River dating from the thirteenth century. He then visited the remains of a Buddhist temple on the River Kharukha (Haruha), the walls of which are still 20 to 40 feet high. At last, on the left bank of the Orkhon, about 30 miles south of its confluence with the Urtu-Tamir, he discovered the far-famed ruins of Kara-Balgassum, where at one time stood the principal residence of the Mongol emperors. The remains of the surrounding city of Karakorum are strewn over a space six miles in circumference, and were connected by canals with the river Jirmanta, where were found the remains of baths near the hot springs. The central buildings stood in 47° 15′ N., 102° 20′ 15″ E., and the remains of ancient Mongol habitations were traced along the whole valley of the Orkhon. Several burial-grounds were found with numerous bas-reliefs, obelisks, Runic inscriptions, and Chinese hieroglyphics. Now the whole region is a lifeless wilderness, as silent as the ancient tombs of the Mongolian nobles strewn over the surface.

In Western Mongolia the chief place is Ulia-sutai, lying on the Kobdo plateau, in a romantic district at the junction of the Ulia-sutai and Bogdo rivers. Ulia-sutai, which is the centre of the military administration, is rather a fortress than a town, being scarcely the third of a mile both ways, and enclosed by high palisades, with a gateway and four towers. Its 4000 inhabitants are nearly all Mongolian officials and troops. Business is transacted in a village about a mile off, where the Chinese traders barter their cotton goods, plush, tobacco, and leather, for sheep-skins, tallow, cattle, horses, buffalo horns, and peltry. This place was first visited in 1868 by the Russian consul Shishmarev, and again in 1872 by Ney Elias, who found the normal temperature in November six degrees below freezing-point. In summer it rises to 110° and 120° F.

Kobdo, which gives its name to the plateau, lies in a wide plain destitute of vegetation, and consists of the government quarters, enclosed by a mud wall, and an open trading quarter. Its trade is said to be more considerable than that of Ulia-sutai, which it exceeds in population. Of its 6000 inhabitants, 3000 are Mongolians and over 1600 garrison troops.

Kirin—Kwan-chang-tsu—Mukden.

Kirin (Girin), the new capital of Manchuria, lies on the Upper Sungari, and is said to have a population of 120,000. Its streets are laid down with planks, and it has a large trade in lumber from the neighbouring wooded uplands. Other important places in Manchuria are

Kwan-chang-tsu, which at the time of Mr. James's visit was the chief trading-place in Manchuria, with a population of about 100,000; Mukden, the old capital, near the Chinese frontier, still the largest city, with an estimated population of 200,000; Petuna (Sin-chung), below Kirin, near the junction of the Nonni; Tsitsihar and Mergen, both in the Nonni valley; Aigun, on the right bank of the Amur, facing Blagoveshchensk, capital of Russian Manchuria; and Ninguta, a flourishing and well-built city on the Hurka (Khulkha), a tributary of the Sungari from the east.

Mr. Wylie, who visited the chief places of Central Manchuria in 1892, describes most of the towns as busy centres of trade for the surrounding populations. "Many of them could boast of very large distilleries, inn-yards of great extent, capable of accommodating hundreds of guests, and oil-works of various kinds, while outside their walls were generally some brick-kilns, brick-works, and lime-kilns. The houses were chiefly built of brick; burnt brick was used for the better houses in the towns, while unburnt brick or mud was used in the country. In some of the cities the shop fronts were quite imposing, being substantially built and lavishly decorated. The streets were generally very wide and level. I saw nothing of the immense activity which characterises these cities during the winter months, when all the inns are full and the shops doing a good trade, when from the south come great numbers of men seeking employment, and the rush of men and the succession of carts is continuous.

The chief crops were the tall millet, four kinds of which are cultivated, the small millet, three kinds; beans of several kinds, hemp of two kinds, cotton in parts south of Kai Yuan, rice chiefly south of Kai Yuan, potatoes more cultivated in the north and to the west of the Sungari. Indigo, wheat, barley, and opium are largely grown, and the great extent to which flour is used and its unusual cheapness testified to the large cultivation of wheat. The hemp chiefly grown is the slender variety, the "linen ma," and of this we saw in many places large fields one after the other; it is grown principally for the sake of the oil expressed from the seeds" (*Geographical Journal*, November 1893).

The Korean towns, which are said to be very numerous, including over 30 of the first rank, were till lately scarcely known even by name. Seul, the capital, and the treaty ports, are described at p. 325.

Topography of China.

In China proper, large cities, some of great antiquity, are even more numerous than in India. There are at least 100 with populations of 100,000 and upwards, while two or three have certainly more than a million inhabitants, although the still higher estimates for Peking and some other places have proved to be gross exaggerations. At the same time, the Chinese cities, which reflect a uniform civilisation of a somewhat stereotyped character, are far from possessing the same antiquarian or architectural interest as those of India, where so many distinct cultures have left their impress on the land. Hence in China the towns seems to be nearly all cast in one mould, the same types of buildings, temples, towers, narrow irregular streets and rickety houses everywhere succeeding each other with oppressive monotony. What variety they may possess is due rather to the nature of the ground on which they stand than to the few slight departures from the standard models which may occur here and there. All are officially grouped in three classes, subordinate to each other. Those of the first rank, distinguished by the title of "fu," have under them several "chau," that is, cities of the second order, under which again come the "hien," or towns of the third rank.

But although these terms are usually added to the names of the towns according to their rank, they really indicate territorial rather than urban administrative divisions. Thus each province is divided into so many *fu* or departments, which are again subdivided into so many *chau* or districts, and these into a number of *hien*, communes or cantons. Hence it is that not one only, as is generally supposed, but several cities in a single province may take the termination *fu*. Such are, for instance, Singan-fu, Hanchung-fu, Tungchau-fu, Yengan-fu, and some others in Shensi, and so on.

Peking.

Above all stands Peking, the present capital of the empire, which lies in the Pei-ho basin, about 40 miles south of the Great Wall, and 100 from the Gulf of Pe-chi-li. It consists of the Chinese or outer quarter and the Manchu or inner quarter, in the heart of which is the imperial palace. The Chinese quarter alone is accessible to foreigners, although an occasional view has been obtained of the Manchu, known also as the "imperial" or "forbidden" city, which seems to differ in few respects from the outer town. They form two parallelograms, so disposed that the shorter side of the Manchu faces the longer side of the Chinese city. A conspicuous object is the Temple of Heaven, which is visited in great state by the Emperor once a year, and never opened except on this occasion. With its various shrines, gardens, courts, moats, etc., it measures altogether two miles in circumference. Attached to it is the Temple of the Seasons, in which four wooden pillars support a cupola, the only

THE TSUNG-LI-YAMEN, PEKING.

structure of the kind said to exist in China (Von Hübner). This sanctuary is entered by the Emperor alone with the imperial princes and their suites.

Other buildings are the great lamaserai or Buddhist temple of Yung-ho-kung, and that of Confucius ("Wen-Miao"), both in the north-east quarter of the Manchu

THE OBSERVATORY ON THE WALL, PEKING.

city. In a little court of the latter are contained all the writings of the sage engraved on black marble tablets for the use of the *literati*. In the Manchu city are also four Roman Catholic churches, and on its east wall stands the famous old observatory with its globes, real masterpieces of science and art, executed by Chinese workers under the direction of the Jesuits. Of the four churches the most interesting is the Pei-tang, or North Church, with which is connected a seminary with a valuable library and a rich cabinet of curiosities. The number of Christians in the vicariate of Peking is at present 27,000, of whom 8000 reside in the capital, including many artisans and all the watchmakers in the place.

Peking produces on the whole an unfavourable impression, caused by the state of the streets, the dilapidated condition of the bridges, many built of marble blocks in the last century, by the neglected appearance of the canals, and the pictures of general decay which meet the eye. The population is far less than was at one time supposed, and is now generally estimated from 800,000 to 1,000,000, while one authority reduces it to about 500,000.

Nan-king—Shanghai.

Much the same ruinous aspect is presented by Kiang-ning, better known as Nan-king, on the Lower Yang-tse, the former southern capital of the empire. A far more important, though scarcely a more inviting place, is the great seaport of Shanghai, a chief centre of British trade in the East, which is situated in a low-lying swampy district on the left bank of the Wu-sung, just above its mouth in the Yang-tse estuary. Shanghai, which has a present estimated population of 700,000, has become the chief seaport in China since 1842, when the first English factories were established at this place. It may be said to stand almost on British territory, for the English "Concession," one of the most permanent results of the first war with the Middle Kingdom, enjoys the immense advantage of self-government. The American and French Concessions, which were afterwards added, have naturally gravitated towards the British settlement. Collectively

NAN-KING.

these quarters extend from the walled Chinese city northwards to and beyond the Suchau, a small creek which here joins the left bank of the Wu-sung, or Hoang-pu ("River of Yellow Waters"), as it is locally called. Despite much dredging and other harbour works, the Wu-sung is continually silting up, and vessels of deep draught are no longer able to ascend the river to Shanghai, or even to cross the bar that has been developed at its mouth in the Yang-tse estuary. Shanghai is thus threatened sooner or later to be cut off from all access to the sea, and this will certainly be its fate unless the imperial government sanctions the engineering schemes that have been proposed to keep open the navigation. The export trade in tea has also fallen off since the rapid development of this industry in India and Ceylon. Shanghai, however, still continues to export large quantities of silk, chiefly to France, and it remains the largest depot for the distribution of European manufactured goods throughout the interior of China.

Ching-kiang—Su-chau.

In the same southern district of the densely-peopled province of Kiang-su are concentrated several other large centres of population, the most important of which are Ching-kiang, east of Nan-king, on the right bank of the Yang-tse-kiang, near the southern extremity of the Grand Canal; and Su-chau (Su-chew), on the canal itself, about 54 miles north-west of Shanghai. Ching-kiang occupies such a commanding position at the head of the Yang-tse estuary, that, although destroyed during the first war with England, and again in 1853 by the Taiping rebels, it is still a flourishing seaport, and a dangerous rival of Shanghai for the foreign trade of the Yang-tse basin. It has a present estimated population of 200,000,

THE BUND, SHANGHAI.

and has already become the second port in the empire for the importation of foreign wares.

Su-chau, which stands at a point where the Grand Canal communicates through several creeks westwards with the Ta-Hu ("Great Lake"), was already a famous place in the time of Marco Polo. The "great and noble Saju," as the Venetian traveller calls it, had at that period a circumference of 60 miles, with six thousand stone bridges high enough for junks to pass through, and a prodigious population numerous enough "to conquer the world." Towards the middle of the present century the "Venice of China," as it is called, had still nearly a million inhabitants; but then came the terrible Taipings, who wasted it with fire and sword, and left most of the space within the walls a heap of smouldering ruins. But, like Ching-kiang, Su-chau rapidly recovered from this disaster, and a population of about 500,000 is supposed to be again crowded within the enclosures, which are now reduced to about 16 miles. Besides its advantageous position for trade at the converging point of many navigable waterways, Su-chau is favoured by a delightful climate, an extremely fertile soil, and easy access both to the coast and to the great inland sea of Ta-Hu. It is also the seat of several long-established industries, such as block-printing, conducted on a vast scale, and brocade and embroidery work famous throughout the whole empire. In Chinese literature it is referred to as a sort of terrestrial Eden. "Above is Paradise, below is Su-chau," say the local poets.

Hankow.

Along the course of the Yang-tse-kiang follow several other large towns and flourishing riverside marts, which being accessible to ocean steamers are virtually seaports.

Such is the great city of Hankow, which, with its wide-spread suburbs and dependencies, forms one of the largest centres of population in the world. Before the district was swept by the Taiping storm, the three contiguous urban groups of Hankow, Wuchang, and Hanyang, about the confluence of the Yang-tse and Han-kiang in the province of Hupeh, had an enormous collective population, estimated by some observers as high as seven or eight millions. But at the time of Blakiston's visit (1861) these multitudes had been reduced to less than a million, and at present (1894) they scarcely exceed 1,500,000. Many of these are permanent residents of innumerable junks, which obstruct the navigation of the main stream, here about a mile wide and accessible to large ocean steamers, with a mean depth of from 40 to 50 feet.

Hankow (" Han-mouth "), which is the administrative capital of Hupeh, occupies probably the most advantageous trading position in the interior of China. Through the Han it communicates with the Hoang-ho basin as far north as the province of Shensi, while the surrounding navigable waters give it easy access to the great Tung-ting-hu lacustrine basin of Hunan and to the great province of Szechuen as far west as the Tibetan frontier. Hence Hankow is the chief emporium of the interior, as well as the centre of the tea trade both with England and Russia. Here are prepared vast quantities of brick-tea for the Siberian and Tibetan markets, and a large business is also done in tobacco, opium, hides, rice, and other local produce. Large steamers make the ascent from Shanghai, a distance of about 700 miles, in three days, and the return journey in two.

The great drawback to the prosperity of Hankow are the tremendous inundations of the Yang-tse and its Han affluent, by which the whole district has at times been submerged to a depth of several feet. In 1869, and again in 1870, the surrounding plains were transformed to a vast inland sea, where the highways were indicated only by long rows of willows and nothing could be seen of the villages except the roofs of the houses. These destructive floods, which coincided with long and severe droughts in the southern province of Hunan, were caused by the heavy rains in the mountainous province of Szechuen in the upper Yang-tse basin. Fires also are much dreaded, and with good reason. In 1850 a conflagration caused by lightning raged with tremendous fury amongst the junks moored in the river between Hankow and Wuchang, and on this occasion some fifty thousand of the floating population are said to have perished by fire and water.

Ching-tu—Chung-cheng.

Ching-tu-fu, capital of Szechuen, the most populous province in the empire, is an historical city situated not on the Yang-tse, but in the valley of its great northern tributary the Min-kiang, at an altitude of about 1500 feet above the sea. In the days of Marco Polo it was already " a rich and noble city," and although since that time frequently wasted and even destroyed, it has still an estimated population of 800,000 including the suburbs, or 350,000 within the enclosures. Over a million of its inhabitants are said to have perished when the place was captured by Kublai Khan, and towards the end of the eighteenth century a great part of the city was destroyed by fire, so that most of the buildings are of recent date. The finest of these are the numerous clubs, which when lit up at night with many-coloured lanterns present quite a fairy sight.

Ching-tu has been called " the Paris of China," and is certainly the pleasantest and best-built city in the empire, with broad, straight, well-paved streets, flanked by houses with graceful, brightly painted façades opening on an inner court planted with shrubs and flowers. The surrounding plain, where ramify the waters of the "four rivers,"—that is, the Min and its tributaries,—is itself a vast garden, one of the best cultivated tracts in the world, growing an abundance of rice, fruits, and vegetables of all kinds. There are numerous local industries, such as weaving, dyeing, and embroidery work; and a considerable transport trade is done in tea, silk, cotton textiles, and other wares, which are forwarded from this distributing centre to Tibet, Tsaidam, and Mongolia.

Even more important as a trading-place than the capital is Chung-cheng, which has been called the "Shanghai of West China," and which is at present probably the largest city in Szechuen, with an estimated population (1894) of 900,000. It stands on the north (left) bank of the Yang-tse-kiang, at its junction with the navigable river Kialing (Pa-ho), which has its farthest sources far to the north in the Kansu uplands. Chung-cheng, with the neighbouring Kiangpeh and an extensive suburb on the right bank of the Yang-tse, covers a vast space on both sides of the main stream and its great affluent. Since Blakiston's time, when the population scarcely exceeded 200,000, it has become the central mart for the local produce and for the manufactured goods forwarded from Shanghai by the Yang-tse-kiang.

Si-ngan.

A much frequented but difficult overland route, leading from Chung-cheng by Han-chung over the Tsing-ling-shan mountains into the Hoang-ho basin, has its terminus at the historical city of Si-ngan-fu, one of the oldest places in China, and for ages the administrative capital of the rich and densely peopled province of Shen-si. Here is still preserved the most ancient memorial of the introduction of Christianity into the Middle Kingdom,—a monument bearing a bilingual inscription in Chinese and Syriac, erected in the eighth century by the so-called "Nestorian" missionaries who were at that time propagating evangelical teachings throughout East Asia. The stone, which bears the date of 781, and the authenticity of which has been placed beyond doubt, was discovered by the Jesuits in 1625, and is now enframed in the wall of a Buddhist temple, where it was placed for safe keeping during the late Muhammadan insurrection.

Si-ngan-fu, which appears to have been the imperial capital for 2000 years (1122 B.C. to 1127 A.D.), is still one of the largest cities in China, containing nearly a million inhabitants within its strongly built walls, which form a vast square facing the cardinal points, and over six miles on all four sides. Each wall, about forty feet high, is pierced by a monumental gateway, from which the four main thoroughfares converge in the heart of the city at the ancient imperial palace, which is at present the residence of the provincial governor, and which is itself enclosed by strong inner walls. Much of the space comprised within the outer enclosures is occupied by gardens, fields, and even waste spaces, where are still to be seen the ruined houses and temples destroyed during the Muhammadan outbreak of 1872.

Si-ngan-fu is a great depot for the tea, sugar, and other produce of the southern provinces forwarded to Kansu, North Tibet, and Mongolia, in exchange for the wool, musk, rhubarb, drugs, and furs imported from those regions. Here is a famous archæological museum containing amongst other treasures the "forest of tablets"; that is to say, a collection of inscriptions and other documents, some of which are 2000 years old. By means of these records antiquaries have been able to reconstruct the sequence of events during several dynasties.

Tung-kwan—Lan-chau.

Si-ngan-fu lies a short distance from the right bank of the Wei, 93 miles above Tung-kwan at its confluence with the Hoang-ho. Although not a large place, the population scarcely exceeding 80,000, Tung-kwan, standing on the Shensi and Hunan frontiers, "is one of the principal gates of China, and has been the scene of numerous dynastic battles, and is on the main line of traffic between east and west and south-west and north-east China" (Colonel Bell). At this point the great Central Asian trade route crosses the Hoang-ho and ascends the rich Wei valley through Si-ngan and Lan-chau-fu to Liang-chau under the Great Wall in Kansu. Lan-chau-fu, which is the capital of this province, stands at an altitude of 5500 feet on the right bank of the Hoang-ho, which is here 250 feet wide, with a deep rapid current. The city is relatively well built, with some fine shops, a flourishing Roman Catholic mission, and about 40,000 houses within the walls.

Direct Russian trade across the Gobi desert reaches this place, which, however, has scarcely yet recovered from the effects of the Dungan insurrection, by which a great part of North-West China was laid waste. "The Muhammadan rebellion lasted for seventeen years, and as its result the whole country traversed from Si-ngan-fu is depopulated and its villages destroyed. A few of the walled towns alone escaped. Out of fifteen millions of inhabitants before it took place, only one or two millions probably now remain. No confidence has yet returned to the people; for it is fourteen years since the rebellion ceased, and the greater part of the land is still untilled. The Muhammadans are still feared; they are braver than the "Heathen Chinee," who is demoralised more or less by opium-smoking. The rebellion was not put down, as is generally supposed, by the prowess of the Chinese soldiery, but by bribery, starving the garrisons out, and the distribution of buttons of rank to the Muhammadan Akhuns or leaders. The missionaries state that the accounts of the rebellion given to them by the Chinese who were partakers in it read like extracts from the book of Joshua: 500 heathens would fly at the sight of one Muhammadan; terror seized them, and they were slaughtered—man, woman, and child. Consequently the only cart line of communication from the Wei valley to Lan-chau, the intermediate base of operations towards Kashgaria, for 450 miles passes through a depopulated, and for several hundred miles an untilled, country, over hills up to 10,000 feet high, having soil of loess, slippery in wet weather, with rivers running in deep gullies, and with a Muhammadan population on either flank holding their heathen brothers in no estimation" (Bell, *loc. cit.*, p. 65).

Tai-yuen—Tien-tsin—Ta-ku.

From Tung-kwan the great Central Asian trade route runs in a north-easterly direction to Tai-yuen-fu, capital of Shansi, but otherwise a place of no great strategic or commercial importance. It is a walled town of about 50,000 inhabitants, standing at an altitude of 2260 feet in the upper valley of the Fuen-ho affluent of the Hoang-ho, 185 miles above Ping-yang-fu.

From this point the great trade route still continues to trend north-eastwards through Pau-ting-fu, administrative capital of the province of Chili (Pe-chi-li) to Peking, capital of the empire. Both of these places lie in the Pei-ho basin, and consequently have their natural outlet in the great northern seaport of Tien-tsin. This place, which has a population (1894) of over a million, stands on a vast plain between lakes Ta-ho and Ta-po, at the junction of the Pei-ho with the Grand Canal. Since 1860, when it was thrown open to foreign trade, it has become the most important place in North China next to the imperial capital, and as a commercial emporium is surpassed only by Shanghai and Canton. Unfortunately the bad state of the bar and the intricate navigation of the Pei-ho render the approaches too difficult for large European vessels, which are compelled to anchor in the roadstead of Ta-ku at the mouth of the river. From this point foreign merchandise is forwarded by barges and junks to the British and other European establishments at Tze-chu-liu, three miles below the city. Formerly a mere village, Tze-chu-liu has become quite a western town, its broad well-kept thoroughfares, substantial buildings, and English residences, presenting a marked contrast to the narrow, unsavoury streets and rickety wooden structures of the Chinese quarter. Here the only noteworthy monument was the fine Roman Catholic Cathedral, whose crumbling ruins now serve only to commemorate the terrible outburst of native fanaticism in 1870, when the church was destroyed and nearly all the French priests and nuns massacred by the infuriated populace.

Since 1886 Tien-tsin, from which the vast and populous provinces of North China draw all their foreign supplies, has been connected by rail with Ta-ku and with the important coal-mining town of Kaiping, the "Newcastle of China." Here quite an English colony has been established, and the great preponderance of British influences in Tien-tsin itself is clearly shown by the subjoined table of the shipping returns for the last two decades :—

	1873.	1879.	1888.	1893.
British .	103,865	194,580	237,175	262,180 tons
American .	99,300	25,475	710	600 ,,
Chinese .	22,000	264,000	165,000	179,000 ,,
Others .	36,000	38,000	35,000	26,000 ,,

As the Pei-ho valley gives direct access to the imperial capital, the Chinese Government has erected some extremely formidable defensive works at Ta-ku, which, following the windings of the river, is distant 60 miles from Tien-tsin. The forts, which stand on both sides of the estuary, were captured by the British fleet in the war of 1858-60; but since then they have been reconstructed and greatly strengthened with guns of the heaviest calibre, and completed by a vast entrenched camp, docks, and other harbour works. It was during the bombardment of these forts that an American commodore came to the aid of the English with the now historical exclamation: "Blood is thicker than water." The approaches to the Pei-ho are further strengthened by the two strongly fortified naval stations of Port Arthur at the extremity of the Liao-ti-shan promontory, and Wei-hai-wei nearly facing it on the south side of the Strait of Pe-chi-li on the north-east coast of Shantung. Both of these strongholds were captured by the Japanese during the war of 1894-5.

Wei-tiẹn—Chefu—Kinfao

In the peninsular province of Shantung, one of the oldest seats of Chinese culture, and birthplace of Confucius, the chief centres of population are the capital Tsinan, the Chinangli of Marco Polo, on the left bank of the Hoang-ho, with a population (1893) estimated at 220,000; Wei-hien, the largest city in the peninsula

THE HARBOUR OF WEI-HAI-WEI.

(250,000 ?); Tsing-chau, a little farther to the south-west, on a coast stream flowing to the Gulf of Pe-chi-li, a great stronghold of Islam in East China; Teng-chau and Yentai, better known as Chefu, the two chief seaports, both on the north-east coast; Taingan-fu, the "city of temples," on the Tawan (Wun) tributary of the Hoang-ho, one of the most frequented places of pilgrimage in the whole of China; and Yenchan-fu, a historical city in a marshy district traversed by the Grand Canal.

About 12 miles west of Yenchan lies the renowned city of Kinfao, "birthplace of Confucius, and still inhabited almost exclusively by his descendants, at least 20,000 of whom bear his name. Although a fine vigorous race, not one of them seems to have distinguished himself in any way during the twenty-four generations which have elapsed since their common ancestor bequeathed his moral code to the empire. The chief temple raised to his memory is one of the largest and most sumptuous in China, and contains a series of inscriptions dating from all the dynasties for the last 2000 years. The accumulated treasures of vases, bronze ornaments, and carved woodwork, form a complete museum of Chinese art. At the entrance of the palace is still shown the gnarled trunk of a cypress said to have been planted by Confucius, while urns, tripods, manuscripts, and other precious objects, said to have belonged to the philosopher, are preserved in the private apartments of the princely head of the family. The domain of this dignitary, who is a direct feudatory of the empire, is no less than 165,000 acres in extent. When Kinfao was seized by the Taiping rebels, they respected the temple, the palace, and all their contents, and even spared the life of the local governor, contrary to their invariable practice. Near the temple is the grave of Confucius, in the centre of a vast space occupied by the family necropolis. Towards the south-west is another cemetery near the small town of Tsin-hien, which for the last twenty-two centuries has received the remains of all the descendants of Mengtze (Mencius), the most renowned disciple of Confucius" (Reclus, viii. p. 177).

Hang-chau—Shaohing—Ning-po—Fu-chau—Amoy.

Before the middle of the present century Che-kiang, northernmost of the seaboard provinces south of the Yang-tse-kiang, and smallest of the great territorial divisions of the empire, was probably the densest peopled region of the whole world. Within an area only 6000 square miles larger than that of Ireland was packed an enormous population officially estimated at no less than 26,000,000, and certainly exceeding 20,000,000, or say from 550 to 600 per square mile. But Che-kiang and the neighbouring provinces in the Lower Yang-tse-kiang basin suffered even more from the Taiping rebellion and its after-effects than did the north-western region from the Dungan revolt. When famine and pestilence, following in the train of the wholesale butcheries on both sides, had done their work, according to Richthofen's estimate, not more than 5,500,000 inhabitants survived in Che-kiang. Since then the empty spaces have been reoccupied by multitudes from all the surrounding lands, and at present (1896) the population probably exceeds ten millions.

Some of the ruined cities have also partly recovered their former prosperity; and one at least, Hang-chau-fu, the capital, is already credited with a population of over 800,000. It stands at the head of a deep and spacious inlet on the left bank of the T'sian-t'ang, largest river in the province, through which a large trade is carried on with the interior. There are some well-stocked shops fine quays, pagodas, towers, and triumphal arches, and the surrounding district is described by the naturalist Robert Fortune as the "Garden of China." Hang-chau is the "Quinsay" which excited the admiration of Marco Polo beyond all other cities in the empire, and of which he spoke in such apparently extravagant terms as to bring upon him the derision of his friends in Europe. Yet his description of a metropolis 100 miles round, with 1,600,000 houses, 3000 baths, 12,000 stone bridges, and so on, was in a measure confirmed by later observers, amongst others Martinus Martini, who assures us that the closely packed houses extended in his time (seventeenth century) without interruption for a distance of 50 li, or about 16 miles.

Shaohing (Show-hing), a little east of Hang-chau, and on the south side of the same bay, is noted for the longest viaduct in the world, which is carried on 40,000 arches for a distance of 86 miles across a formerly marshy tract eastwards to Tsinhai at the mouth of the Yung-kiang. But Shaohing itself, which 2000 years ago was the capital of a great empire in South-East China, is now a decayed place known chiefly for the preparation of the so-called "Shaohing wine," a rich and highly appreciated cordial distilled from rice.

Ning-po, which stands on the left bank of the Yung-kiang at its junction with the Yuyao, is now the most flourishing centre of trade in Che-kiang. It owes this pre-eminence partly to the advantages it enjoys as one of the treaty ports, partly to its position at the confluence of two navigable rivers and at the converging point of several canals ramifying throughout Che-kiang and the conterminous province of Kiang-si. Ning-po is also a great military stronghold famous in the annals of Chinese warfare; it was captured during the "Opium War" by the British, who made it the base of their operations

FU-CHAU.

against Nan-king. But in recent years much of its foreign trade has been diverted to Shanghai and to the deep and sheltered port of Tinghai, capital of the neighbouring Chusan Archipelago.

Below Ning-po follows Wen-chau, another treaty port, which, however, is no longer frequented by British shipping. Beyond this place South Che-kiang merges almost imperceptibly in the neighbouring province of Fu-kien, the capital of which, Fu-chau, on the left bank of the Min, is the largest seaport between Shanghai and Canton, with an estimated population (1894) of 700,000, including the extensive suburb of Nantai on the opposite side of the river. In this suburb reside the European merchants, chiefly English, who deal principally in the "Bohea" and other famous black teas grown in the upland districts of Fu-kien. The great arsenal, the largest in China, established by the imperial government on the Min a short distance below Fu-chau, was destroyed by the French during the hostilities which were carried on in 1884 without any formal declaration of war on either side.

South of the capital follows the busy seaport of Amoy, which stands on the island of Hia-men at the head of a spacious and well-sheltered bay, affording excellent anchorage to shipping. Amoy, which has a population of about 300,000, has had commercial relations with Europe for several centuries. But the port was closed to all except the Spaniards from 1730 to 1842, when it was thrown open to the trade of the world by the Anglo-Chinese treaty of Nan-king. The surrounding district is densely peopled, and the population increases so rapidly that large numbers annually seek employment abroad. A considerable proportion of the Chinese emigrants, and especially of the coolies employed on the plantations, are drawn from this part of Fu-kien. Ku-cheng (Kow-chang), a walled city two

days' journey from Fu-chau, was the scene of an outbreak of fanaticism in August 1895, which led to the massacre of the English missionaries, Mr. and Mrs. Stewart, their family, and some young women engaged in mission work.

On the coast between Fu-chau and Amoy lies the former provincial capital Tswan-chau, commonly called Tsaetung, which by some authorities has been identified with Marco Polo's Zayton (Zaïtun), said at one time to have been the largest seaport in the world. Both Arabs and Italians (Genoese) traded with this place, which was famous for its velvets and other costly stuffs, and from which, according to Ibn Batuta, the word *zaituniah* ("satin") takes its name. Tswan-chau lost its trade by the gradual silting up of its harbour, when most of the traffic was diverted north to Fu-chau and south to Amoy.

Canton.

But all these places have in recent times been outstripped by Canton, properly Kwang-chau-fu, capital of the province of Kwang-tung, and for many years the chief emporium of South China, as well as one of the largest cities in the world, with an estimated population (1893) of 1,800,000. The "picturesque charms" of Canton, with its narrow, tortuous lanes, rickety houses, quaint gables, many-coloured façades, gaudily-painted sign-boards, and streamers fluttering in the breeze, groups of motley crowds encumbering the unpaved, ill-kept thoroughfares, are often dwelt upon in rapturous language by passing European observers. But such attractions are dearly paid for by perhaps the greatest aggregate of human squalor, filth, and misery in the whole world. Canton is in truth little better than one vast unsavoury slum, where a certain stage of outward culture has been reached by a people ignorant of the primary conditions of true civilisation. Such is the contempt of all sanitary regulations, that the whole place is a seething mass of material and moral corruption, in which the bubonic pestilence, most horrible of all contagions, is always lurking, and occasionally breaks out with great virulence, as in the year 1894, when it even spread to the neighbouring British settlement of Hong-Kong. In Chicago courageous strangers visit the huge shambles where thousands of animals are "packed" in a few hours; in Canton their eyes are feasted with the indescribable horrors of Chinese courts of justice, prisons, and execution grounds.

Nevertheless Canton is so favourably situated for trade that it must always remain a great commercial mart, in which the ravages of vice and disease are repaired by a continuous flow of immigration from the thickly-peopled rural districts. It stands at the head of the intricate system of ramifying channels which form the common delta of three great navigable rivers, the Si-kiang and Pe-kiang in the west, and the Tung-kiang in the east. But most of these channels are constantly shifting their beds, hence are unnavigable except by junks and barges. Even the eastern or largest branch, the "River of Canton" as it is called in a pre-eminent sense, is accessible to vessels of average size only as far as Hoang-pu (Whampoa), eight miles below Canton, while ships of heavy draught are obliged to stop much farther down, outside the bar, which has a depth of scarcely 13 feet at low water. This branch, the Chu-kiang, or "Pearl River" of the natives, is defended at the head of the estuary by two forts crowning two rocky cliffs, which have been compared by the Chinese to the jaws of a tiger; hence the European expression *Bocca Tigris* ("Tiger's Mouth"), familiarly the "Bogue," which is commonly applied to the narrows at this part of the Canton River.

Canton, which lies about 40 miles above the entrance to the estuary, is one of the oldest places in the south of China. Some 2000 years ago it bore the name of Nanwuching, which was afterwards changed to Yang-ching, the present official designation, Canton being simply a European corruption of Kwang-tung, the name of the province of which it is the administrative centre. For ages it has traded directly with Formosa, the Philippines, Malaysia, and Indo-China, and the huge, unwieldy junks still engaged in this coasting traffic, though now largely replaced by steamers, stand out conspicuously amid the multitudes of smaller craft crowding all the channels of the delta, and affording accommodation to a permanent floating population of some 300,000 or 400,000 souls. The foreign trade, which is carried on chiefly under the British flag, ranks in importance next to that of Shanghai, the total annual exchanges averaging about £6,000,000. Tea and silk are the two chief items of the export trade, the more important imports being rice, opium, raw silk, cotton and wool, and textile fabrics of all kinds.

Canton was long the only Chinese seaport open to foreign traffic. But between the first Anglo-Chinese war of 1841-42 and 1889 as many as twenty-one other places, the "treaty ports" as they are called, were successively thrown open to the trade of the world. On the Yang-tse-kiang there have also been established six stations, where steamers are permitted to stop for the purpose of landing and shipping goods and passengers, and recently Chungking in Se-Chuen towards the Tibetan frontier has been added to the list of riverside free ports. At this point, 1600 miles from Shanghai, and twenty-six to over thirty days by boat above Ichang, the Yang-tse is already a majestic stream considerably

CANTON, SHOWING RIVER AND FOREIGN SETTLEMENT.

over a mile wide at high water. Chungking was visited in 1889 by Mr. A. E. Pratt, and officially by Mr. Hobson, Commissioner of Customs, for the purpose of opening it to foreign trade. The party conducted by Mr. Hobson took thirty-three days to ascend from Ichang through the gorges and rapids to Chungking; but the pilot of the expedition was of opinion that the place might be reached for eleven months in the year by steamers drawing 5 feet. The rapids vary greatly with the rise and fall of the river, many disappearing during the floods, while new and more formidable ones come into existence. The famous Ichang gorges are described as of a highly romantic and somewhat gloomy aspect, varied with lofty and inaccessible mountains, lovely glens and charming scenery, while presenting no great difficulties to the navigation of light craft. The Yeh-tan reefs, most formidable of the rapids, run obliquely athwart the stream for two-thirds of its width, leaving the remainder of the channel clear, but swept by a seething, swirling mass of water. The extent of poppy culture on the riverside tracts is astounding, apparently enough to supply a great part of China were the Indian opium trade suppressed.

Kwei-yang—Wu-chau—Yun-nan—Tali—Momein.

In Kwei-chau, which from the geographical standpoint may be regarded as a southern extension of Sze-chuen, the only important centres of population are the two administrative cities of Kwei-yang-fu and Chen-yuan-fu. Kwei-yang-fu, which is the provincial capital, stands on a small head-stream of the Wu-kiang (Kung-tan), and also communicates by navigable waters with the great lake Tung-ting. In the vicinity are seen the remains of some old temples, and of a palace said to have at one time been a royal residence. But the whole region suffers from malaria, and from the raids of the unreduced Miao-tse wild tribes, and even before the destructive Taiping wars the province was one of the least thickly peopled in China, with an estimated population of little over 5,500,000.

This remark applies also in great measure to the two conterminous southern provinces of Kwang-si and Yun-nan, the former with little over 5,000,000 inhabitants, the latter despite its vast size (108,000 square miles), with a population reduced from under 12,000,000 before the wars to about 6,000,000 in 1894. Kwei-ling-fu, capital of Kwang-si, lies in the hilly north-eastern district, about 250 miles north-west of Canton, with which it communicates through the Kwei-fung affluent of the Si-kiang. Near the confluence a canal branching off from the main stream gives access through the Siang (Heng-kiang) to Lake Tung-ting at the foot of the Nan-shan range.

But at present the most important place in Kwang-si is Wu-chau, which is probably the largest city in South-West China, with an estimated population (1893) of about 200,000. Wu-chau has the advantage of standing on the north (left) bank of the Si-kiang below the Kwei-ling confluence, and is consequently accessible to large river craft from Canton. It was even reached by the allies during the Anglo-French expedition of 1859, and since then it has become a flourishing depot for the salt and manufactured goods forwarded by water from Canton, and taken in exchange for the timber, cabinet-wood, rice, copper ores, and other produce of the conterminous province of Yun-nan.

Yun-nan-fu, administrative centre of Yun-nan, lies in that part of the province which drains to the Yang-tse-kiang, and is one of the few places in this district which escaped the widespread ruin caused by the terrible Panthay insurrection of 1855. It stands near the north bank of a large lake, the so-called "Sea of Tien," which sends its overflow through the Pulu-shing northwards to the Yang-tse-kiang. Although its present population scarcely exceeds 50,000, Yun-nan-fu, probably Marco Polo's Yashi, is a very ancient city which cannot fail soon to recover all its former prosperity. It abounds in natural resources of all kinds, an exceedingly fertile soil yielding heavy crops of corn, tobacco, flax, fruits, and opium, and the richest copper mines in the whole of China. A neighbouring hill is crowned by a temple constructed of this metal, the current price of which is regulated by the Yun-nan-fu market. Formerly the district produced great quantities of wax; but the bees are said to have been all killed by surfeiting on the poppy flower since the introduction of the opium industry.

Another noted historical place in Yun-nan is Tali-fu, which may be said to lie within the Mekhong Catchment basin. At least it stands at an altitude of over 6500 feet on the west side of the great lake Tali, which during the floods rises to a height of nearly 20 feet, and sends a copious discharge through the Yanghi-kiang to the left bank of the Lan-tsang, as that section of the Mekhong is called which traverses the province of Yun-nan. In the days of Marco Polo, Karayang, as Tali-fu was then called, ranked as one of the chief cities of South China, and is referred to by the great traveller as the "capital of seven kingdoms." During the Panthay revolt it was also chosen as the residence of "Sultan Soliman," otherwise known as King Tu-wen-hsia, and this honour brought upon it the wrath of the imperial government. Its capture by the Chinese troops in 1873 was followed by the massacre of about half of its 50,000 inhabitants, and when the place was visited by Gill, four years later, it was still in ruins.

But Tali-fu lies on the direct trade route between Upper Burma or Sze-chuen, and as it is also favoured by an excellent climate and abundant natural resources, the effects of the recent calamities will probably soon be obliterated. The surrounding district is extremely fertile, the neighbouring hills are rich in minerals of all kinds—the precious metals, salt, fine marbles—and the lake itself teems with fish, which are captured by aquatic birds trained for the purpose. This great lacustrine basin develops a gentle crescent of some 30 miles in the direction from north to south, with an average breadth of 6 or 8 miles, and an extreme depth of over 300 feet. The navigation is rendered dangerous, especially in winter, by the sudden squalls which descend the slopes of the Tien-tsang range, skirting the west side at an elevation of 9000 to 10,000 feet above the level of the lake, or about 16,000 feet of absolute altitude. As in the Yun-nan-fu district, poppy culture has in recent years acquired a great development on the surrounding plains, and the opium produced at Tali-fu is esteemed the best in Yun-nan. This increase of the opium industry goes hand-in-hand with the decline of the foreign tea trade, and while India (with Ceylon) is rapidly taking the place of China, as the great producer of tea, China promises soon to become independent of the Indian market for its supply of opium. Thus a balance of commercial interests may be brought about, without the financial disorder likely to follow the Utopian schemes of sentimental political economists.

In that section of Yun-nan which is comprised within the Irawadi basin, the chief place is Momein, properly Tengyueh-ting, on the Ta-ho (Shin-cha-ho) affluent. Although a town of scarcely 5000 inhabitants,

Momein is regarded by the English as the future gateway to the rich provinces of South-West China. In all the proposed railway schemes its name comes to the front, and its political and commercial importance must certainly increase with the development of the French aggressive policy in the Mekhong basin. Before the late Panthay insurrection, Momein, which dates from the fourteenth century, had already become an important market for the mineral products of Yun-nan, and here were worked the jade from Mogoung, the amber from the Hu-kong valley, the copper of Yun-nan-fu, the gold, silver, tin, lead, and other metals widely diffused throughout the province. But these industries have languished since the reoccupation of the place by the imperialists in 1873, and at present Momein depends chiefly on its cattle and rice trade. According to Anderson it is regarded as the headquarters of the confederation of the nine Shan states, known before the recent political changes by the name of Koshan-Pri.

9. *Highways of Communication.*

Roads as understood in the West can scarcely be said to exist anywhere in Central and East Asia. Even in China proper, settled as it has been by a practical and intelligent people for ages, the main highways are little more than beaten tracks left to take care of themselves. Morrison tells us that on one occasion in 1880 Mr. Wylie's carter preferred to deviate from the roadway into the neighbouring fields, where walls 18 inches high had to be crossed.

But such as they are, well-known routes have for thousands of years intersected the empire in all directions, marking the lines followed at all times by trade, at intervals by migrating hordes and conquering hosts, or occasionally by such illustrious explorers as the pilgrim Hwen Tsang, and the famous traveller Marco Polo.

China proper is intersected in every direction by 2000 imperial highways, which with the great number of navigable streams, and the extensive system of canalisation, render the country one of the richest in means of communication in the whole world. Unfortunately the State has neglected to keep either roads or canals in repair, or protect them from the wear and tear of time and weather, so that they are now partly impracticable. Morrison gives a deplorable account of the present condition of the Grand Canal, on which Peking largely depends for its supplies. In 1880 some parts were in such a ruinous state that the boats could not pass through, and portages were formed at the sides for discharging and reshipping farther on.

In Tibet the chief road connecting Gartok and Lassa leads also to Rudok, Leh, and India, on the one hand, and on the other to Batang, Litang, and China. This highway is fairly well maintained with "tarjums," or stations, at intervals of 20 to 70 miles, where horses, yaks, and coolies are kept in readiness for the Government messengers and officials. The messengers are never allowed to stop, except to eat and change horses. To prevent them from taking off their clothes, the breast fastening of the overcoat is sealed, no one being allowed to break the seal except the official to whom they are sent. At the end of their 800 miles' ride they look haggard and worn, their faces cracked, their eyes bloodshot and sunken. The entire distance is covered in about twenty-two days.

Next in importance is the northern route between Lassa and Peking, running by Nak-chu-ka to Kuku-nor, Sining, Mongolia, and the capital. This route is generally followed by the Kutukhtu of Urga on his pilgrimage

LITANG, ON THE BATANG ROAD.

to Lassa, and was also traversed by Huc and Gabet in 1845, and partly by Prjevalsky in 1872. From Nak-chu-ka two other routes are said to run north-west to Polu and across the Sok country to Lob-nor. In 1874 the Pundit followed a track from Lake Pang-gong and Noh across grassy plains and lacustrine valleys by Tengri-nor to Lassa. A path also leads from Noh near Rudok north to Khotan across the west end of the Kuen-lun and through Polu. Although at present less used than that over the Karakorum, this route may some day acquire great importance, as it seems to afford the most direct and easiest means of communication between India, Tibet, and Kashgaria, *via* Rudok, the Bogo-la Pass, Totling, and the Niti Pass.

But the most convenient routes at present between Tibet and India are—1. That by Lake Palti over the Pari-jong Pass, down the Chumbi valley, and over the Yelap Pass to the Tista valley and Darjiling, partly followed by Bogle, Turner, and Manning; 2. Nain Sing's route over the Karkang-la Pass to Tawang, whence three roads lead to Odalguri and Dewangarhi in Assam.

But the whole line of the Himalayas is pierced by numerous passes giving access from Tibet to India, the chief of which are the Niti, between Garhwal and Gartok; the Kirong, from Katmandu; the Nilam route leading through the Butia Kosi Gorge, where the stream is crossed fifteen times, and where the path for one-third of a mile is formed by stone slabs from 9 to 18 inches wide, supported by iron pegs driven into the face of the cliff 1500 feet above the torrent; lastly, the Karkang-la, crossed by the Lassa-Tawang route.

The Nan-lu is connected with the Pe-lu, or corresponding northern route, by several passes leading north and south over the Tian-shan, between Zungaria and the Tarim basin. Of these the westernmost is the Muzart, or "Ice Pass," between Aksu and the Ili basin, which leads over a huge glacier close to the Tengri-khan, and which is extremely difficult, although kept open by gangs of labourers. Farther east are the routes between Turfan and Urumchi, explored by Dr. Regel, and between Hami and Barkul, where the Tian-shan would seem to gradually merge in the Altai system.

Between the Pe-lu and Nan-lu, consequently in the heart of the Tian-shan, there is a route, followed in 1876 by Prjevalsky, leading from Kulja in the Ili valley over the Narat Pass (9800 feet) and across the Lesser Yulduz plateau to Lake Bagrach and Korla, and thence by the Tarim valley to Lob-nor and the Altyn-tagh range. Here the "Kalmuk Road" leads south to Tibet, while another track is said to run along the north foot of the Altyn-tagh to Sa-chu, and thence across the Gobi north to Ngasi-si-chan and Hami. Hami is also reached from Su-chau (Su-chow) at the western extremity of the Great Wall by a path which forms a section of the great historical highway between China and the West.

At present the chief trade route between China and Russia runs farther east from Peking *via* Kalgan to Urga and Maimachin, close to Kiakhta on the Siberian frontier. South of this the more direct road runs from Hankow through North Kansu and across the Gobi to Hami, and thence by the Pe-lu route through the Zungarian depression, or the alternative Irtish valley, to Orenburg. The Russians are beginning to see that the future trade route must follow this line, which is practicable for carriages throughout the whole distance of 2580 miles from Zaisan to Hankow, except a section of 160 miles, which presents no difficulties to pack animals. It can be traversed in 140 days, whereas by the far more difficult Kiakhta road, which is 1800 miles longer, it takes 202 days to reach Peking.

From Peking to Kashgar the Great Central Asian Trade Route, traversed from end to end by Colonel Bell in 1887, has a total length of nearly 3500 miles. It passes as a cart-road through the province of Shansi to Si-ngan-fu, capital of Shensi on the Wei; thence through Kansu by Lan-chau and Liang-chau to Su-chau, and so on across the Gobi north-west to Hami; thence by the Pe-lu to Urumtsi, and from this place by the Nan-lu to Kashgar. The three main sections are: Peking to the Wei valley, 770 miles; Wei valley to Hami, 1322; Hami to Kashgar, 1347; total, 3439. "It is for very many miles suited for a single line of traffic only, passing through deep and narrow gullies in loess hills; unmetalled throughout, becoming impassable after heavy rain or snow, and even light rains halving the rate of progress over it. It passes for hundreds of miles between the Wei valley and Hami, to a great extent through a depopulated and uncultivated country, partly peopled by Chinese Muhammadans inimical to the Chinese Government" (Bell, *loc. cit.*).

Meantime there is only one railway in the empire. The short line, 9 miles long, opened by an English company in 1876, from Shanghai to Wusung on the Yang-tse estuary, had a brief career of sixteen months, when it was bought up and destroyed by the Government. But since then a line has been constructed from Tien-tsin to Tongku near the Taku forts, a distance of 27 miles, and for 67 miles beyond that point to the Tungshan and Kaiping coalfields. Lately this line, intended mainly for the transport of minerals, has been extended 40 miles farther on in the direction of the Manchurian frontier, and it is now (1894) reported that it is to be immediately pushed on through Manchuria to the Amur River. The urgency on military grounds of a northern line to meet the Russian Trans-Siberian railway seems

home for the study of the physical sciences. Hence the majority are satisfied with committing to memory antiquated formulas and axioms of doubtful practical value. Each degree is denoted by a distinctive academical head-dress, and the various public appointments are awarded to the lettered classes in proportion to their literary qualifications.

Great improvements, however, have in recent years been introduced into the educational system, thanks especially to the action of Mr. Robert Hart, Inspector-General of Chinese Customs. A series of elementary scientific works have been translated for circulation amongst all classes, and the *Tungwen,* or Foreign College of Peking, already embraces a scientific curriculum superadded to the study of European languages, which was its original aim. It has now several foreign professors, and over a hundred students, and among the subjects taught are chemistry, natural history, mathematics, physiology, and astronomy. A complete course lasts for eight years, and successful students are appointed to high offices in the various official departments.

The administration of State affairs is, in the first instance, entrusted to a Privy Council composed of four members—two of Manchu, and two of Chinese nationality. Two representatives of the "Kang-ling," or "Great College," take part in the sittings, their duty being to see that no measures are adopted at variance with the fundamental laws as contained in the writings of Confucius. The members of the Council and their associates are called "Ta-hyo-si," or Ministers of State, and control the ministerial functions. Amongst these are—1. The Ministry of the Civil Officials, which watches over the conduct of these officers; 2. The Ministry of Finance; 3. The Ministry of Customs and Ceremonies, whose province it is to see that the national laws and usages are observed by the people; 4. The Ministry of War; 5. Public Works; 6. The Supreme Military Tribunal.

Theoretically above all these departments is the "Tu-khe-yi-veng," or Board of Public Censors, consisting of forty or fifty members, with a Chinese and a Manchu President. According to ancient usage, each member of this Council has the right of personally directing the Emperor's attention to anything in the State calling for censure. At the deliberations of the various ministries one of the censors is always present, though not entitled to vote. Others visit the various provinces and inquire into the administration of the viceroys.

But with all this censuring the country continues subject to much maladministration. Under the ruling dynasty the public functionaries are understood to be often wanting in probity. In an absolute yet democratic system, in which knowledge and personal merit are the only legal avenues to distinction, a universally high moral tone can alone ensure respect for the laws and administration of justice. But these qualities began to lose their efficacy as soon as the State became tainted with corruption. In a patriarchal form of government, imposing on the authorities no check beyond the duty to act in a paternal spirit, the absence of morality had necessarily evil results for the nation, which thus became, to some extent at least, the victim of venality. Those political institutions of the *literati*, which, seen from a distance, are apt to excite the admiration of enthusiastic believers in the higher theories of a freedom and equality limited only by intelligence and morality, have here produced a depressing thraldom. The lower mandarins are chiefly occupied in raising money, either to redeem the pledges made to those of higher rank to whom they owe their appointments, or to procure still higher honours by fresh payments. All the authorities are surrounded by satellites attached to them as bodyguards, police agents, or executioners. These men are charged with the faults usually characteristic of pretorian guards. And the administration is far from being conducted with the integrity and efficiency which would be expected in a European State.

The imperial forces consist of two distinct branches, one belonging to the nationality of the present Manchu dynasty, the other composed of Chinese and other races. The first, on which the chief reliance is placed, is organised in eight "banners," and supplies the garrisons of the large towns, fortresses, and other important positions throughout the empire. Hitherto their defective equipment and rudimentary instruction have rendered these troops quite incapable of contending with European forces greatly inferior in numbers. Of late years, however, a marked improvement, especially in the artillery service, has been effected by the English instructors employed to train the men.

The second branch consists of about 1,000,000 Chinese and other races distributed over the empire, who reside with their families, support themselves by their own labour, and assemble only on special occasions in the capitals of their several provinces.

Before the war of 1894-5 China also maintained a powerful fleet, which, under European officers, had been brought to a state of considerable efficiency, but which was destroyed by the Japanese during the course of hostilities. Captain Lang, R.N., who was in command between 1883 and 1890, speaks well of both services at that time. "There is the making of anything out of the Chinese. They are well-trained and excellent marksmen, and the discipline is very good. As compared with the Japanese, the Chinese navy is about equal. Probably the Japanese have more dash and go,

A TATAR GENERAL. PRINCE CH'ING. LI-HUNG-CHANG.

THREE YELLOW JACKETS.

and there is among them more *esprit de corps*. But when I was in command the Chinese were well to the fore in steam tactics, and their drills were excellent and unsurpassed anywhere. All depends, however, on how they are led. If the men have confidence in their leaders they will face anything. The Northerners especially, the Shantung and the Tien-tsin men, never fear death. In a service like the Chinese, however, there are many different types of men, and between the people of Fuchu and Tien-tsin, for instance, there is as much difference as between Englishmen and Frenchmen. The great difficulty in China is that the services are looked down upon by the civilian classes, while in Japan men of princely birth serve in the army and navy. The Chinese ironclads are very good, though some of their guns are becoming obsolete. The vessels are kept in good order, and, contrary to popular belief, are perfectly clean. The guns also, Krupps and Armstrongs, are in excellent order. The strength of the army on paper is 1,750,000, but I do not believe they could put into the field more than 200,000 men. There are not many European-drilled troops, but these are exceedingly well trained and are wonderful marksmen. At rifle firing at 500 or 600 yards no one can touch them. When under arms, however, one half of the Chinese army is made up of savages."

Enjoying the advantage of a vast seaboard, intersected by numerous streams, estuaries, and canals, abounding in cereals, tea, silk, cotton, flax, hemp, sugar-cane, indigo, tobacco, coal, and iron ores, China possesses perhaps the largest material resources of any country in the world. For the last half century the West has paid in cash fully three-fourths of the value of the exports in silk and tea alone. And if we consider the amazing industry of the Chinese, the energy and endurance of the working classes, their thrift and frugal lives, their respect for precedent,

TEA EXPORTED TO GREAT BRITAIN.[1]

Year.	Quantity.	Value.
1888	98,098,000 lbs.	£4,017,000
1890	68,551,000 „	2,616,000
1891	57,024,000 „	2,204,000
1892	50,575,000 „	1,820,000

REVENUE (1892).

Land tax	£4,000,000
Maritime customs under foreign supervision	4,750,000
Native customs, maritime and inland .	1,280,000
Transit levies, foreign and native . .	2,260,000
Salt, taxes and levies . . .	1,980,008
Rice tribute	570,000
Licences	410,000
Total . . .	£15,250,000

External debt (1892) estimated £5,000,000 ; 1896 (since the war with Japan), estimated £65,000,000.

SHIPPING (1892).

	Vessels (mostly steamers) entered and cleared.	Tonnage.
Great Britain . . .	18,973	19,317,000
China	14,532	6,561,000
Germany	2,016	1,466,000
Japan	719	631,000
United States . . .	111	62,000
France	144	253,000

ARMY (BEFORE THE WAR OF 1894-95).

Manchuria, two Army Corps . .	70,000
Centre, peace footing . . .	50,000
Turkestan, „	40,000 (?)
Territorial Militia, peace footing . .	200,000

NAVY (1892).

Battleships—1 first-class, 1 second-class, 3 third-class.
Cruisers—9 second-class, 47 third-class.
Torpedoes—2 first-class, 26 second-class, 15 third-class.
Port Defence Vessels, 9 ; Gunboats, 11.

[1] This Table shows the steady decline of the China tea trade, owing to the development of tea culture in India and Ceylon.

Railways (1895)—None open for general traffic; one short mineral line.

Telegraphs—System well developed; since 1892 connected through Siberia with Europe; most of the treaty ports on frontier stations connected.

Foreigners Resident in the Treaty Ports (1894).

British Subjects	4163	Portuguese	410
Americans	1336	Spaniards	357
Japanese	1070	Sundries	1200
Frenchmen	786		
Germans	770	Total	10,092

Emigration.

To United States since 1820	230,000
To Peru, 1860 to 1894	120,000
To Singapore last 50 years	130,000
To Dutch East Indies last 50 years	350,000
To Australia last 50 years	? 30,000

Hong-Kong.

Area in sq. miles, 29; Population (1893), 221,440.
Revenue, £466,000; Expenditure, £488,000; Debt, £200,000.
Imports, £1,800,000; Exports, £837,000.
Registered tonnage, 28,000; Tonnage entered and cleared, 10,294,000.

CHAPTER VI

JAPAN

1. *Boundaries—Extent—Area—Name.*

In its widest sense the Japanese Empire comprises the Kuriles, Yesso (Yezo), Nip-hon (Japan proper), the Liu-kiu islands, the Bonin Archipelago, and, since 1895, Formosa with the adjacent Pescadores group. It thus stretches in an almost continuous chain across 28 degrees of latitude (50°-22° N.) for over 1900 miles from the southern extremity of Kamchatka to the Philippine Islands. But in the narrower sense Japan proper consists of the four large islands of Yesso, Hon-do (Hon-shiu), Kiu-shiu, and Shi-koku, together with about 3850 circumjacent rocky islets lying between 30° 35′ to 45° 30′ N., and 129° to 146° E., with a total area of 147,655 square miles, and a population (1892) of 40,719,000. It is washed east and south by the Pacific Ocean, and separated on the north by La Perouse Strait from Sakhalin, on the west by the Strait of Korea and the Sea of Japan from Korea and Russia. Between Yesso and Hon-do flows the narrow Sugaru (Sangar) Strait, while Kiu-shiu is separated by the Bugo Channel, Suwo Sound, and Shimonoseki Strait, from Shi-koku and Hon-do. Lastly, between Shi-koku and Hon-do flows the marvellous Inland Sea (Seto-Uchi), which, with its countless islets, bluffs, headlands, and inlets, its limpid waters and glorious sub-tropical vegetation, presents a varied panorama of almost unrivalled natural scenery. At the same time, the navigation of these waters is everywhere endangered by innumerable rocks, shallows, reefs, and intricate channels, while the more open Sea of Japan is swept by fierce cyclones accompanied with waterspouts, whirlpools, and rapid currents.

The term Japan, applied in various modified forms to this great Archipelago, is not merely synonymous, but absolutely identical with the corresponding native name Nippon. The history of this word Nippon, which is by European writers wrongly restricted to the large central island of Hon-do, is extremely curious. The original Chinese form was Nit-pon, meaning the Orient, from *nit* = sun, and *pön* = origin. The word was in this form adopted about the seventh century of the Christian era by the Japanese, who soon assimilated the *t* to the *p*, whence Nip-pon, Nip-hon, and even Nif-hon. But in China the *t* was first dropped, whence Ni-pon or Ni-pen, and the initial *N*, through Mongolic influence,[1] afterwards changed to *J*, whence Ji-pen, the form current in the time of Marco Polo, whose Venetian Zipangu appears to derive from it, and to be the parent of all the European varieties of the word *Japan*.[2] This word was, as stated, from the first applied to the whole Archipelago, and not exclusively to the large island, for which the Japanese had no general name till that of Hon-do—that is, Original or Main Division—was introduced some years ago. Hence in our maps *Nip-pon* ought to be either altogether suppressed or extended to the whole group—that is, made

[1] During the Yen or Mongol dynasty (1260-1366), the mandarin or court language was greatly influenced by the Mongol phonetic system.

[2] The attempt made by Mr. George Collingridge (*Geo. Jour.*, May 1894) to show that Marco Polo's Zipangu refers not to Japan but to Java, scarcely calls for serious refutation.

synonymous with *Japan*, both being varieties of the common prototype *Nit-pon*.[1]

2. *Relief of the Land: Highlands—Volcanoes, Fuji-yama, Aso-san, Asama-yama.*

Till recently the Japanese Archipelago was commonly supposed to consist mainly of igneous ejections accumulated over a fissure in the ocean-bed, and forming a connecting link in the long chain of volcanic formations encircling the Pacific Ocean. But this view can no longer be accepted, or accepted only to a limited extent, since the geological surveys of Dr. Edmund Naumann, continued over a period of five years (1881-85), and extended to nearly every part of the empire except Yesso, have shown that the Archipelago is on the contrary "an advanced frontier of Asia," consisting for the most part of very old sedimentary rocks, deposited, like the mainland itself, in deep water in palæozoic times, and upheaved, like so many other mountain systems, by lateral pressure due to the gradual shrinkage of the earth's crust through secular cooling. Doubtless extensive longitudinal fissures were left, through which igneous matter was ejected in later ages. But although most of the loftiest summits are extinct craters, volcanic agencies have on the whole played a relatively small part in the geological history of Japan. If the Archipelago be compared, with the old geographers, to garlands of flowers, then the volcanoes may be likened to small pearls threaded among these garlands.

Amongst the fossiliferous rocks occurring on the uplands is a "Radiolarian slate," as Naumann calls it, the composition of which shows it to be of "a formation

[1] The expression Dai-Nippon—*i.e.* "Great Nippon"—also occurs, and answers to our "Great Britain," as applied to all the British Isles.

corresponding to that of the mud from the deepest part of the ocean-bed. We learn from these slates that the Japanese chain, or a large part of it, was submerged deep beneath the ocean surface during some portion of the palæozoic era. The Radiolarian slate is a deep-sea sediment, and perhaps the oldest sediment of this kind known" (*Geo. Proc.*, 1877, p. 89).

The neighbouring Pacific waters are the deepest that have yet been anywhere sounded; but they shoal somewhat gradually towards the east coast, while the incline is still more gentle in the comparatively shallow sea of Japan on the west side. Above these waters rise the Japanese uplands, which cover the greater part of the surface, and which, viewed as a single orographic system, are found to consist of a long series of folds running normally in the direction of the main axis of the Archipelago. But towards the central and widest part of Hon-do a great transversal cleft, Naumann's *Fossa Magna* ("Great Ditch"), marks off an area of profound disturbance between the northern and southern sections of the system. For some distance north of this cross fissure, above which rises Fuji-yama (Fujinoyama), culminating point of the Archipelago (12,425 feet), the folds curve round so as to run for the most part transversely to the insular trend, but resume the normal direction about 38° N. lat., between Sado Island and Sendai Bay. The folds themselves appear to be the result of a horizontal thrust or pressure acting in the direction of the Pacific Ocean, but arrested and deflected sideways at the point where it came in contact with the submerged Shichito Chain, stretching from Tokyo Bay southwards to the Bonin Islands. At least the Fossa Magna, where the dislocations begin, is disposed in a line with this marine ridge.

During its long life above the marine waters, the

original structure of the Japanese highlands has been somewhat obliterated by weathering, denudation, erosive action, mechanical pressure, and igneous agencies. Nevertheless, three primitive zones—an outer towards the Pacific, a median, and an inner facing the mainland—may still be distinguished, and are somewhat clearly marked, especially in the southern section south of the Fossa Magna. Here the outer zone traverses the islands of Kiu-shiu and Shikoku, and the Kii and Akaishi districts of Hon-do, rising to a height of over 7700 feet in Shikoku, and to about 10,000 near the transverse fissure. Beyond this point it is continued at intervals by the Quinto, Abukuma, and Kitakami mountain masses.

In the south the median zone is now represented by the innumerable rocky islets of the Inland Sea, a vast flooded depression disposed in the normal direction between the outer and inner zones. North of the Fossa Magna this basin is continued by a median range with crests 6000 feet high, extending to Awomori Bay at the northern extremity of Hon-do, and bearing numerous igneous cones. Both in the north and in the extreme south (Kiu-shiu) the median zone is the chief sphere of volcanic activity in the Archipelago, and here are accumulated enormous masses of erupted rocks.

Lastly, the inner zone, skirting the shores of the Sea of Japan, is of a more fragmentary character, its most salient feature being isolated volcanoes rising above circular basins formed by abrupt depressions. Such are the Sanpei and Daisen basins in the south facing the Oki Islands, in the north those of Gassan, Chokai, Moriyoshi, and Iwaki, extending from near the parallel of Sado Island to Sangar (Sugara) Strait, between Hondo and Yesso.

In Yesso (Yezo) exploration has been greatly retarded by the absence of roads through the trackless forests covering the greater part of the interior. The island, however, was crossed by Mr. B. S. Lyman in 1874, and in 1891 by Professor J. Milne and Mr. A. H. Savage Landor, by a route running at right angles to that followed by their predecessor. The whole surface is hilly and in parts mountainous, the highest peaks being Shribetsi in the south (7874 feet); Unabetsu in the north-east (5039); Ofuyu in the west coast (6000); Ishikari (7710) and Tokachi (8200) near the centre. Both old and recent eruptive rocks occur, as in Hon-do; but sedimentary formations seem to predominate, developing numerous ridges of moderate elevation. The narrow intervening valleys are watered by small streams, which do not converge in any large fluvial basins, but for the most part find their way in independent channels to the coast. These rivers teem with salmon, while the immense forests contain much valuable timber—oak, elm, walnut, birch, and maple—which might be exported at a profit. Coal abounds, and the mines opened at Sorachi are now connected by a railway with the coast. Yesso, which has an area of 86,880 square miles, and a population (1891) of 314,000, bears a curious resemblance in outline to a skate-fish, with head facing north-east and tail curving round Volcano Bay in the direction of Hon-do.

In Hon-do the main axis towards the middle of the island recedes somewhat from the east coast, where is developed an alluvial lowland district watered by numerous streams, and occupied by Tokyo (Yedo), capital of the empire. But west and south of this district the hills attain their greatest elevation in Mounts Nantai (8195 feet), Asama (8260), Haku (9185), and the magnificent snow-capped cone of Fuji-yama (12,400), culminating point of the Archipelago. Fuji-yama, which rises in solitary grandeur some 70 miles south-west of Tokyo, is visible in clear weather for a distance of nearly 100 miles. It has been quiescent since the year 1707. But although the highest, Fuji-yama is not the largest volcano in Japan. This honour is claimed by Aso-san, in Kiu-shiu, 20 miles from Kumamoto, the crater of which is said by Milne[1] to be 12 miles in diameter, and consequently

FUJI-YAMA.

larger than Maunaloa, hitherto supposed to be the largest in the world. It was visited in 1880 by Milne, who found it occupied by several villages.

The Asama-yama, which occupies a somewhat central position to the north-west of Tokyo, was ascended in 1873 by the French traveller G. Bousquet. From its crater, 1000 feet across, this volcano emits constant volumes of smoke and vapour, and from its summit a magnificent prospect is commanded of the surrounding country.

[1] *Popular Science Review*, Dec. 1880.

Other superb cones are Chokai-san on the north-west coast of Hon-do, and Tateyama, one of the most conspicuous and loftiest peaks (nearly 10,000 feet) of the Shinano Hida range, towards the northern extremity of the Fossa Magna. Like Fuji-yama, near the southern extremity, both are famous places of pilgrimage, and both command magnificent prospects of the surrounding lands and seas. Except a few hot-springs at the foot of the mountain, there are no traces of recent volcanic action on Chokai-san; but on the western slopes of Tateyama is seen the largest and most interesting solfatara in the Archipelago. "The Japanese call it Figoku, 'Hell,' and no place in the whole world could remind one more of the infernal regions. From hundreds of openings steam is emitted with a shrill, hissing noise, and sulphurous vapours belch forth in large volumes At the edge of the solfatara I found some small mud volcanoes in regular action. In some of the openings grew graceful flower-like cups of a beautiful yellow colour, formed of minute and glittering crystals of sulphur, one of which was about 6 feet high" (Naumann).

Although many of the volcanoes have been in eruption during the historic period, nearly all are now extinct, or at least quiescent. But in 1878 Naumann witnessed a tremendous outbreak on the island of Oshima at the entrance of Tokyo Bay. From a small cone springing from the floor of a huge circular crater a column of fire was projected into the air to a height of 1000 feet, while masses of molten lava streamed down the slopes.

But if eruptions are rare, earthquakes are all the more frequent, one might almost say of daily occurrence, although seldom of a violent character.

Since the opening of the country to foreign intercourse, however, two very destructive disturbances have been recorded, that of the Tokyo district on 22nd February 1880, and the still more violent convulsions of the Mino and Owari provinces, which began on 28th October 1891, and continued till the end of March 1892, as many as 2588 shocks being felt at Gifu, and 1495 at Nagoya. This event, which has been carefully studied by Professor B. Koto of the Imperial University,[1] was felt over an area of 50,000 square miles, or 60 per cent of the Japanese Archipelago, and in the central parts its effect was to greatly modify the topography of the country, rendering existing surveys quite useless in some districts. On the plain near Nagoya the ground was riven with myriads of fissures, small mud volcanoes being thrown up along the Shonaigawa River, where a bamboo grove slid 60 feet back, the trees remaining upright. Gifu was nearly ruined, and every house overthrown in the continuous street, 20 miles long, running thence to Nagoya. Several other places shared the same fate, and even greater havoc was wrought in the hilly Mino district, traversed for 40 miles by a new line of fault, where everything lying near the great throws of shale was destroyed. The solid ground became for a time like a liquid sea of waves, the destruction being complete in the epicentric district, 4200 square miles in extent. Near Kimbara, in the Neo basin, the sides of the valley slid into the river, and in the upper reaches a great part of the mountain slopes glided down to the lowlands. One result of the earthquake was the formation of a huge fissure, which was traced for over 40 miles through the Neo valley from Katabira to Fukui in Echizen, cutting across hills and paddy-fields, and raising the soft earth into a ridge, like the track of a gigantic mole. "The old Japanese idea that earthquakes are caused by the burrowing of a gigantic insect

[1] "The Cause of the Great Earthquake in Central Japan, 1891," in *Journal of the College of Science, Imperial University*, 1893.

might well be suggested by such a phenomenon." The Seismological Society, founded at Yokohama, issues regular *Transactions,* which have already thrown much light on the nature and cause of these phenomena.

3. *Hydrography: Table of Rivers above* 50 *miles long—Lake Biwa.*

The lofty range stretching southwards from Mount Asama forms the water-parting between the Pacific and the Sea of Japan. But owing to the disposition of the mountain system, covering probably nine-tenths of the whole surface, no room is left for the development of large rivers. Those that do exist bear somewhat the character of mountain torrents with very rapid courses, and liable to sudden and disastrous floodings in their lower reaches. Hence they are almost more damaging than beneficial even for irrigation purposes. To navigation they are not merely useless, but a positive hindrance, owing to the large quantities of sedimentary matter which they bring down, and with which some of the best harbours in the country have been gradually filled in. Such has especially been the fate of Osaka and Niigata harbours, formerly accessible to the largest vessels, but which can now be approached only by small craft. In Japan "a river-bed is a waste of sand, boulders, and shingle, through the middle of which, among sand-banks and shallows, the river proper takes its devious course. In the freshets, which occur to a greater or less extent every year, enormous volumes of water pour over these wastes, carrying sand and detritus down to the mouths, which are all obstructed by bars. Of these rivers the Shinano, being the biggest, is the most refractory, and has piled up a bar at its entrance through which there is only a passage 7 feet deep, which is perpetually shallowing."[1]

Subjoined is a table of all the Japanese rivers above 50 miles in length:—

Name.	Source.	Outlet.	Length. Miles.
Shinano	E. Shinano	Niigata	180
Tone	N. Kodsuke	Gulf of Tokyo, and Pacific	170
Kitakami	N. of Rikuchiu	Ishinomaki and Murohama, E. coast Rikuzen	140
Ishikari	N. prov. Ishikari, Yesso	Ishikari, W. coast	130
Tenriu	Lake Sua	Pacific	120
Kiso	S.W. Shinano	Pacific	115
Sakata	S. of Usen	Sakata, W. coast Usen	110
Okuma	S.W. of prov. Iwaka	Watari, E. coast Iwaki	110
Noshiro	W. of prov. Rikuchiu	Noshiro, W. coast of Rikuchiu	100
Akano	Lake Inawashiro in prov. Iwashiro	Near Niigata, W. coast of Echigo	90
Sumida	E. Musashi	Gulf of Tokyo	90
Toshima	S.E. of prov. Ugo	Kubota, W. coast of Ugo	70
Fujii	Koshiu	Pacific	70
Yodo	E. Iga	Bay of Osaka	70
Baniu	Yamanaka (Koshiu)	Pacific	60
Oi	N. Koshiu	Pacific	55

Of the few lakes none are of any size except Biwa, a magnificent sheet of water some 45 miles long, with a mean breadth of about 10 miles. Biwa, which is traversed by the River Yodo, lies within 8 miles of Kioto, the ancient capital of the Mikados, who usually spent the summer months with their suites on its romantic banks. It is closed north and west by lofty forest-covered mountains, and elsewhere skirted by an open, highly-cultivated country dotted over with numerous villages and tea-houses, the resort of pleasure-seekers from all parts. Its clear waters, which abound

[1] *Unbeaten Tracks*, i. p. 212.

in fish, are enlivened by fleets of tiny craft, including probably one hundred small steamers always crowded with passengers.

LAKE BIWA FROM MÜDERA.

4. *Natural and Political Divisions: Yesso, Hon-do, Kiu-shiu, Shi-koku.*

The four great islands, forming so many main natural divisions, were formerly divided into nine "do," or departments, which comprised as many as 85 "kuni," or provinces. But these old divisions have, since the revolution of 1868, been suppressed, and the empire is now, for administrative purposes, divided into three "fu" or urban districts, and 60 "ken" or "prefectures." Yesso (Yezo), however, forms with the Kuriles a so-called "Hok 'kaido" or "Northern Sea Circuit," administered by the "Kaitakushi," or Colonisation Department. The aspect of this little-known island, which is considerably larger than Ireland, is described as peculiarly romantic. The southern peninsula, which was visited in 1878 by Miss Bird, everywhere abounds in magnificent woodland scenery, commanded towards the east by the volcano of Komaga-take, whose cone rises 3830 feet above a charming lake at its base. The crater has fallen in on the east side facing seawards, and it is everywhere intersected by yawning chasms emitting dense volumes of steam. Here the prospect over land and sea is superb, the middle distance and background being occupied by wooded hills, whose deep green foliage merges imperceptibly in a violet tint. A long chain of volcanoes stretches at the foot of the spectator, with the quiescent Mount Ussu bounding the distant horizon.

The shores of Volcano Bay are extremely rich in minerals, including gold, silver, lead, iron, rock oil, and coral. At Yesan, in the south-east corner of Yesso, there is a "solfatara" with sulphur works, and another at Mount Iwauno-bori, the central cone of a triple group of volcanoes. The interior of Yesso, which is said to be crossed by a series of mostly extinct volcanoes, still remains to be explored, the expeditions hitherto undertaken having been mainly restricted to the neighbourhood of Volcano Bay and the adjoining peninsula. The colonisation of Yesso by the Japanese, begun in 1869, has already made considerable progress, especially in the southern districts about Esashi and Hakodate, and on the arable land round Sapporo and Mombetsu. In 1890 nearly 50,000 acres had already been brought under cultivation, yielding fair crops of rice, wheat, millet, sorghum, hemp, buckwheat, maize, oats, and potatoes, besides apples, grapes, and other fruits. The Japanese settlers have increased from 49,000 in 1869 to nearly 300,000 in 1890, while the irreclaimable Ainu aborigines are disappearing. In 1890 they were estimated at not more than 14,000, exclusive of half-breeds.

The great central island of Hon-do, with an area of 92,000 square miles, although thickly peopled and generally well cultivated, is by no means such a naturally rich land as is generally supposed. The soil is sandy and only moderately fertile, except in the Tokyo district, which is covered with a rich black loam. Here the people are prosperous and comfortable, but elsewhere the rural districts have often a poverty-stricken air. The wretched houses serve to shelter man and beast alike, the villages are in some places indescribably filthy, and the natives clothed in rags or covered with loathsome sores. Miss Bird speaks of the painful sight presented by the importunate crowd pressing upon one another.

Yet "their industry is ceaseless, they have no Sabbaths, and only take a holiday when they have nothing to do. Their spade husbandry turns the country into one beautifully-kept garden, in which one might look vainly for a weed. They are economical and thrifty, and turn everything to useful account. They manure the ground heavily, understand the rotation of crops, and have little if anything to learn in the way of improved agricultural processes. The appearance of poverty may be produced by apathy regarding comforts to which they have not been accustomed. The dirt is preventible, and the causes of the prevalence of cutaneous diseases among children are not far to seek."[1]

A glance at the map of Japan will show that the three large islands in the south almost touch each other at certain points, where they are separated only by narrow channels. Of all the large islands, Shi-koku, with an area of 7000 square miles, is the least volcanic.

[1] *Unbeaten Tracks*, i. p. 166.

Even in Kiu-shiu the sandstone formation prevails over the igneous rocks. Like all the other islands, it abounds in grand natural scenery, and especially romantic is its fiord-like and rugged iron-bound south coast. The narrow Shimonoseki Strait separating it from Hon-do, and giving access to the Inland Sea, affords the shortest and pleasantest route by steamer from Nagasaki on its west coast to Yokohama.

Outlying Dependencies: Liu-kiu, Bonin, Formosa.

The Japanese Archipelago properly so called is continued by the Liu-kiu chain and Formosa southwards to the Philippine Islands, and by the less important Bonin chain south-eastwards in the direction of the Marianas. Liu-kiu (Lu-cheu in Chinese, Riu-kiu in Japanese, Du-kiu of the natives) describes a curve of about 600 miles between Kiu-shiu and Formosa (30°-24° N. lat.), and comprises altogether four distinct groups:—

I. SITSI-TO (Linschoten), which with the neighbouring Tanega, Yokuno, and a few other islets, has a total area of about 460 square miles. This group, which is mainly volcanic, belongs geographically and administratively to Kiu-shiu, hence is not usually included in the Liu-kiu system. II. HOKUBU-SOTO (*northern group*), also administratively dependent on Kiu-shiu, includes the large islands of Oshima (Oho-shima), with area 320 square miles, and population over 50,000, and Tokuno, highest of the non-volcanic islands (2200 feet), and population 30,000. III. TSIUBU-SOTO (*central group*), including Okinava, largest of the whole Archipelago (500 square miles), with population about 130,000, and summits 1400 to 1500 feet. IV. NAMBU-SOTO (*southern group*), including Iriomote (120 square miles), Isigaki (100), and Mivako (60).

Since 1879 groups III. and IV., which have a collective area of nearly 950 square miles, and a population (1892) of over 400,000, form a separate administrative division, the Ken (prefecture) of Okinava, with capital Siuri (Siuli), on the south-east coast of Okinava, 3 miles from Nafa (Nava), chief seaport of the Archipelago. Both groups consist of granite, schist, sandstone, calcareous and coralline masses, and, like the non-igneous parts of Japan, belong geologically to the Asiatic mainland. The coral-builders still work in the surrounding waters, whose temperature is raised by the warm Kuro Siwo current. But they are also swept by terrific hurricanes (typhoons), which greatly endanger the navigation, and which tended to keep the Archipelago isolated from its powerful Chinese and Japanese neighbours till comparatively recent times.

Down to the second half of the fourteenth century Siuri was a royal residence, capital of an independent insular state, which appears to have attained a considerable degree of culture under Chinese influences. Regular relations had already been established with China early in the seventh century, and soon after with Japan; but the islanders enjoyed absolute autonomy till the year 1372, when they voluntarily recognised the suzerainty of the Middle Kingdom. But the Japanese were even then the virtual masters, for the powerful and wealthy daimios of Satsuma enjoyed a monopoly of the trade of the Archipelago. The presents offered to his representatives gradually assumed the character of a tribute, to enforce which an expedition was despatched in 1609, resulting in the formal recognition of Japanese supremacy. Thus the little state found itself bound to allegiance to both its neighbours; but China was relatively distant, while the Satsuma prince, representative of the Mikado, was so to say in possession, and after the successful

Japanese expedition to Formosa in 1874, the king was called upon to break off all relations with Peking. This was followed by the formal annexation of the islands in 1879, when the northern groups were attached administratively to Kiu-shiu, and the southern constituted a separate prefecture.

Both in physique and speech the Liu-kiu natives betray their Japanese origin. The language resembles the Satsuma dialect, and fully-developed beards, suggestive of a remote Ainu strain, are met perhaps more frequently in Liu-kiu than in Japan itself. But there is also an unmistakable Chinese strain, especially in Okinava, where the Imperial Government established a colony from Fo-kien towards the close of the fourteenth century. The natives have all the courtesy and geniality of their Japanese kindred, while they have carried the Chinese veneration for the dead to a degree probably unapproached by any other people in the world. "It were scarcely too much to say that if the living dwell in hovels, the dead dwell in palaces, so imposing are the vaults, of which each family, even the very poorest, possesses one. The roofs of these burial-vaults may be seen from a considerable distance at sea, on account of the dazzling white plaster that distinguishes them from the surrounding vegetation. On the occasion of a death, the corpse is conveyed to the family vault in solemn procession, a Buddhist priest leading the way, hired mourners following with bitter wails, and the kinsmen of the dead bringing up the rear. The religious rites duly concluded, the body is left shut up for two years. Then the family again assemble for the purpose of washing the bones and depositing them in their final resting-place, an earthenware urn, which is lifted on to one of the numerous shelves that run round the interior of the vault. The name of the dead and the date are inscribed in Chinese characters on the front of the urn in a space left for that purpose."[1]

The same authority states that Buddhism as a religion and rule of life is practically extinct, having been killed by Confucianism. He speaks of the natives as one of the most civilised peoples in the world, possessing an ancient history, a system of farming which would put European agriculturists to shame, and a skill in diplomacy which till lately preserved the national independence from the encroachments both of China and Japan.

BONIN, on the contrary, has neither history nor indigenous inhabitants. Its very existence was unknown till the year 1675, when the northern clusters, now known as *Kater* and *Parry*, were discovered by some Japanese fishers, and named Monin-Shima, the "Desert Islands." Then the group was forgotten till 1823, when another fisher, Coffin, an American whaler, determined the position of the southernmost cluster named from him. Next year, Ebbet, also an American, discovered, or rather re-discovered, the fourth cluster, later called *Peel Islands*, both Peel and Coffin having already been sighted and correctly mapped in 1639 by Matthys Quast and Abel Tasman during their expedition in quest of the legendary Rica de Oro and Rica de Plata Islands.

The whole Archipelago, which lies about 500 miles from the nearest part of the Japanese coast, stretches for a distance of 86 miles nearly north and south between 26° 30′ and 27° 45′ N. lat., and lies just beyond the 140th meridian. Most of the islands, which have a collective area of 380 square miles, appear to be high, ranging from a few hundred to 1200 or 1300 feet. They are mainly of igneous origin, abounding in old lavas and basalts, with even a few cones terminating in craters,

[1] Basil Hall Chamberlain, *Journal of the Anthropological Institute*, August 1894, p. 58.

although no traces of recent volcanic action have been observed. The forests clothing many of the slopes comprise the areca, pandanus, and other palms, the sago plant, tree ferns, and a species of mulberry which grows to a great size.

The temporary settlements made by the Americans at one or two points in the south, and by the English in Peel, most important of the four groups, were all abandoned, some before, some after the year 1861, when the claim to their possession was decided against England in favour of Japan. The centre of administration has been established at the old English station of Port Lloyd, now Oho Minato, in Peel, the only permanently inhabited member of the Archipelago.

One of the chief advantages secured by Japan on the conclusion of her successful war with China was the large island of FORMOSA, which, with the little Pescadores group off the west coast, was ceded to her in perpetuity by the Treaty of Shimonoseki, April 1895. Formosa (Taiwan), which is separated from Fu-kien by the broad Fu-kien or Formosa strait, is about 240 miles long north and south, with an extreme breadth of 80 miles and a total area of nearly 15,000 square miles.

Formosa—that is, the "Beautiful"—received this name from the early Spanish navigators, struck by the charming aspect of its magnificent wooded heights. A volcanic range with long extinct craters crosses the interior, merging westward in an extensive fertile plain watered by numerous limpid streams. This inviting region has attracted many industrious colonists from Fu-kien, who have brought most of the land under cultivation. But the eastern districts, where the hills reach down to the coast, afford little room for tillage, and have consequently been left to the wild native tribes of Malay stock. In the central range the culminating points are

Mount Morrison (10,850 feet), Mount Sylvia (11,300), Ta-shan (12,000 ?), and an unnamed peak near Sylvia (12,800).

A trip along the east coast presents a panorama of the loveliest scenery imaginable. Above the highly-cultivated lowlands, the advanced ridges rise to a height of 5000 feet, beyond which the horizon is bounded by the central range with a mean elevation of 10,000 feet. The outlines of these highlands are as fantastic as they are beautiful—domes and slender towers, curious jagged cliffs and steep rocky ramparts towering everywhere above the soft grassy slopes, and sending down mountain torrents, which merge lower down in sparkling silvery streams. Here and there the brighter hues are toned by the darker shadows of the native hamlets clustering in the rocky gorges. These villages are grouped in two classes, the independent and the "Peppohoan," which before the cession to Japan acknowledged the authority of the Chinese, intermarried with them, and adopted their customs, while retaining their own Malay speech. These subject tribes serve to promote intercourse with the independent natives of the interior. They are a people of much promise, amongst whom Christianity seems to be gaining ground.

The wild tribes tattoo the face, and, according to Captain Box, construct elegant huts of bamboo and palm leaves. Over the doorway are often suspended, as trophies, the skulls of wild boars, deer, and apes, and a more than usually vainglorious savage made a display of a tuft of six pigtails detached with his own hand from the heads of his Chinese victims.

Amongst the natural products of Formosa, which is about the size of Sardinia and Corsica rolled into one, are sulphur, petroleum, coal, and camphor.

Much valuable information on Formosan geography

and natural history is embodied in Mr. Hosie's official report for 1893. The railway, begun some years ago to connect the western districts with the port of Kelung in the north-east, is now completed, and has been extended to the more productive southern districts. Immense coal-fields extend over the whole island, and much sulphur is derived from the numerous sulphur springs of the north-east. Geysers even occur on the Kelung branch of the Tamsui River, and in the south coral limestone with oyster shells, at an elevation of 2000 feet, point to volcanic upheaval in comparatively recent times. The west coast is fringed with mud-banks deposited by the numerous mountain streams. The land is thus continually encroaching on the sea on the west side, while the rock-bound east coast remains stationary.

The only large river is the Tamsui in the north, which owing to a shallow sand-bar is inaccessible to large vessels, except at high water. Nevertheless the harbour of Kelung, north-eastern terminus of the railway, is the best in the island. Takon harbour in the south-west is now silted up, and has been abandoned for An-ping, the port of Tai-nan-fu, farther north, which, however, is exposed to typhoons.

The vegetation is more tropical than on the neighbouring Chinese mainland, and includes the rattan, betel-nut, palm, and some other plants not found in China. These may have been brought by the Kuro-Siwo (Kuro-Shimo), the "Gulf Stream" of the Pacific, which appears to have brought the rattan to Hai-nan, midway between Formosa and the Malay Archipelago, where it is indigenous. The insects also resemble those of Hai-nan and Malaysia, their distribution being probably due to the prevailing winds. The temperature ranges from about 40° F. in February to a maximum of 100° F. in June and July, while the annual rainfall appears to average 75 or 76 inches. The present Chinese and mixed settled populations are estimated at from two to three millions; but no estimate can be formed of the number of the savage tribes of the interior, who are of a Malayan type, with traces of intermixture with the Japanese, especially in

MALAY VILLAGE IN FORMOSA.

the northern districts. Here the men are tattooed, with a single verticle blue line down the middle of the forehead, and a similar line from the centre of the under lip to nearly the middle of the chin; the women present a grotesque appearance, having tattooed designs on both upper and lower lips, extending across the cheeks to the ears in a narrowing blue line. These aborigines appear to have been carried by storms from various islands of Malaysia, their language showing marked Malayan affinities, while head-hunting and other primitive Malay usages are still practised. Even quite recently a native of the Philippines was drifted by a storm to the north-east coast of Formosa, where he landed and settled down, thus showing the probability of similar involuntary migrations in prehistoric times.

The soil of Formosa is extremely fertile, yielding all sorts of vegetables in unparalleled profusion. There are two annual harvests of rice and sweet potatoes, these being the chief food products. There are also numerous textile, oleaginous and medicinal plants, though hitherto little attention has been paid to these resources of the Formosan woodlands. The main industries are tea-growing, sugar-raising, camphor and sulphur extraction. Besides coal, traces of gold have been discovered in various districts; but no auriferous reefs have yet been struck, and all the gold is derived exclusively from washings. The trade of the island, which might be greatly developed, is chiefly with the United States, which takes nearly all the tea, with Japan, to which more than half of the sugar is sent, with the surrounding Chinese ports, and with Hong-Kong, to which all the camphor is forwarded for Europe.

5. *Climate.*

Japan lies entirely within the temperate zone, and, thanks to the proximity of the warm Pacific currents, it enjoys a far milder climate than the neighbouring Asiatic mainland lying under the same parallels of latitude. It may be described on the whole as equable and healthy, without any great extremes of temperature, and with a copious rainfall distributed over the whole year, but heavier in summer than winter. The summer heats are tempered by the cool northern sea-breezes prevalent during that season, while in winter the eastern seaboard is greatly influenced by the warm equatorial current, known as the Kuro Siwo or "Black Stream." The coasts washed by this current, which presents many remarkable analogies to the "Gulf Stream" of the Atlantic, enjoy an exceptionally mild climate, while the less-favoured northern sections are relatively colder than the parts of Europe crossed by the same parallels. In latitudes corresponding to those of Marseilles and Gibraltar it freezes hard in winter; and Yesso, lying between 42° to 46° N., is altogether much colder than Scotland, lying between 55° to 59°.

The southern islands are also naturally affected by their low latitude (30° to 34° N.), and here the glass rises normally in summer to 96° and 98° Fahr. in the shade, while in winter there is little snow except on the highest peaks. Between these extremes lies the great island of Hon-do, where the climate of the east coast is so genial and healthy that it has become the chief sanitarium for Europeans settled on the less salubrious and more relaxing Chinese seaboard. The temperature at Yokohama ranges between 20° and 90° F., averaging about 60°, with not more than eight or ten snowy days in winter, and a correspondingly short period of oppressive summer heats. But the west side of the island, at least as far north as about 38° lat., being less affected by the Black Stream, is somewhat colder, with an average snowfall of thirty-two days at Niigata. Here the "canals and rivers freeze, and even the rapid Shinano sometimes bears a horse. In January and February the snow lies 3 or 4 feet deep, a veil of clouds obscures the sky, people inhabit their upper rooms to get any daylight, pack-horse traffic is suspended, pedestrians go about with difficulty in rough snow-shoes, and for nearly six months the coast is unsuitable for navigation owing to the prevalence of strong, cold north-west winds. And all this in latitude 37° 55′—three degrees south of Naples. . . . Europeans and their children thrive well in all parts of the Empire."[1]

Basing his conclusions on the publications of the Imperial Meteorological Observatory of Tokyo, Herr J. Hann remarks that in general temperature increases rapidly with decreasing latitude on both coasts, but more in the north than in the south, and more on the east than on the west side, except in spring on the west coast. Up to about 36° lat. the west is a little cooler than the east, but north of 38° lat. it is much warmer. This is due to the trend of the long axis of Hon-do, which about 36° lat. turns nearly due north and south, the consequence being that the cooling influence of the mainland is lessened in the north by the increased breadth of the Sea of Japan, while the warm Kuro Siwo turns eastwards at 38° lat., and the east coast north of 36° lat. comes within the influence of the cold Kurile current extending southwards to Cape Daihosaki. On the other hand, all the west side is washed by a branch of the Kuro Siwo, and also exposed to the prevailing warm west and north-west winds. As regards the rainfall, this observer finds that it is heaviest in the south-east, especially the Sikoko and Kiu-shiu coastlands, and also the middle portion of the west coast of Hon-do, the mean annual discharge in all these districts being from 90 to 100 inches. In general, the autumn and winter rains are more copious on the west than on the east coast, in consequence of the prevailing moist winds on that side.

[1] *Unbeaten Tracks*, i. p. 219.

6. *Flora and Fauna: Tea-culture—The Crows of Yesso.*

With an abundant rainfall, a moderately fertile soil, and a temperature ranging from the almost Siberian winters of Yesso to the tropical heats of Kiu-shiu, the vegetation of Japan could not fail to be one of the richest and most varied on the globe. Notwithstanding its long settlement and comparatively dense agricultural population, the forest area is still about four times more extensive than the portion brought under cultivation. This is due to the relief of the land, in which the highland formation, ill suited for tillage, and here well adapted for the growth of timber, so greatly prevails over the plateau and lowland formations. Nevertheless sufficient space is left on the plains and on the more gently-sloping hill-sides for the development of an extensive and skilful agricultural system yielding results of the most abundant and varied character. The vegetation of the lowlands, offering a striking contrast to that of the higher grounds, presents a magnificent flora, including many rare specimens which have already contributed to enrich the European botanical collections. The fruits are naturally of large size, and have been further developed by careful and systematic culture. Rice, which, with wheat, millet, fish, and vegetables, forms the staple of food, arrives at great perfection, one variety being fully equal to the best kinds produced in India or America. Wheat, barley, millet, and buckwheat are also extensively grown on the more hilly districts above the level of the rice zone; but the crops are of inferior quality.

Next in importance to rice are the mulberry and tea plantations, which cover large tracts in almost every part of Hon-do and the southern islands. "The tea-plant was introduced from China about the beginning of the ninth century. . . . The plant is pollared to render it more branchy, and therefore more productive, and must be five years old before the leaves are gathered" (Mossman).

Of the forest trees the most prevalent are the cypress, firs, and other conifers, including the "umbrella pine" (*Koya Maki*), a magnificent variety often growing to a height of 100 feet, with a dense foliage of broad leaves

OLD PINE TREE AT BIWA.

in shape somewhat like an umbrella. Still larger are the cedars, which tower to a height of 200 feet, with trunks 18 feet in girth. But of all forest growths, the most remarkable is the Camphor Tree, which has a trunk over 50 feet round. The camphor is extracted from the stem and roots cut into small pieces, by a simple process of decoction. There is also a large species of oak yielding an edible acorn, which is boiled and much esteemed for its nutritious properties by the peasantry. But economically a far more important tree is the mulberry, of which there are several varieties. Besides supplying the natural food of the silkworm, the bark of this tree is also used in the manufacture of paper, cordage, and coarse dress materials. Yet even the mulberry must yield in importance to the bamboo, which is as indispensable in Japan as it is in China. It grows very rapidly in dense thickets to a height of 50 or 60 feet, and the uses to which it is put are endless. The framework of the houses and most of their contents, sails of junks, screens, mats, paper, pipes, walking-sticks, are amongst the innumerable objects made from this most useful of plants.

The wild Mammalia of Japan includes no large animals except the bear, wolf, and wild boar. Amongst the smaller ones are the fox, red and black badger, the monkey, marten, otter, and squirrel, all closely allied to the corresponding species on the mainland, and pointing at a former connection of the Archipelago with the Asiatic continent (Pumpelly).

Very remarkable is the paucity of domestic animals, which are limited to the ox, the horse, two varieties of the dog, the cat, and poultry. Neither sheep, goats, nor the ass seem to be anywhere indigenous in these islands. Economically the silkworm is the most valuable animal, supplying the material of one of the staple industries. From the Japanese cocoons the deteriorated stock in Italy and other Western States has been largely renovated in recent years. "One has to walk warily in many villages lest one should crush the cocoons which are exposed upon mats, and look so temptingly like almond comfits." [1]

The avifauna includes few birds of bright plumage, and scarcely any songsters. But the eagle, hawk, heron, quail, stork, and pheasant are numerous, while the crow forms a salient feature of the landscape, especially in

[1] *Unbeaten Tracks*, i. p. 257.

Yesso. Here "there are millions of them, and in many places they break the silence of the silent land with a babel of noisy discords. They are everywhere, and have a degree of most unpardonable impertinence, mingled with a cunning and sagacity which almost put them on a level with man in some circumstances. Five of them were so impudent as to alight on two of my horses, and so be ferried across the Yurapugawa. In the inn-garden at Mori I saw a dog eating a piece of carrion in the presence of several of these covetous birds. They evidently said a great deal to each other on the subject, and now and then one or two of them tried to pull the meat away from him, which he resented."[1] The Japanese is much larger and stronger than the European species, or about the size of our ravens, and fully a match for small dogs.

The fauna no less than the flora of Yesso differs in many respects from that of the neighbouring Hon-do. While conifers prevail in the latter, the forests in the former consist mainly of hard wood. The jays and woodpeckers are of different species; the Yesso birch grouse is not found in Hon-do, while the Hon-do ptarmigan and pheasants, as well as a sheep-faced antelope or goat, a monkey and a black bear, are all confined to the southern island. This, combined with many symptoms of upheaval, especially on the west and north coasts, points to a time when the narrow Sugaru Strait, now only 11 miles wide, formed a broad marine channel between Yesso and Hon-do.

7. *Inhabitants: Ainus and Japanese—The Shinto and Buddhist Religions—Christianity.*

From prehistoric times Japan has been exclusively occupied by the Ainus and Japanese, two races funda-

[1] *Unbeaten Tracks*, ii. p. 149.

mentally differing from each other in physique, speech, and culture. The Japanese are obviously a branch of the Mongolian family, but a branch separated from the parent stock for a far longer period than is commonly supposed. The Aiñus, whose affinities will be elsewhere discussed, are at present entirely confined to the northern island of Yesso, the southern districts of Sakhalin, and some of the neighbouring Kurile group. But there can be little doubt that they were at one time in possession of the whole of Nippon, or at all events of Hon-do, whence they were gradually driven northwards, or else partly exterminated or absorbed by the Japanese intruders from the mainland.

Milne concludes that in former times Ainus lived in the districts of Hon-do, where the kitchen middens and other archæological remains are now found, as they still live in Yesso, where similar remains are still found. And this is confirmed by tradition and history, according to which the present Japanese, on arriving in Nippon, "found it tenanted by Ebisu or barbarians, whom they recognise as the ancestors of the modern Ainus. Year by year the aborigines were driven step by step towards the north. About the year 800 they were struggling near Morioka, and by the year 1200 they seem to have been practically exterminated from Nippon, and those who remained or had taken refuge farther to the north in Yesso were completely subjugated."

The Ainus.

The traces of the partial fusion of the two races are most conspicuous in the northern districts of Hon-do, where the "Ebisu" held out longest and came most frequently into contact with the Japanese. They undoubtedly occupy, both socially and intellectually, a very low position, and some writers have even denied them the capacity of improvement. But Captain Blakiston and Miss Bird (Mrs. Bishop), both careful observers of the habits of this race, assure us that they possess many excellent qualities, and gladly take advantage of every opportunity to better themselves. Yet, like so many

AN OLD AINU.

other primitive peoples, they seem incapable of enduring the contact of a higher culture, in the presence of which they slowly disappear. Blakiston, Brandt, Pumpelly, and others are unanimous on this point; and if the process of extinction has lasted for perhaps thousands of years, the reason probably is because the Ainus found room to withdraw continually northwards, and thus escape destruction. They are now reduced to about 10,000 in

Yesso, where alcohol, small-pox, and the gradual reclamation of the land, must eventually cause them to disappear. Further fusion with the ruling race, who regard them with infinite contempt, seems now no longer possible.

The Ainus are described by Miss Bird as "about the middle height, broad-chested, broad-shouldered, very strongly built, the arms and legs short and muscular, the hands and feet large. The bodies of many are covered with short, bristly hair. I have seen two boys whose backs are covered with fur as fine and soft as that of a cat. The foreheads are very high, broad, and prominent, and at first sight give one the impression of an unusual capacity for intellectual development. The nose is straight but short, the cheek-bone low, the eyebrows full, forming a straight line nearly across the face. The eyes are large, tolerably deep-set, and very beautiful, the colour a rich liquid brown, the expression singularly soft, the skin of an Italian olive tint, mostly thin, and light enough to show the changes of colour in the cheeks." [1]

The usual dress of both sexes consists of a deer-skin garment reaching below the knees and girdled round the waist. The women are tattooed on the forearm, about the lips, and on the lower part of the cheeks, the pattern representing upturned mustachios of a light-blue shade, and producing an ugly effect. The practice has recently been prohibited by the Government.

Yet their religious notions are of the very vaguest, and their gods mere wooden sticks and posts, whittled so as to let the shavings fall down in curls, and stuck about in their houses, on the hill-tops, by running waters, on precipices and mountain passes. But, as amongst some of the neighbouring continental tribes, "their chief divinity seems to be the bear, who, however, is eaten as well as worshipped. A young bear, captured in the early

[1] *Unbeaten Tracks*, ii. *passim.*

spring and confined in a cage, is kept in the chief's house, where it is suckled by a woman, and played with by the children until it gets strong and dangerous, when the great Bear Feast is celebrated.

"Truth is of value in their eyes; infanticide is unknown, and aged parents receive filial reverence, kindness, and support, while in their social and domestic relations there is much that is praiseworthy."[1]

The Ainus are entirely subject to Japan, and are now chiefly occupied with hunting and fishing. They are also employed to collect seaweed, and at one spot Blakiston even found them engaged in tillage, and raising a few crops of millet, potatoes, and turnips.

The Japanese.

During many ages of isolation in their insular homes the Japanese people have developed many characteristic physical and mental traits, by which they may easily be distinguished from the neighbouring continental races. Nevertheless their slightly oblique eyes, small nose, black lank hair, sparse beard, salient cheek-bones, and yellowish complexion, still attest their relationship with the great Mongolian family. But from what branch of that stock they more immediately descend, whether from the Chinese, Manchu, Korean, or Mongol proper, is a question which is still far from being determined. Its ultimate solution will probably depend upon a more exhaustive study of the Japanese language.

But the race has been so long separated from the parent stock that its linguistic affinities have become obscured almost beyond the hope of recovery. The Chinese elements in the national speech are all of comparatively recent date, and directly introduced since the

[1] *Unbeaten Tracks*, ii. *passim*

dawn of the historic period. They lie entirely on the surface, and in no way affect the inner structure of the language, which has had time to become differentiated into a very distinct and at present completely isolated form of speech. It is an extremely soft and musical tongue, being in this respect fully on a level with the Italian, especially when spoken by ladies of the upper classes.

Both the pronunciation and the writing system present some very formidable obstacles to the student. Thus the Russian traveller Golovnin, who was for a long time a captive in Japan, tells us that, for instance, the word *fi* (= fire) seems to oscillate between the sounds of *fi, hi, psi, fsi.*

Although the standard speech is everywhere understood and spoken with considerable uniformity, every province has its peculiarities, increasing in the ratio of the distance from the capital. The best Japanese is perhaps spoken in Niigata, and by the very highest circles in Tokyo. The men and women have also a different pronunciation, to which the ear takes some time to get accustomed. As in English, the absence of verbal, nominal, and other inflexions gives at first sight a false impression of ease and simplicity. But as our studies progress, we come upon idiomatic forms, shades of expression, and modifications, which more than outweigh the apparent advantage secured by the absence of those grammatical elements.

Still more intricate is the writing system. The Japanese are the only people known to history who have spontaneously and without outward pressure adopted their characters, literature, philosophy, and moral code from a foreign nation. All these things they have borrowed from the Chinese, retaining the cumbrous Chinese ideographic system, even after they had themselves developed a true phonetic syllabary. This "I-ro-ha" system, as it is called, from the names of the three first letters, consists of forty-seven syllables, of which there are two distinct forms, the Kata-kana and Hira-gana. Unfortunately the matter is still further complicated by the great variety of forms assumed by the Hira-gana according to the whim of the writer. Hence the student is compelled first to master the Chinese ideographic signs, then the two phonetic alphabets, and lastly a general mixture of all three—this last being at present the ordinary style. Of the Chinese hieroglyphics there are further three distinct forms—one restricted to printed works and poetry; the second to official documents, indentures, and records of all sorts; the third, or "grass" character, confined to correspondence and everyday use.

Although based on that of China, the Japanese culture has always presented many striking analogies to the European standard, to which it is now being altogether assimilated. This extraordinary and surprisingly rapid social revolution must be mainly attributed to their capacity for at least imitating the features of foreign institutions. Possessed of considerable mental endowments and quickness of apprehension, the Japanese are pre-eminently distinguished by their love of knowledge and their appreciation of the higher interests of humanity. Although on the whole of a kindly and lovable disposition, especially when compared with the Chinese and other branches of the Mongolian family, their character is the reverse of childlike or puerile. Beneath many genial and amiable qualities there is often betrayed a spirit of treachery, suspicion, and revenge, which will for years pursue its victim under the cloak of the most seemingly cordial friendship. A mercenary disposition and unbridled licentiousness are also amongst the darker shades of a picture which is nevertheless apt, by its cheerful and brighter aspects, to beguile the unwary stranger; for the Japanese of all classes are highly courteous and obliging, personally brave and proud of their forefathers' great deeds, altogether a warlike people, distinguished beyond all others for their contempt of

A JAPANESE GIRL.

death and for an almost morbid sense of honour. This latter sentiment, leading often to duels, has also given rise to the peculiar institution of the hara-kiri,[1] or "happy despatch," an abnormal and cruelly refined method of self-immolation indigenous in Japan from time im-

[1] From *hara* = stomach, and *kiru* = to cut; but the more usual native name is seppuku, which is in fact the Chinese reading of the same characters.

memorial, but now happily fallen into desuetude, and abolished as an official punishment.[1]

All travellers describe the first appearance of Japan as the embodiment of some Eastern legend, full of surprises and arousing a sense of bewilderment in the stranger. He rubs his eyes, asks himself—Can all this be actually true, and not rather some day-dream or tale out of *The Thousand and One Nights?* Here everything has a smiling outward appearance—the bright skies, the sparkling waters, the glorious vegetation, man himself with his parti-coloured dress and sprightly humour. The natives seem to be always gossiping, free of cares, almost frivolous, but still courteous, and this is as true of the sturdy porter as of the pretty waitress in the tea-house. Even the mendicant seeks rather to create interest by means of all manner of professional jokes and buffoonery. The whole family, down to the third generation, may be seen solemnly amusing themselves flying fantastic kites or firing off rounds of squibs and crackers in broad daylight! A common species of fireworks consists of rockets and bombshells, which when shot up let off, not fire, but smoke of the brightest and most varied tints, or else figures of paper and wire, which unfold in mid-air and flutter slowly to the ground.

In all cities there is the so-called *Joshiwara,* or quarter entirely set apart for pleasure, and enlivened by troupes of clever acrobats, strolling players, jugglers, and clowns, all performing in the open streets. The numerous theatres may also be visited for a trifling fee, and here the audience remains squatted in family groups for hours together.

Wrestling is even a more popular entertainment than the playhouse, often exciting a degree of frantic enthusiasm resembling that of the Spaniards at a bull-fight.

[1] A full and accurate description of a "suicide to order" of this description may be seen in Mitford's *Tales of Old Japan.*

The Japanese are altogether a gay, pleasure-seeking people, usually devoting the whole evening to relaxation, which is always begun with the inevitable bath.

It may be a doubtful question whether the social vices are much more prevalent in Japan than in the West. The ideas associated with love and marriage may be of a simpler, less restrained character, but they are quite as rigidly adhered to. Marriage itself was formerly a mere formality—a mutual written engagement—the husband arranging matters with the parents and bringing his bride home without the intervention either of church or state. She then became the housewife, as in Europe, directing all domestic affairs and admitted to her husband's confidence. Her very first act was the sacrifice of her beauty. She made herself old and repulsive, shaved her eyebrows, and applied a black enamel to her teeth. Although the Japanese are monogamists, separations were not unfrequent, the wife being put away after the Jewish fashion by a writing of divorcement. Infidelity was punished with great severity. But now all this is changed. Marriage is recognised as a civil contract.

In Japan the women are fond of dress. The hitamono, or silken under-garment, is generally of a bright red colour. Over it, according to the season and occasion, are worn two or three, and even as many as five or six, flowing robes, the so-called "kimono," which fall down over the feet. These also are mainly of silk or crape, those underneath of a light, the others of a dark colour, and all are girdled round the waist by the "obi," six or eight feet long, a foot wide, generally of satin or some heavy silk material. The ends of this girdle are tied into a large square bow behind. Umbrellas and fans are worn by both sexes, but beyond this, about every third man you met formerly wore merely a wrap round the loins, and some elegant tattoo patterns in red and blue on various parts of his body. These were mostly porters, fishermen, "bettos" or grooms, and others of the lower classes, all of whom are now obliged to go decently clothed, while tattooing is forbidden. Men of patrician rank, princes or daimios, and high functionaries, wear silk, and formerly carried two swords in their girdle. They had also the exclusive right of riding, and letting the hair grow long; but this privilege has recently been extended to all classes, who have also largely taken to the European dress.

JAPANESE TATTOOING.

The houses are mostly one-storied, of bamboo, with projecting roofs and very large rooms, but little furniture to speak of—neither chairs, tables, nor bedsteads, articles which are all replaced by the indispensable mats and quilting. This is the case with all classes, the only perceptible difference consisting in the more or less costly material of the matting.

The Japanese bed consists of the so-called "futon," a capacious wadded garment with sleeves, into which the sleeper creeps and then draws another coverlet over him. The "makura," or pillow, consists of a little wooden box 8 inches long, on which is laid a paper cushion, renewed every night. The wealthy classes alone have fireproof stone houses, the outer wall in all the others consisting merely of a bamboo grating, so that the inner court is visible through four or five rooms from the street. Oiled paper is often substituted for bamboo, an arrangement which explains the fact that Tokyo is burned down and renewed about every seven years. Paper is also universally used instead of glass for lamps and lanterns, and the risk of fire is increased by the recklessness with which the children play with the dangerous element. The towns are divided into five districts, regularly visited during the night by the "Kanobo," or firemen, and each district has a well-organised "shikashi," or fire-brigade, an observatory, guard-house, and the usual appliances for extinguishing fires.

A singular object, which never fails to rivet the attention of strangers on their first arrival, is the "jinrikisha," or hackney-coach, a sort of go-cart on two wheels, drawn by the "ninsoku," or native porters. The streets are crowded with these vehicles, which are by no means uncomfortable, and move along with amazing rapidity.

Of the many religious sects three only have acquired a special prominence. The oldest of these is the so-called Shinto religion, the cult of the "Kami-no-michi," or spirits, one of whom is supposed to be represented in the flesh by the reigning sovereign.[1] The followers of this creed talk vaguely about a sublime being diffused throughout the universe, far too holy and ethereal to be directly addressed in prayer. In its oldest and simplest

[1] Kami-no-michi is the Japanese reading of the two Chinese signs for Shin-to = the way of the gods.

form the Shinto religion identifies the "tenka," or heavens, with the deity, of which they are the abode. It seems also to have believed in the immortality of the soul, and in everlasting rewards and punishments in the after state. But these notions have become antiquated, or completely forgotten. Material objects of its worship

JINRIKISHAS.

are the heavenly bodies, the elements as well as the powers of nature, which had, at a very early period been already conceived as self-existing personalities, and addressed in prayer as spirits, or "kami." Of the kami there are "eight millions"—that is to say, they are innumerable. But the whole system, which is exceedingly barren and empty, resolves itself into a vague reverence for ancestry, and especially for that of the Mikado, who reigns in virtue of his divine origin. For the present Mikado is the 121st lineal descendant of Jimmu Tenno (660 B.C.), fifth in descent from Amaterasu, the Sun-Goddess, and the great divinity of the Shinto religion.[1] Hence Shintoism has become a convenient instrument in the hands of the government for enforcing obedience to the reigning dynasty.

On this system was based the Shinto theocracy, which usurped the place of the military authority. But it was

INTERIOR OF THE IYEMITSU TEMPLE AT NIKKO.

subsequently compelled to yield to Buddhism, which was introduced about 550 A.D., and struck deep root, although not without undergoing profound modifications. The Buddhist shrines and temples, which cover the land, are of unique structure and often of vast size, and are filled with all sorts of idols, the abortions of the most distorted fancy, and far more calculated to strike terror than to

[1] "The reigning house of Japan descends from the Sun-Goddess Amaterasu" (*Japan nach Reisen und Studien*, by J. J. Rein, Leipzig, 1881, i. p. 245).

inspire confidence in their votaries. On all festive occasions the people troop in crowds to these places. To all the larger temples are attached regular spectacles, playhouses, panoramas, besides lotteries, games of various sorts, including the famous " fan-throwing," and the shooting galleries, where the bow and arrow and the blowpipe take the place of the rifle.

But since the recent revolution, Buddhism, like all institutions formerly encouraged by the Shogun, has fallen into disfavour. The accumulated wealth of the priests has been confiscated, the temples spoiled of their artistic treasures, the monks driven from their monasteries, and many of these buildings converted into profane uses. Countless temple bells have already found their way to America, or have been sold for old metal.

The third religious phase of thought is the so-called " Siza," a philosophic system, which is a feeble imitation of the ethical teachings of Confucius, but which has never found much favour with the ruling classes. It seems to be a sort of refined materialism, such as in fact underlies the whole religious thought of the nation. It relies altogether on moral truths and axioms. While outwardly conforming to the Kami religion, most of the *literati* and upper classes have hitherto been adherents of this school. But the Chinese is now being supplanted by what is called the " English Philosophy," represented in Japan chiefly by Buckle, Mill, Herbert Spencer, Darwin, Huxley, many of whose works have already been translated into Japanese.

Christianity was introduced by the Jesuit missionary Francis Xavier, about the middle of the sixteenth century. An impartial study of the early records shows that the Roman Catholic priests were sometimes preparing the minds of the people for the subversion of the native rule, and subjugation of the country to Spain or Portugal.

Hence arose the persecution of 1596, which ended in the extirpation of Christianity and the expulsion of the Portuguese from Japan.

But since the restoration of the Mikado to supreme power, perfect liberty has been extended to the Christian religion. No obstacles are now thrown in the way of proselytism, and native preachers are allowed to proclaim their doctrines throughout the land, although foreign missionaries are still restricted to the Treaty ports.

Of the foreign missionaries the Russian "pope," Father Nikolai, attached to the Muscovite embassy in Tokyo, seems to have hitherto had the greatest success. The Greek church is always crowded, and the people in the neighbourhood are selling off their shrines and idols, fearing they may become a drug in the market, so "fashionable" is the "orthodox" rite becoming. The Roman Catholic clergy are said to be very successful in the rural districts.

Meantime the Buddhists, alarmed at the progress of their rivals, have been stimulated to fresh exertions.

In Japan there exist properly speaking two nations side by side—the nobles and the commons. The nobles, who have been stripped of most of their prerogatives by the recent changes, were divided into two classes—the "Kuge" or "Court Nobility," and the "Buke" or "Aristocracy of the Sword"—of which there were again various degrees with six main divisions.

The whole people have from the earliest times been grouped into seven or eight classes according to their various social ranks and occupations:—1. The Daimios, the highest nobles of the state, originally powerful feudal lords, standing in somewhat the same relation to the Mikado that the imperial princes and counts did in mediæval Germany to the emperor. Their number corresponded at first to that of the "Koku," or provinces, so that there were originally eighteen "Kokushiu Daimios." 2. The hereditary nobles, from which class were chosen the provincial governors, the generals and Government officials. In this class are included the "Hattamotto," or lower aristocracy. 3. The Shinto and Buddhist clergy. 4. The military class of the "Yakonins," and "Samurai." These were the four higher classes, who were privileged to carry two swords and to wear loose trousers. 5. The upper middle classes — physicians, officials, notaries, etc. 6. Merchants and traders, whatever their income. 7. Retail dealers, small shopkeepers, pedlars, hawkers, artists, painters, artisans. 8. Sailors, fishermen, peasants, day labourers.

There were also the so-called "yeta," or "yetori," a sort of pariah class numbering from 250,000 to 300,000, and scattered all over the empire. They were tanners, skinners, workers in leather, and public executioners and cremators. They were not permitted to dwell with the other classes in towns and villages, nor were they included in the census returns. They were also excluded from inns and tea-houses, and when travelling were compelled to eat from separate dishes in the open air.

But all these classifications have now been merged in the following three:—1. The Kwazoku, or nobles; 2. Shizoku, or old military class; 3. The Heimin, or commoners. All social privileges have been abolished, and every man is now equal before the law of the land.

8. *Topography: Chief Towns—Hakodate.*

In Yesso the only places that can be called towns are Sapporo, the capital, on the west coast; Matsumai and Hakodate, both on Sugaru Strait over against Hondo. Hakodate is probably the most important northern seaport, with a spacious and safe harbour and excellent anchorage. The town, in which is concentrated more than one-fourth of the whole population of the island, is a very flourishing place, with a large export trade in fish, skins, trepang, and other local produce. Here are several missionary stations, seventeen schools, several large public buildings, and a British Consulate. But the direct foreign trade has almost entirely ceased, foreign merchandise being now imported by native merchants in native vessels, while the exports are also forwarded in native vessels to Hon-do and China.

Sendai—Niigata.

In Hon-do the most noteworthy places north of the imperial capital are Sendai on the east, Toyama and Kanazawa on the west coast. Sendai, which had a population of 64,000 in 1892, stands 12 miles from its port of Sivogama on the Sendai Ura ("Sendai Bay"), 190 miles north by east of Tokyo, and about the same distance from Aomori at the northern extremity of Hon-do. With this place it is connected by the Omiya-Aomori railway and with Sivogama by a short branch running from the main line to the coast. Sendai is the largest market town in North Japan, and does a considerable trade in fish and salt. It took an active part in the civil war of 1868, siding with the Daimios, which was in accordance with the historic traditions of the place.

Amongst the "three wonders" of the Japanese world is reckoned the shallow Matsusima inlet at the head of Sendai Bay. This secondary basin, which has scarcely 20 feet at high water, is studded with innumerable little wooded limestone islets, as many as 808 according to the local report, ranging from about 30 to 300 feet in height, the whole presenting a lovely panoramic view, as seen at sunrise from the neighbouring forest-clad

Miura heights. Many members of the Archipelago have been hollowed into caves or carved into arcades by the natural action of the waves; the hand of man has excavated fantastic grottoes in the steep sides of others; nearly all are crowned with clumps of cryptomeria or other leafy vegetation, while the summit of one has been sculptured to the likeness of a gigantic Buddha.

Niigata, capital of the province of Echigo, stands over against the large island of Sado on the west side of Hon-do, about 20 miles south of the parallel of Sendai. Although its harbour is no longer accessible to large vessels, Niigata derives some importance from the fact that it is the only Treaty port on the west coast between Hakodate and Nagasaki, a distance of 1100 miles. It is a flourishing place, with some large public buildings, hospitals, barracks, normal schools, a geological museum, a missionary station, and a population (1892) of 47,000, but no direct foreign trade. The Teremachi, or "Temple Street," as one of the thoroughfares is called, is lined on one side for nearly its whole length by dens of vice, on the other by Buddhist temples, grounds, and priests' houses. The temples are mostly large handsome structures, with splendid high altars, candelabra, bronzes, and statues of Buddha, "with glories round their heads in gorgeous shrines, looking like Madonnas" (Miss Bird, now Mrs. Bishop).

Toyama—Kanazawa.

South of Niigata, but still north of the parallel of Tokyo, follow the two prosperous coast towns of Toyama and Kanazawa, the former on the north, the latter on the south side of the peninsula of Noto. Toyama, which gives its name to the spacious Toyama Ura ("Toyama Bay"), stands 35 miles east by north of Kanazawa, on the Dzindzu-Gawa, 8 miles above its mouth. The place might almost be called an oriental "Apothecary Hall," such is the number of hands here employed in the preparation and sale of Chinese "patent medicines." In winter they are occupied in the concoction of an endless variety of drugs and nostrums, which they retail in the fine season throughout the whole of Japan. The industry, however, has suffered much in recent years, partly from the progress of education, partly from the heavy taxes imposed by the imperial government on such wares.

Toyama occupies a geographical position on the west coast curiously analogous to that of Tokyo on the east side, the chief difference being the reversed direction of their respective peninsulas, Noto projecting northwards into the Sea of Japan, Kadsusa southwards into the Pacific Ocean. The expression "Little Yedo," formerly applied to Toyama, had reference to this topographical similarity, and not to any political or social resemblance. In these respects there is no room for comparison, and Toyama, despite its historic memories, never was a very large place. In 1892 the population scarcely exceeded 60,000, while its neighbour Kanazawa numbered as many as 94,000 inhabitants.

Kanazawa (Isikawa) stands on the two little rivers Sai and Asano, 5 miles from the west coast and 186 miles west by north of Tokyo. It enjoys some reputation as a manufacturing centre, and is specially noted for its chased bronzes, painted porcelain ware, and silk stuffs. Takamats, some 5 miles to the west, is the outport of Kanazawa and of the surrounding industrial district.

Tokyo—Yokohama—Kama-kura.

Tokyo, formerly Yedo, the imperial capital, lies about the middle of the east coast at the head of Yedo Bay, in 35° 50′ N. Although Yedo Bay, now renamed the Gulf of Tokyo, is one of the largest and finest in the world, Tokyo itself enjoys few of the advantages of a seaport, even the smallest vessels being obliged to lie some two or three miles off the shore. Like so many other Jàpanese ports, it has been partly obstructed by the alluvia of the River Tone, the southern branch of which enters the head of the gulf close to the capital.

The city is spread over an undulating plain, enclosed on the south by the bay, on the east and north by the broad and picturesque River Sumida, westwards partly by flat paddy fields, partly by a ridge of low hills planted with conifers and bamboo thickets. East of the Sumida River lies the large suburb of Honjo, continued southwards by the village of Shinagawa, which is itself an extension of the suburb of Takanawa.

The old divisions of Tokyo were Shiro, Soto-shiro, Mitsi, and Honjo. On a wooded eminence in the centre of Shiro—that is, in the very heart of the capital—formerly stood the palace occupied till 1868 by the Shogun, but destroyed by fire in 1870. The park-like grounds were enclosed by lofty fortified walls and a broad moat flooded with running water. Round the Shiro spreads the Soto-shiro quarter, which is again surrounded by a wide ditch connected with the palace moat. This district is everywhere intersected by countless canals communicating both with the inner circular moats and with the sea. Here resides the commercial community, and here are situated the finest warehouses and public buildings.

Towards the north-east another wooded height was formerly covered with splendid Shinto and Buddhist temples, as well as with the tombs of several Shoguns of the last usurping family. But nearly all the temples were destroyed during the troubles of 1868-69, and this

TOKYO.

hill is now laid out as a public park, with a permanent exhibition building in the centre.

But the great feature of old Tokyo were the numerous "yashiki," or palaces of the feudal lords. Some of these establishments occupied extensive wooded enclosures, and consisted of clusters of houses surrounded by one-storied and whitewashed outhouses occupied by the Samurai, or "two-swordsmen," and other retainers of the Daimios. But since their owners have been stripped of their privileges, these strongholds have been dismantled and mostly destroyed.

The industries of Tokyo are chiefly limited to artistic and fancy goods, such as bronzes and lacquer ware. The old law obliging the Daimios to reside for six months of the year in the capital, to which their families were confined all the year round, tended greatly to the development of the city. But since the repeal of that law in 1868 the population has fallen from about 1,500,000 to 1,160,000 in 1892.

On the south side of the Gulf of Tokyo lies the city of Yokohama, which, with the neighbouring port of Kanagawa, is open to Europeans. Thanks to its commodious harbour in the vicinity of the imperial capital, Yokohama has become the chief centre of the trade with the West, and the headquarters of the European establishments in the empire. It has been connected since 1872 with Tokyo by a well-constructed railway 18 miles long, the first opened in the empire.

Near the north-east side of Sagami Bay, about 11 miles from Yokohama, stand the ruins of the historical city of Kama-kura, which was the capital of the Shogun Yoritomo and of his immediate successors in the twelfth and thirteenth centuries. The extensive remains of palaces, gateways, stone bridges, tombs, and of over a hundred temples still attest the former size and splendour of this princely residence. One of the temples, which has been rebuilt, is dedicated to Hatsiman, whose worship is associated with the glorious epoch of the Empress Okinagatarasihimé (Zingu Kogo), during whose reign (second century of the new era) Korea was first conquered by the Japanese.

Another famous monument is the tomb of the 8300

MAIN STREET, YOKOHAMA.

legendary heroes, all of whom committed the happy dispatch in compliance with the rigid code of honour of mediæval Japan. Here is also a "Dai-buts" ("Great Buddha"), a colossal bronze statue of the god in the usual attitude of contemplation, 40 feet high, mounted on a granite pedestal and enclosing a little shrine. There are two other Dai-buts in Japan, one near Kioto said to be 53 feet high, the other in the great temple at Nara, also over 50 feet and weighing 450 tons.

On the east side of Hon-do south from Yokohama

follow the busy towns of Nagoya and Yamada, the former of which ranks for population (180,000 in 1892) as the fourth city in Japan, being exceeded only by Tokyo,

DAI-BUTS, KAMA-KURA.

Osaka, and Kioto. Nagoya, officially Aitsi-ken, stands on a fertile plain 3 miles from the northern extremity of Mia-ura or Owari Bay, 32 miles east of Kioto, on the route between that city and the capital. It is a great industrial centre, noted especially for the production of the highly-prized Owari ware, and of innumerable little fancy objects remarkable for their endless variety of form and colour. Nagoya is also the chief market for the porcelains of the neighbouring town of Seto, where this industry is said to have been carried on without any interruption for the last 2000 years. Like some other Japanese cities, Nagoya is laid out on the chess-board plan so common in the New World. All the streets are perfectly straight, disposed at right angles and lined with interminable wooden houses of two stories with overhanging tiled roofs. Of the religious edifices the most remarkable is the convent of Kenchiu-ji, containing the tombs of the powerful feudal lords of Owari, one of the three great families that have supplied Shoguns to Japan. Nagoya was the imperial capital under Ota Nobunaga, the powerful protector of St. Francis Xavier and of the Christian party during the religious troubles of the sixteenth century. But owing to the silting up of Owari Bay it has long ceased to have direct access to the sea, and all merchandise is now forwarded through Atsuda to the port of Yokka-itsi 21 miles farther down on the west side of the bay.

On the peninsula enclosing the south side of Owari Bay stands the city of Yamada, with a population of little more than 30,000, but renowned throughout the empire in connection with the neighbouring temples of Ge-ku and Nai-ku. These centres of the Shinto religion, which are visited by constant streams of pilgrims from all parts of Japan, are reputed to be 3000 years old, and the original structures may possibly date from the beginning of the new era. But the present temples are mere "facsimiles" of the old temples, being pulled down and re-erected every twenty years on the exact lines of the first buildings. In their reconstruction timber of the same species is used, the same straw thatching is adhered to, and all the arrangements and fittings of the interior are faithfully preserved.

The most interesting city in the interior is undoubtedly Miako, or Kioto, on a branch of the Yodo River about 8 miles west of Lake Biwa. Kioto is the ancient religious capital of the empire, and former residence of the Mikado. It is said to contain 93 Shinto and 935 Buddhist temples, amongst which are two conspicuous above all the rest for their size and splendour. According to the native accounts, one of these enshrines 333,333 idols, while the other contains the famous colossal figure of Buddha seated on the lotus. The Japanese look upon this city as a sort of Athens, from time immemorial the centre of learning and literature, where the most classic form of speech is current. It is still the seat of the book trade and a focus of native literature, while at the same time constantly crowded with Buddhist pilgrims from all quarters. Its principal industries are porcelain and bronzes. But although recently renamed Saikio, or the western capital, Miako is now in no sense a capital, the residence of the Mikado and the whole administration having been definitely removed to its great rival Tokyo, the "eastern capital."

Kioto is connected by the River Yodo with Ohosaka or Osaka, the "Venice of Japan," and the queen of Japanese cities. It is intersected in all directions by innumerable canals which are crossed by no less than 3500 bridges, and the people live here altogether more on the water than on land. But since the opening of so many Treaty ports Osaka has lost much of its former wealth and importance, and its harbour is now so obstructed with sandbanks that the whole of its foreign trade and shipping has been gradually passing northwards to Yokohama and Tokyo.

KYOTO.

Facing Osaka, on the coast of the Inland Sea, are the seaports of Hiogo and Kobe, which, strictly speaking, form but one city, always spoken of by the natives under the latter, by Europeans under the former name. Hiogo, which is now connected by rail with Osaka and Otsu on Lake Biwa, commands the markets of central Japan, and forms an important connecting link between the various ports thrown open to trade along the east coast.

Hirosima.

Other flourishing seaports in the Inland Sea are Wakayama, at the eastern entrance south of Osaka, and Hirosima (Hiroshima) on the north coast. Hirosima, which in its general aspect somewhat resembles Osaka, is next to that place the most important commercial centre in the Inland Sea. But in the eyes of the natives its chief distinction arises from its association with the neighbouring Itsku-sima, "Island of Light," dedicated to the goddess Bentin, and regarded as one of the three "wonders" of Japan. The soil of this hallowed spot is never disturbed by the peasant's hoe, and provisions are daily brought to the inhabitants from Miyasima and some other places on the mainland. The population consists exclusively of priests, innkeepers, fishers, and wood-carvers, who retail the products of their art to the thousands of pilgrims constantly resorting to the sacred island. The chief temple, built in an extremely simple style, stands on the beach, and dates from the year 587 of the new era. It contains some curious wood-carvings representing historic events, portraits of celebrities, inscriptions, and many of the numerous votive tablets also possess historic interest.

Nagasaki—Kumamoto—Kagosima—Fukuoka.

Nagasaki, southernmost of the Treaty ports, stands on the west coast of Kiu-shiu. It occupies a convenient position at the head of a romantic inlet, which here forms a magnificent landlocked harbour about 500 miles from Shanghai, and 600, by the Inland Sea, from Yokohama.

TAKOBOKO, NAGASAKI.

Here is the rocky islet of Takoboko (" Papenberg "), from which native Christians are said to have been hurled into the sea at the close of the persecutions in the sixteenth century. Not far off is the artificial island of De-shima, occupied by the factory where the Dutch monopolised the trade of Japan for two hundred years. But all restrictions are now removed, and the barriers between the station and the mainland have recently been cleared away. Thanks to its genial climate, Nagasaki has become a sort of sanitarium for the European traders settled on the Chinese seaboard.

In the interior of Kiu-shiu, the largest place is Kumamoto, which had a population of 56,000 in 1892. It stands on the right bank of the Sira-kava (Midori-gava), 53 miles east of Nagasaki and not far from Simabara Bay. Kumamoto is the capital of one of the ten military divisions of the empire, and here is a formidable stronghold built of huge blocks in the old Japanese style, with walls inclined inwards and shaded with camphor trees. Although near the coast, Kumamoto is essentially an inland city, the neighbouring waters being quite shallow and accessible only to flat-bottomed craft.

On the other hand, the spacious and deep inlet of Kagoshima-ura, on the south coast of Kiu-shiu, is accessible to the largest vessels. The seaport of Kagoshima, which gives its name to the inlet, and which was formerly capital of the powerful daimio of Satsuma, was bombarded and nearly destroyed by the British fleet in the year 1863. But it has already recovered from this disaster, and is now a prosperous industrial city of nearly 60,000 inhabitants, engaged in cotton-weaving, the manufacture of small arms and of porcelain, more especially a skilful imitation of "old Satsuma ware." In the Satsuma district is the superb Kaimon-ga-tate volcano, the Mount Horne of the marine charts, rising to a height of about 3000 feet at the entrance of the Kagoshima Gulf.

Fukuoka, on the north-west coast of Kiu-shiu, about 60 miles north by east of Nagasaki, scarcely yields in commercial activity to Kagoshima at the opposite extremity of the island. With Hakata, from which it is separated only by a narrow coast stream, it forms a large centre of population (55,000 in 1892) engaged in trade and several local industries. Of these the most important is silk-weaving, and Hakata is specially noted for its stout silk fabrics. Fukuoka was formerly the residence of the powerful daimio of Chikuzen, and is associated with many stirring events in the history of mediæval Japan. Of the numerous temples dotted over the district, the most conspicuous was that of Yeiyas, which crowned an eminence within the city limits, but which was destroyed during the revolution of 1868. It was dedicated to the founder of the princely House of Tokugawa, in which the office of Shogun had become hereditary during the last 300 years of its existence.

Tokusima—Kotsi—Taiwan.

In the mountainous island of Shikoku the largest place is Tokusima, which had a population of over 60,000 in 1892. It stands at the north-east extremity of the island close to the Kii Channel, that is, the eastern entrance to the Inland Sea, and is distant about 310 miles west by south from Tokyo. Lying near the mouth of the Yosinogawa, largest river in Shikoku, Tokusima enjoys the advantage of water communication with the interior.

Nevertheless this former residence of the Hatsizuka daimios is outstripped in industrial activity by the busy town of Kotsi, where are centred the largest paper-mills in the whole of Japan. Kotsi, which lies on the south coast at the narrowest part of the island, was also a princely residence in feudal times when it was the chief seat and stronghold of the Toza daimios.

Taiwan, capital of Formosa, to which it gives its native name, lies in somewhat shallow waters on the south-west coast. It is a Chinese city of the ordinary type, with a closely-packed population of over 350,000.

9. *Highways of Communication—Railways.*

In Hon-do the three great historical highways are:—1. The Oshiu-Kaido, running from Tokyo for 444 miles northwards to Aomori on Sugaru Strait; 2. The Tokai-do, skirting the south-east coast for 507 miles between Tokyo and Kioto; 3. The Nakasendo (Kishu-Kaido), running through the heart of the country for 323 miles, also between Tokyo and Kioto. This alternative route is much more difficult than the coast road, and is often blocked by snow in winter.

Niigata on the west coast is connected with Tokyo by two routes, 264 and 225 miles long respectively, both running for a part of the distance along the Nakasendo line. From Tokyo the Koshiu-Kaido, 77 miles long, leads to Kofu in the province of Kai, whence it is continued for 32 miles to the Nakasendo road at Shimono Suwa.

None of these highways are macadamised, and those of Shikoku and Kiu-shiu especially are said to be in a very bad state. But even in Nippon many of the roads in the interior are about as bad as they well can be. "The road from Kurumatoge westwards is so infamous that the stages are sometimes little more than a mile. Yet it is by it, so far at least as the Tsugawa River, that the produce and manufactures of the rich plain of Aidzu with its numerous towns, and of a very large interior district, must find an outlet at Niigata. At present it is a perfect quagmire, into which great stones have been thrown, some of which have subsided edgewise, and others have disappeared altogether."[1]

The first railway, a line 18 miles long between Tokyo and Yokohama, was opened in 1872, and since then the system has acquired considerable development. In 1892

[1] *Unbeaten Tracks*, i. p. 186.

the State railways had a length of over 55 miles, while thirteen private companies owned lines with a total length of 1166 miles, and several others were in progress in various directions. Some radiate from the capital towards Fuji-yama and Asama; Hiogo is connected with Lake Biwa by a substantial line running by Osaka to Kioto and Otsu, and this romantic district may also be reached by rail from Wakasa Bay on the west coast.

Steam navigation has also made great progress, and several lines of steamers now ply regularly between all the chief seaports of the empire, as well as with China and Korea. Many of these vessels have been built in Japan, and are worked entirely by native hands.

10. *Administration: The Mikado and Shogun—The Revolution—Army and Navy—Education—Art—The New Ideas.*

The peculiar political and social institutions of the Japanese people, which are now rapidly yielding to the influence of Western ideas, were slowly and independently developed during the long ages that they have been in possession of their secluded island empire. How long they may have here been settled it is now hard to say, their historic records dating no further back than the third century of our era. These records embrace two clearly-defined epochs, the "oshei" and the "hashei," as they are called. The first, reaching to the year 1192 A.D., covers the period of the absolute authority of the Mikados, or hereditary emperors of divine origin, who claimed direct descent from Jimmu, reputed founder of the present dynasty in 660 B.C.[1]

[1] *Mikado* (from *mi*=sublime, and *kado*=gate) has etymologically the same meaning as the "Sublime Porte." The idea seems to be that "the Mikado is too sublime a being to be spoken of directly or otherwise than

The second period corresponds with the growth of the power of the Shoguns, or military rulers. This is the mediæval period, which was not brought to a close till 1868, when the Revolution restored the Mikados to the supreme authority. The power of the Shogun, the true significance, of which had long been misunderstood, was not fully and legally recognised till the beginning of the sixteenth century, when the office became hereditary in the family of its founder, Tokugawa. The popular idea was that there were two Emperors, the temporal Shogun residing in Yedo (Tokyo), and the spiritual Mikado residing in Miako (Kioto). But in point of fact the Shogun was never anything more than the military vicegerent of the Mikado, without any legal claim to joint sovereign right.[1] In course of time, however, he largely usurped the control of state affairs, retaining the Mikado in a sort of honorary imprisonment; he also enjoyed absolute executive power over the daimios, or territorial lords, who, like the feudal barons of mediæval Europe, governed their own territories under the Shogun. But the Shogun himself was always legally subject to the Mikado, who was at all times recognised both as a sacred being and as the supreme ruler of Japan. Consequently the Revolution which crushed the feudal lords merely put an end to the Shogun's military despotism, and reinstated the Mikado in the full exercise of his sovereign rights.

This reaction was followed by perhaps a still more important event in 1873, when the Mikado voluntarily promised to his subjects a representative constitution.

in a figurative sense" (Rein, *Japan*, i. 245). The title, however, is somewhat antiquated, and the head of the State is now more commonly known as the *Kotei* or "Emperor."

[1] From *Sho*=general, and *gun*=army; hence *Shogun*=commander-in-chief. The early European writers spoke of the Shogun as the Tykoon, from the Chinese *Tai-kun*="Great Lord."

The promise was followed in 1875 by the appointment of a Senate, and of the "Dai-shin-in," consisting of the Supreme Council (Shoin), and the College of Ministers (Uin). The Scheme of representation was further developed by the establishment in 1878 of the provincial and departmental Assemblies, and was completed by a full National Assembly, which, by the imperial decree of October 12, 1881, met for the first time in the year 1890.

The original nine divisions of the empire, which had long ceased to answer to the requirements of the executive government, were replaced in 1871 by the three "fu," or urban districts, and the 66 "ken" or prefectures, which have since been reduced to 60. Each ken is governed by a "ken-rei," directly dependent on the Ministry of the interior, and its name corresponds in almost every case with that of its chief town.

According to the law of 1873 regulating the conscription, all Japanese subjects are liable to serve from their twentieth year, and must pass three years either in the army or the navy. They then pass into the army of reserve for four years, after which they form part of the *landwehr* for another five years. Behind the landwehr is the *landsturm*, which comprises every male from 17 to 40 not in the active or reserve forces, and which is liable to be called out in case of national emergency. The Japanese troops, who have at all times been distinguished by extreme bravery, are now disciplined after the European method, although no foreigner receives a permanent appointment in the army. Most of the war materials are prepared in the arsenals of Tokyo and Osaka, and even the army rifle, the Murata, is a Japanese invention.

For naval defence Japan is divided into two districts, each under a vice-minister subject to the naval minister at Tokyo. Here is the naval school; but the chief dockyard is at Yokosuka. The fleet includes several powerful ironclads, some built in the country, others in Europe. The crews are noted for their skill, courage, and coolness, and the thorough efficiency of both the naval and military services was proved beyond question by the events of the war of 1894-95 with China.

Special attention has of late years been paid to the question of public instruction. For educational purposes the country has been divided into seven circles, each with a "dai-gak'ko" or High School in the central town. Elementary education is compulsory, and most of the primary and higher schools are supported by government and local rates.

Japan has also joined the International Postal Union, and already possesses 17,000 miles of telegraphic lines.

The chief branches of local industry resemble those of China, from which country Japan has received its culture. The ancient and world-renowned bronze, lacquer, and paper wares still continue to be produced, as well as the porcelains and works in enamel, an art introduced some three centuries ago. The chief centres of the porcelain industry are Satsuma, Hizen, Kioto, and Kaga. In Hizen there are inexhaustible deposits of kaolin, or china clay, and here is produced the so-called "egg-shell" ware.

Extremely interesting are the paintings on paper or silk, and the books illustrated with woodcuts. Strips of tissues several yards long and about a yard wide are painted with figures in a somewhat conventional style, and with plants, in which a happy mean is observed between mannerism and the realistic style. Great skill and exuberance of fancy are displayed in the numerous illustrated works, which include many sketch-books with rich and varied landscapes, studies of men and animals, genre scenes and the most extravagant caricatures.

To the local arts have in recent years been added lucifer-match making and some other western industries, in which the Japanese now compete with the European producers, not only in Japan itself, but also in India and other Asiatic markets.

Meantime the revolution of 1868 forms the starting-point of a new era; and although it has been followed by many sweeping changes and reforms of a fundamental character, the country must still pass through a restless and anxious period of transition before the new ideas can become thoroughly assimilated. The nation has been suddenly awakened, as it were, from sleep, and seized all at once with a passionate desire to share with the peoples of the West in all the intellectual and material triumphs of a slowly-matured culture. This movement constitutes an almost unique feature in the history of Asiatic nations. But in the eagerness to appropriate the results of such a culture, mistakes must be expected. Fleets and armies have been improvised, and magnificent buildings, such as the Mint at Osaka, erected at a needless expenditure of vast sums that might have been better applied. Costly embassies are maintained in Washington and at the chief European Courts, while large sums are devoted to the education of several hundred students in Europe and America. Every inducement has been held out to foreign professors to settle in the country; orders of chivalry have been instituted on the European model; the natives have been encouraged to adopt the European dress; and the idea has even been broached of abolishing the national speech and replacing it by the English language and letters. According to the most judicious opinions, however, the crisis cannot be said to have been passed; errors springing mainly from excessive zeal have yet to be repaired. Many enthusiastic observers may hope that the collective wisdom of the nation will guide the Japanese race to a bright goal. Still, prudent judges will suspend their opinion regarding the moral and intellectual future of this interesting race.

11. *Statistics.*

AREAS AND POPULATIONS (1892).

Divisions.	Area in sq. miles.	Population.
Northern Nippon . . .	30,204	6,261,150
Central Nippon . . .	36,600	15,912,791
Western Nippon . . .	20,681	9,345,388
Shikoku	7,031	2,903,332
Kiu-shiu	16,840	6,326,905
Hokkaido (Yeso and Kuriles)	36,299	340,374
Formosa with the Pescadores ceded by China (1895) .	15,000	3,000,000 (?)
Total . .	162,655	44,089,940

GROWTH OF POPULATION.

1874 . .	33,623,000	1889 . .	40,072,000
1880 . .	35,925,000	1890 . .	40,454,000
1887 . .	39,069,000	1891 . .	40,718,000
1888 . .	39,607,000	1896 . .	44,089,940

CITIES WITH OVER 40,000 INHABITANTS (1892).

Tokyo . . .	1,180,000	Hakodate . . .	60,000
Osaka . . .	479,000	Kumamoto . . .	59,000
Taiwan . . .	360,000	Toyama . . .	59,000
Kioto . . .	308,000	Fukuoka . . .	56,000
Nagoya . . .	185,000	Wakayama . . .	56,000
Kobe . . .	148,000	Kagoshima . . .	55,000
Yokohama . .	143,000	Okayama . . .	50,000
Kanasawa . .	92,000	Niigata . . .	49,000
Hirosima . .	90,000	Sakai	46,000
Sendai . . .	70,000	Matsuye . . .	45,000
Nagasaki . . .	63,000	Naha	44,000
Tokusima . .	62,000	Fakui	42,000

FINANCE (1893).

Revenue.		Expenditure.	
Land tax	£7,700,000	Public Debt charges .	£4,000,000
Saké, malt, and soy tax	3,300,000	Ministry of War .	2,500,000
Post and telegraphs .	1,200,000	" Marine .	1,100,000
Customs	900,000	" Finance .	810,000
State services . .	710,000	" Justice .	700,000
Tobacco-tax . .	360,000	" Education	180,000
Other Inland revenue	570,000	" Trade and Agriculture	190,000
Licenses, fees, etc. .	400,000	" Post and Telegraphs	1,140,000
Income-tax . .	200,000	Civil List, etc. . .	640,000
Interest on deposits .	185,000	Provincial Government	980,000
Forests	200,000	Hokkaido Government	300,000
Miscellaneous . .	160,000	Redemption of paper .	200,000
Bank licenses, stamps	260,000	Cabinet, Diet, etc. .	140,000
Temporary revenue .	550,000	Sundries . . .	2,900,000
Total . .	£16,715,000	Total . .	£15,780,000

Public Debt (1894)	£55,000,000
Paper currency	5,000,000

TRADE RETURNS.

Year.	Imports.	Exports.	Total.
1880	£7,250,000	£5,510,000	£12,760,000
1884	5,764,000	6,603,000	12,367,000
1890	16,000,000	11,000,000	27,000,000
1894	18,000,000	19,000,000	37,000,000

CHIEF EXPORTS AND IMPORTS (1892).

Exports.		Imports.	
Silk, cocoons, etc. .	£7,800,000	Sugar	£1,900,000
Silk textiles . .	1,600,000	Cotton yarn . .	1,400,000
Tea	1,700,000	Cotton piece goods .	580,000
Rice	830,000	Calico	340,000
Coal	900,000	Metals	1,000,000
Copper	490,000	Wool and woollen goods	1,110,000
Porcelain, bronze, etc.	460,000	Drinks and provisions	1,200,000
Dried fish . . .	440,000	Petroleum	1,060,000
Camphor . . .	250,000	Machinery, ships, etc.	590,000
Drugs, chemicals .	180,000	Drugs	470,000

CHIEF CUSTOMERS OF JAPAN (1894).

	Exports to.	Imports from.
Great Britain	£790,000	£5,600,000
United States	5,500,000	1,220,000
China	1,500,000	3,400,000

	Exports to.	Imports from.
France	£3,900,000	£660,000
East Indies, Siam . .	490,000	1,700,000
Germany	270,000	1,400,000
Korea	260,000	380,000
Australia	180,000	64,000
Russia	113,000	370,000
Italy	260,000	14,000

SHIPPING (1892).

	Cleared.	Tonnage.	Entered.	Tonnage.
Japanese steamers .	403	346,000	401	338,000
„ sailing vessels .	855	29,500	762	28,700
Foreign steamers .	897	1,277,000	1017	,424,000
„ sailing vessels .	82	85,000	78	80,000
Total . .	2,237	1,737,500	2,258	,870,700

British shipping, 582 of 967,500 tons entered and cleared.

AGRICULTURAL RETURNS (1892).

	Acres.	Bushels.		
Rice .	6,752,000	205,240,000	Tea . .	75,000,000 lbs.
Wheat .	1,064,000	15,247,000	Sugar . .	109,700,000 „
Barley .	1,601,000	33,774,000	Silk cocoons	7,900,000 bushels
Rye .	1,593,000	30,043,000	Silk, raw .	12,100,000 lbs.

LIVE STOCK (1893).—Cattle, 1,095,000 ; Horses, 1,554,000.

MINERAL PRODUCTS (1890).

	Government Mines.	Private Mines.
Gold . .	630 lbs.	141 lbs.
Silver . .	16,413 „	98,066 „
Copper . .	78,300 „	35,700,000 „
Iron . .	6,690,000 „	5,000,000,000 „
Lead . .	...	1,320,000 „
Coal . .	184,000 tons	2,500,000 tons
Antimony .	4,000,000 lbs.	416,000 lbs.
Sulphur .	...	36,320,000 „

ARMY (1894).

	Men.	Horses.	Guns.
Imperial Guard . .	6,200	560	40
Line (six divisions) . .	66,000	8,800	470
Total peace footing .	72,200	9,360	510

First Reserve, 100,000 ; Landwehr, 99,000.

Navy (1894).

	No.	Tonnage.	Horse-Power.	Guns.
Ironclads . .	5	10,810	15,560	65
Cruisers and others .	17	36,000	85,400	247

Education (1892).

Schools.	No.	Teachers.	Attendance.
Elementary . .	25,374	69,608	3,154,000
Kindergarten . .	147	317	8,660
Lower Middle . .	57	815	14,380
Higher Middle . .	7	314	4,442
High Girls' . .	29	332	2,768
Normal . . .	49	669	5,354
Technical . . .	88	1,683	18,447
Special . . .	1,682	4,300	85,806
Universities . .	3	248	8,662

Libraries, 20; Volumes, 182,000. Books published (1891), 22,568; Periodicals, 716. Periodicals circulated, 199,168,000.

Crime.

	1887.	1889.	1891.
Serious offences . .	4,397	2,431	3,260
Minor „ . .	79,723	86,555	154,087

Postal Service.

Year.	Letters.	Parcels, Books, etc.	Income.	Outlay.
1894	255,000,000	62,000,000	1,120,000	910,000

	Miles.	Miles of Wire.	Dispatches.
Telegraphs (1894) . .	9,100	24,970	6,497,000

	Miles open.	Passengers.	Income.	Outlay.
Railways (1894) .	1,870	27,000,000	£1,980,000	£893,000

Telephones (1894), 4356 miles of wire; offices, 28; subscribers, 2672.

INDEX

END OF VOL. I.